AF389331

MEMS Packaging Technologies and 3D Integration

MEMS Packaging Technologies and 3D Integration

Editor

Seonho Seok

MDPI • Basel • Beijing • Wuhan • Barcelona • Belgrade • Manchester • Tokyo • Cluj • Tianjin

Editor
Seonho Seok
C2N (Center for
Nanosciences and
Nanotechnologies)
University-Paris-Saclay
Palaiseau
France

Editorial Office
MDPI
St. Alban-Anlage 66
4052 Basel, Switzerland

This is a reprint of articles from the Special Issue published online in the open access journal *Micromachines* (ISSN 2072-666X) (available at: www.mdpi.com/journal/micromachines/special_issues/MEMS_Packaging_Integration).

For citation purposes, cite each article independently as indicated on the article page online and as indicated below:

LastName, A.A.; LastName, B.B.; LastName, C.C. Article Title. *Journal Name* **Year**, *Volume Number*, Page Range.

ISBN 978-3-0365-4258-4 (Hbk)
ISBN 978-3-0365-4257-7 (PDF)

Cover image courtesy of Hyungdal Park

Contents

About the Editor

Seonho Seok

Seonho Seok received the M.S. and Ph.D. degrees in electrical engineering from Seoul National University, Seoul, Korea in 1999 and 2004, respectively. He was a postdoctoral researcher at Center for Advanced Transceiver Systems (CATS) in Seoul National University and at IEMN, CNRS in France from 2004 to 2007. He worked as a permanent researcher of CNRS at IEMN from 2007 to 2014. Starting from March 2014, he is working at micro-nano systems and biotechnologies team in C2N (Center for Nanosciences and Nanotechnologies)-University Paris-Saclay.

His current research interests are wafer-level (vacuum) packaging, transfer technique based on adhesion engineering, 3D (heterogeneous) integration, MEMS and sensors, 3D antennas, energy harvesting, and related multi-physics modellings, etc.

Preface to "MEMS Packaging Technologies and 3D Integration"

MEMS packaging is an essential technique for successful commercialization of MEMS products as MEMS has moving parts and an application-specific nature. A classic approach of MEMS packaging is to bond silicon or glass cap wafers to MEMS wafers. Therefore, it is typically implemented under high pressure and high temperature conditions. Advanced approaches use a thin-film deposition technique and then a cavity for MEMS is realized via sacrificial etch through access holes at the thin film cap. The packaging cap transfer technique is a compromise between the two approaches, since it makes it possible to bond and transfer a thin packaging cap to the released MEMS device. MEMS devices and IC are being integrated in a 3D fashion to achieve a better performance, and implantable devices need special packaging techniques. Thus, this Special Issue gathers research papers, short communications, and review articles that focus on MEMS packaging technologies and related integration methods. This Special Issue will help you to catch the trend of the recent MEMS or electronic device packaging and 3D integration technologies.

Seonho Seok
Editor

micromachines

Editorial

Editorial for the Special Issue "MEMS Packaging Technologies and 3D Integration"

Seonho Seok

Center for Nanoscience and Nanotechnology (C2N), University-Paris-Saclay, 91400 Orsay, France; seonho.seok@u-psud.fr

Citation: Seok, S. Editorial for the Special Issue "MEMS Packaging Technologies and 3D Integration". *Micromachines* 2022, 13, 749. https://doi.org/10.3390/mi13050749

Received: 29 April 2022

Accepted: 30 April 2022

Published: 9 May 2022

Publisher's Note: MDPI stays neutral with regard to jurisdictional claims in published maps and institutional affiliations.

As fabrication technologies advance, the packaging of MEMS device is being developed in two main directions: MEMS device packaging and MEMS or sensor system integration. MEMS device packaging is an essential technique for successful commercialization of MEMS product, as MEMS devices inevitably have moving and fragile parts. MEMS devices tend to be packaged in chip-scale, or, namely, zero-level packaging as fabrication technologies advance in place of conventional package of a commercial product such as DIP (Dual Inline Package). For certain MEMS products, vacuum packaging has been achieved with zero-level packaging approaches [1,2]. Such a MEMS packaging has mainly been realized by bonding techniques with joint materials such as metals, polymers, etc. A packaging cap with housing cavity is bonded to a sealing ring surrounding MEMS devices. The lid wafers are generally fabricated with so-called hard materials such as silicon, glass, and so on. For certain MEMS devices, different cap materials, for example, thin films or polymer films, have been adopted for high frequency devices and inertial sensors [3–7]. Thin film packaging is fabricated by conventional semiconductor process and sacrificial etch and thus may have advantages of small package sizes and low costs due to its high throughput. The bonding techniques typically used for MEMS device packaging are interfacial bonding and intermediate layer bonding [8–12]. The interfacial bonding depends on the chemical reaction between two joint materials, while the intermediate layer bonding needs additional materials as adhesive layers. Anodic bonding and silicon fusion bonding are frequently used interface bonding techniques. The interfacial bonding requires high surface cleanness as well as high surface flatness, and it is carried out under high temperature and high applying pressure conditions. Thus, the interfacial bonding has certain constraints for temperature sensitive MEMS devices. Intermediate layer bonding needs good adhesion materials with associated substrates to avoid unwanted delamination of the sealing layers. As the intermediate layer material determines the bonding condition such as temperature, it can be implemented at relatively low temperature. For the packaging based on bonding technologies, attention should be paid to thermal expansion coefficient difference among the associated materials because it would cause undesired high packaging stress. Packaging stress is a principal cause of its reliability, and thus modeling and simulation of electronic and MEMS packages have been frequently performed through FEM (Finite Element Method) to understand mechanical behavior of the packages [13–17]. To obtain reliable simulation results, material properties and their behavior depending on temperature or external load should be well characterized. As an alternative packaging approach, thin film encapsulation integrates the packaging process with the MEMS device process on the same wafer. MEMS structures covered by an additional sacrificial layer are first released by sacrificial etching through channels or holes, and then the access holes are sealed by depositing an overcoat material. The thin-film packaging materials should be deposited or formed without degrading or changing the properties of MEMS structure, and it takes a longer time to release overall packaging cap including the packaged MEMS devices via the accesses of etching solution or gas [18]. MEMS packaging has application-specific features. In other words, it has different approaches depending on application. For example, high-performance inertial sensors may need vacuum packaging fabricated with solid metal-to-metal bonding, but it could not be applied to RF-MEMS packaging because such a metallic sealing ring may create additional loss or undesired parasitic effect at higher frequencies. Another aspect of MEMS product packaging is integration and packaging with integrated circuit (IC) chips, as MEMS output signals should be processed for operation. The integration of MEMS and ICs is implemented through a hybrid multi-chip solution or an SoC solution. SoC solutions have the fabrication and integration of MEMS and IC components implemented on the same substrate, and the fabricated chips are separated only at or near the end of the fabrication process. The hybrid integration of MEMS and IC technology has been implemented by 2D integration approaches. MEMS and IC wafers are fabricated independently and placed on a common printed circuit board (PCB) and then interconnected with wire-bonding. Multi-Chip-Module (MCM) is an advanced integration technique of hybrid integration; MEMS and IC chips are placed side-by-side in a common package and interconnected at the package level, typically via wire and/or flip-chip bonding [19]. This approach has evolved into system-in-packages, also referred to as vertical or stacked multi-chip modules, consisting of chips that are attached on top of each other and interconnected via wire and/or flip-chip bonding, either directly or through additional re-distribution layers. The main benefits of these 3D stacked approaches are their higher integration densities, shorter signal path lengths, and smaller package footprints/volumes in comparison with the 2D multi-chip modules. Moreover, package-level system-on-packaging can integrate optics, wireless communication, and power module along with MEMS and IC on a common package. For the packaging and integration explained earlier, interconnection techniques have been important as the packaging size determines the cost of the final product [20–22]. Interconnect techniques frequently used for MCM packaging are wire-bonding, solder balls, metal stud bumps, and ACF (Anisotropic Conductive Film). Most of the interconnected techniques require certain amount of pressure at elevated temperature for efficient bonding between materials in joint. In case of metallic joints, such as solder balls and metal stud bumps, a relatively thin adhesion or UBM (under bump metallurgy) layer would be critical for the package reliability, as the adhesion layer could be delaminated or disconnected due to intermetallic diffusion. In certain cases, the length of metallic interconnect determines the life-time of the package due to shear stress limit [23].

Due to the emergence of novel electronic devices such as flexible electronics or implantable medical devices, the Si IC should be integrated with flexible substrate to comply with new applications. Two-dimensional-material-based circuit approaches are attractive for flexible electronics, but advances in key areas such as robust manufacturing and reliable mechanical characterization are still necessary [24]. The integration of existing IC and novel biocompatible polymeric devices is highly demanded for new implantable medical devices, for example, neural prosthetics [25,26]. Such medical devices should be encapsulated in a biocompatible way for human body implantation. The reliability and life-time of the implant system are highly dependent on both packaging materials and technology. In general, implantable device packaging houses the electronic or mechanical system through the polymer encapsulation [27], welding, or bonding of metal [28,29] and ceramics [30]. Materials of the polymer encapsulation package include epoxies, silicones, polyurethanes, polyimides, silicon-polyimides, parylenes, polycyclic-olefins, silicon-carbons, benzocyclobutenes (BCB), and liquid crystal polymers. Furthermore, the packaging size of the implant should be minimized to avoid unwanted foreign body reaction (FBR) during implantation. The biocompatible packaging has been shown in different implantable medical devices, for example, a cardiac monitoring system implemented with commercial three-axis accelerometer, pressure sensor device mounted on a stent graft, implantable retina stimulator implemented by MFI (MicroFlex Interconnection) technology, thin-film interconnect for 1000-electrode retina prothesis, etc. [31].

In conclusion, MEMS packaging has been evolved from MEMS device packaging to MEMS system packaging as the application of MEMS devices has been widely extended.

Innovative and efficient packaging technologies becomes more and more important as well as new packaging materials. Concerning heterogeneous integration highly demanded for new applications, the interconnection between different material such as silicon and polymers should be adapted in order to reduce mechanical and electrical optimization. This Special Issue presents 12 research papers and 1 review article on recently developed MEMS packaging technologies and 3D integration. It will serve to elucidate the need for new packaging technologies and its recent research trend.

Conflicts of Interest: The author declares no conflict of interest.

References

1. Lee, B.; Seok, S.; Chun, K. A study on wafer level vacuum packaging for MEMS devices. *J. Micromech. Microeng.* **2003**, *13*, 663–669. [CrossRef]
2. Cherniak, G.; Avraham, M.; Bar-Lev, S.; Golan, G.; Nemirovsky, Y. Study of the Absorption of Electromagnetic Radiation by 3D, Vacuum-Packaged, Nano-Machined CMOS Transistors for Uncooled IR Sensing. *Micromachines* **2021**, *12*, 563. [CrossRef] [PubMed]
3. Santagata, F.; Zaal, J.J.M.; Huerta, V.G.; Mele, L.; Creemer, J.F.; Sarro, P.M. Mechanical Design and Characterization for MEMS Thin-Film Packaging. *J. Microelectromech. Syst.* **2012**, *21*, 100–109. [CrossRef]
4. Zekry, J.; Tezcan, D.S.; Celis, J.-P.; Puers, R.; van Hoof, C.; Tilmans, H.A.C. Wafer-level thin film vacuum packages for MEMS using nanoporous anodic alumina membranes. In Proceedings of the 16th International Solid-State Sensors, Actuators and Microsystems Conference (Transducers), Beijing, China, 5–9 June 2011; pp. 974–977.
5. Lee, B.-K.; Choi, D.-H.; Yoon, J.-B. Use of nanoporous columnar thin film in the wafer-level packaging of MEMS devices. *J. Micromech. Microeng.* **2010**, *20*, 045002. [CrossRef]
6. Seok, S.; Rolland, N.; Rolland, P.-A. Packaging methodology for RF devices using a BCB membrane transfer technique. *J. Micromech. Microeng.* **2006**, *16*, 2384–2388. [CrossRef]
7. Kim, J.-G.; Seok, S.; Rolland, N.; Rolland, P.-A. Polymer-based zero-level packaging technology for high frequency RF applications by wafer bonding/debonding technique using an anti-adhesion layer. *Int. J. Precis. Eng. Manuf.* **2012**, *13*, 1861–1867. [CrossRef]
8. Bower, R.W.; Ismail, M.S.; Roberds, B.E. Low temperature Si3N4 direct bonding. *Appl. Phys. Lett.* **1993**, *62*, 3485–3487. [CrossRef]
9. Bourim, E.-M.; Kang, I.-S.; Kim, H.Y. Investigation of Integrated Reactive Multilayer Systems for Bonding in Microsystem Technology. *Micromachines* **2021**, *12*, 1272. [CrossRef]
10. Lee, J.-H.; Li, P.-K.; Hung, H.-W.; Chuang, W.; Schellkes, E.; Yasuda, K.; Song, J.-M. Geometrical Effects on Ultrasonic Al Bump Direct Bonding for Microsystem Integration: Simulation and Experiments. *Micromachines* **2021**, *12*, 750. [CrossRef]
11. Yamamoto, M.; Matsumae, T.; Kurashima, Y.; Takagi, H.; Suga, T.; Takamatsu, S.; Itoh, T.; Higurashi, E. Effect of Au Film Thickness and Surface Roughness on Room-Temperature Wafer Bonding and Wafer-Scale Vacuum Sealing by Au-Au Surface Activated Bonding. *Micromachines* **2020**, *11*, 454. [CrossRef]
12. Seok, S.; Fryziel, M.; Rolland, N.; Rolland, P.-A. Enhancement of bonding strength of packaging based on BCB bonding for RF devices. *Microsyst. Technol.* **2012**, *18*, 2035–2039. [CrossRef]
13. Seok, S. Fabrication and Modeling of Nitride Thin-Film Encapsulation Based on Anti-Adhesion-Assisted Transfer Technique and Nitride/BCB Bilayer Wrinkling. In Proceedings of the 2016 IEEE 66th Electronic Components and Technology Conference (ECTC), Las Vegas, NV, USA, 31 May–3 June 2016; pp. 1301–1307.
14. Ayhan, A.O.; Nied, H.F. Finite element analysis of interface cracking in semiconductor packages. *IEEE Trans. Components Packag. Technol.* **1999**, *22*, 503–511. [CrossRef]
15. Wang, P.-H.; Huang, Y.-W.; Chiang, K.-N. Reliability Evaluation of Fan-Out Type 3D Packaging-On-Packaging. *Micromachines* **2021**, *12*, 295. [CrossRef] [PubMed]
16. Lee, Q.-Y.; Lee, M.-X.; Lee, Y.-C. A Hybrid Fuzzy Decision Model for Evaluating MEMS and IC Integration Technologies. *Micromachines* **2021**, *12*, 276. [CrossRef]
17. Jiang, B.; Huang, S.; Zhang, J.; Su, Y. Analysis of Frequency Drift of Silicon MEMS Resonator with Temperature. *Micromachines* **2021**, *12*, 26. [CrossRef]
18. Seok, S.; Rolland, N.; Rolland, P.-A. A theoretical and experimental study of the BCB thin-film cap zero-level package based on FEM simulations. *J. Micromech. Microeng.* **2010**, *20*, 095010. [CrossRef]
19. Fischer, A.C.; Forsberg, F.; Lapisa, M.; Bleiker, S.J.; Stemme, G.; Roxhed, N.; Niklaus, F. Integrating MEMS and ICs. *Microsyst. Nanoeng.* **2015**, *1*, 15005. [CrossRef]
20. Liu, Z.; Fang, M.; Shi, L.; Gu, Y.; Chen, Z.; Zhu, W. Characteristics of Cracking Failure in Microbump Joints for 3D Chip-on-Chip Interconnections under Drop Impact. *Micromachines* **2022**, *13*, 281. [CrossRef]
21. Wang, M.; Ma, S.; Jin, Y.; Wang, W.; Chen, J.; Hu, L.; He, S. A RF Redundant TSV Interconnection for High Resistance Si Interposer. *Micromachines* **2021**, *12*, 169. [CrossRef]

22. Roshanghias, A.; Dreissigacker, M.; Scherf, C.; Bretthauer, C.; Rauter, L.; Zikulnig, J.; Braun, T.; Becker, K.-F.; Rzepka, S.; Schneider-Ramelow, M. On the Feasibility of Fan-Out Wafer-Level Packaging of Capacitive Micromachined Ultrasound Transducers (CMUT) by Using Inkjet-Printed Redistribution Layers. *Micromachines* **2020**, *11*, 564. [CrossRef]

23. Wu, C.; Liu, J.; Yeung, N. The effects of bump height on the reliability of ACF in flip-chip. *Solder. Surf. Mt. Technol.* **2001**, *13*, 25–30. [CrossRef]

24. Glavin, N.R.; Muratore, C.; Snure, M. Toward 2D materials for flexible electronics: Opportunities and outlook. *Oxf. Open Mater. Sci.* **2020**, *1*, itaa002. [CrossRef]

25. Park, H.; Choi, W.; Oh, S.; Kim, Y.-J.; Seok, S.; Kim, J. A Study on Biocompatible Polymer-Based Packaging of Neural Interface for Chronic Implantation. *Micromachines* **2022**, *13*, 516. [CrossRef] [PubMed]

26. Seok, S.; Park, H.; Kim, J. Characterization and Analysis of Metal Adhesion to Parylene Polymer Substrate Using Scotch Tape Test for Peripheral Neural Probe. *Micromachines* **2020**, *11*, 605. [CrossRef] [PubMed]

27. Hassler, C.; von Metzen, R.P.; Ruther, P.; Stieglitz, T. Characterization of parylene C as an encapsulation material for implanted neural prostheses. *J. Biomed. Mater. Res. Part B* **2010**, *93*, 266–274. [CrossRef]

28. Kramar, T.; Michalec, I.; Kovacocy, P. The laser beam welding of titanium grade 2 alloy. *GRANT J.* **2012**, *1*, 77–79.

29. Schuettler, M.; Ordonez, J.S.; Santisteban, T.S.; Schatz, A.; Wilde, J.; Stieglitz, T. Fabrication and test of a hermetic miniature implant package with 360 electrical feedthroughs. In Proceedings of the 2010 Annual International Conference of the IEEE Engineering in Medicine and Biology, Buenos Aires, Argentina, 31 August–4 September 2010; Volume 2010, pp. 1585–1588.

30. Chlebowski, A.L.; Chow, E.Y.; Ellison, C.; Irazoqui, P.P. Integrated LTCC packaging for use in biomedical devices. *Bio-Med. Mater. Eng.* **2012**, *22*, 361–372. [CrossRef]

31. Seok, S. Polymer-Based Biocompatible Packaging for Implantable Devices: Packaging Method, Materials, and Reliability Simulation. *Micromachines* **2021**, *12*, 1020. [CrossRef]

micromachines

MDPI

Article

A Study on Biocompatible Polymer-Based Packaging of Neural Interface for Chronic Implantation

HyungDal Park [1,2], Wonsuk Choi [1,3], Seonghwan Oh [1,3], Yong-Jun Kim [2,*], Seonho Seok [4,*] and Jinseok Kim [1,*]

[1] Center for Bionics, Korea Institute of Science and Technology (KIST), Seoul 02792, Korea; hyungdal@kist.re.kr (H.P.); wonsuk@kist.re.kr (W.C.); emart11@kist.re.kr (S.O.)
[2] School of Mechanical Engineering, Yonsei University, Seoul 03722, Korea
[3] Department of Biomedical Engineering, Korea University, Seoul 02841, Korea
[4] Center for Nanoscience and Nanotechnology (C2N), University-Paris-Saclay, 91400 Orsay, France
[*] Correspondence: yjk@yonsei.ac.kr (Y.-J.K.); seonho.seok@u-psud.fr (S.S.); jinseok@kist.re.kr (J.K.)

Abstract: This paper proposed and verified the use of polymer-based packaging to implement the chronic implantation of neural interfaces using a combination of a commercial thermal epoxy and a thin parylene film. The packaging's characteristics and the performance of the vulnerable interface between the thermal epoxy layer and polyimide layer, which is mainly used for neural electrodes and an FPCB, were evaluated through in vitro, in vivo, and acceleration experiments. The performance of neural interfaces—composed of the combination of the thermal epoxy and thin parylene film deposition as encapsulation packaging—was evaluated by using signal acquisition experiments based on artificial stimulation signal transmissions through in vitro and in vivo experiments. It has been found that, when commercial thermal epoxy normally cured at room temperature was cured at higher temperatures of 45 °C and 65 °C, not only is its lifetime increased with about twice the room-temperature-based curing conditions but also an interfacial adhesion is higher with more than twice the room-temperature-based curing conditions. In addition, through in vivo experiments using rats, it was confirmed that bodily fluids did not flow into the interface between the thermal epoxy and FPCB for up to 18 months, and it was verified that the rats maintained healthy conditions without occurring an immune response in the body to the thin parylene film deposition on the packaging's surface.

Keywords: polymer packaging; neural interface; chronic implantation

Citation: Park, H.; Choi, W.; Oh, S.; Kim, Y.-J.; Seok, S.; Kim, J. A Study on Biocompatible Polymer-Based Packaging of Neural Interface for Chronic Implantation. *Micromachines* **2022**, *13*, 516. https://doi.org/10.3390/mi13040516

Academic Editor: Nam-Trung Nguyen

Received: 23 February 2022
Accepted: 24 March 2022
Published: 26 March 2022

Publisher's Note: MDPI stays neutral with regard to jurisdictional claims in published maps and institutional affiliations.

1. Introduction

Until recently, implantable bioelectronic devices have been developed in various forms for the purpose of monitoring the conditions of diseases through measurements of biological or neural signals and treating or suppressing specific diseases through electrical/optical/magnetic/ultrasound-based stimulations. Basically, implantable bioelectronic devices have been packaged with a material that can minimize the occurrence of an immune response without being damaged or leaking, even when surrounding muscles move after implantations. Most recently, commercialized bioelectronic medical devices have used Ti packaging that can minimize deformations from external forces, such as muscle movements in the main body, and the outside of such packaging is encapsulated with a silicone elastomer that satisfies biocompatibility. At the same time, the packaging has a small number of electrical channels which have millimeter-level diameters, so an individually sealed connector is used to connect it to the main body of implantable bioelectronic devices [1–3]. In addition, since most Ti packaging has relatively bulky dimensions compared to neural electrodes, the implantable site is limited to the subcutaneous area of the chest. For the above reason, the neural electrode and the Ti-packaged main body are connected through a wire while maintaining a certain distance or more. Such a configuration does not cause serious problems when the number of channels for neural signal monitoring or stimulation signal transmission is small. However, if the number of channels increases, there is a high possibility of problems due to the plurality of lines connected between the neural electrodes and the main body. In order to prevent failures or the twisting of each wire, a method of winding each line in a circle several times in the body is also applied [4,5]. However, this approach is difficult to utilize as an ultimate solution because it has no choice but to increase the length of the wire inserted into the body and increases the area of the location where the immune response in the body occurs. Recently, demand for a multi-channel neural interface with a high neural signal selectivity and fine local stimulation has increased remarkably. This multi-channel neural interface is being used as an advanced approach to implementing a neural signal-based neuroprosthesis for the daily life recovery of amputees and for electroceuticals, which is rising as a new treatment technology for chronic diseases [6–14]. The multi-channel neural interfaces basically have diverse designs depending on the implant sites (brain, spinal cord, peripheral nerve, and vagus nerve), and each neural electrode is commonly connected to a pre-amplifier device regardless of whether it is wire- or wireless-driven [11–14]. For the connection between the neural electrode and the pre-amplifier device, it is common to use wire bonding or an unsealed connector integrated on a PCB and to mainly use the form of encapsulating the connection part and pre-amplifier device with a biocompatible polymer [15–19]. However, the reason to use the small packaging configuration with the biocompatible polymer can be inferred from the spatial limitation of the implantation site. Moreover, in clinical trials, the system is generally removed again with a surgical procedure after a certain period of an experiment [20–22]. Not only has a finite-element analysis study been based on the characteristics of polymer materials applied to polymer-based packaging and materials used as a diffusion barrier but also a verification study of a lifetime has been reported based on a bodily fluid penetration between the polymer insulation layer and the metal electrodes [23,24]. Consequentially, research on polymer-based packaging methods and characteristic verifications are insufficient, even though the purpose of the neural interface is the capability of chronic implantation with the continuous monitoring of the disease condition and delivering stimulation signals.

As shown in Figure 1, in this study, we introduce the biocompatible polymer-based packaging approach for chronic implantation with the results of mechanical property evaluations and acceleration experiments between two kinds of polymers (packages and neural interface materials). Finally, verification results of the lifetime of actually developed polymer-based packaging are discussed with the results of in vivo experiments.

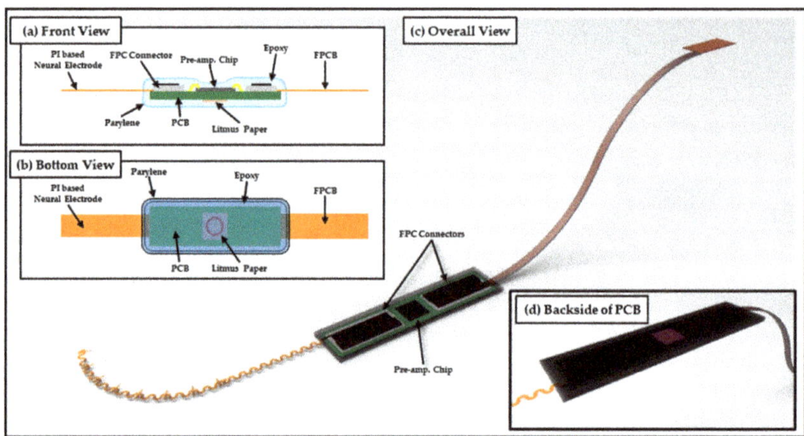

Figure 1. Scheme of polymer-based packaging for neural interface system. (**a**) Front view and (**b**) bottom view show details of packaging inside, (**c**) overall view, and (**d**) backside view.

2. Materials and Methods

2.1. Design and Fabrication of Polymer-Based Neural Interface Packaging Prototype

The neural interface proposed in this study used SPAE (Spiral Probe Array Electrode) [25], a photosensitive polyimide-based neural electrode developed through previous studies; the neural interface system is constructed by connecting PCB integrated with pre-amplifier (RHS2116 Stim/Amplifier Chip, Intan Technologies, Los Angeles, CA, USA) and FPCB (Flexible Printed Circuit Board).

As shown in Figure 1, the proposed neural interface system was encapsulated using thermal epoxy (EPO-TEK® 302, Epoxy Technology, Billerica, MA, USA) from the neural electrode interface to the FPCB interface to connect the PCB substrate and the external system. During the process, square-shaped litmus paper (KA.22-93A, DOOSAN Scientific, Seoul, Korea) responsive to bodily fluids was added to the backside of the PCB in order to enable an intuitive visual leakage check. Thereafter, a thin parylene layer was deposited using commercial parylene deposition equipment (VPC-500, Paco Engineering, Incheon, Korea) to satisfy the biocompatibility of the outer surface of the packaging.

The detailed process of polymer-based packaging of the neural interface is shown in Figure 2a. The fabricated neural electrode and FPCB are individually connected with the PCB through FPC connectors (Molex 502078-3310 and Molex 502078-3760, Molex, Lisle, IL, USA), and the two FPC connectors are electrically connected to the pre-amplifier integrated on the PCB for each channel. In addition, a peeler gauge with a thickness of 50 µm was attached to the backside of the PCB for the convenience of visually checking leakage through litmus paper without any difficulty caused by the contrasting effect of red and green complementary colors (Figure 2a(i)). After integrating each component, encapsulation was performed using thermal epoxy in a state in which a rectangular parallelepiped PDMS (Polydimethylsiloxane, Dow Corning, Midland, MI, USA) structure, which is manufactured with a mixture ratio of 10:1 and a curing condition of 65 °C/1 h and was attached to the back of a PCB (Figure 2a(ii)). The recommendation of the curing condition of thermal epoxy by the manufacturer was 2 h at room temperature. However, it is possible to improve the adhesion between the thermal epoxy and polyimide by adjusting the curing conditions. It should be noted that all thermal epoxy curing processes have been done at 65 °C for 2 h. (Related information can be found in the Results in Sections 3.2 and 3.3) PDMS has low mechanical strength and adhesion due to weak interactions between molecular chains and a lack of polar groups [26]. Various methods, such as phosphorus and boron synthesis, ultraviolet-ozone treatment, and an oxygen plasma treatment for improving adhesion, have been studied [27–29], but in this study, it is possible to easily remove the PDMS structure after curing the thermal epoxy due to the low adhesion of PDMS. Thereafter, square-shaped litmus paper with a circle pattern with an aqueous marker was attached at the corresponding position, and additional encapsulation was carried out using thermal epoxy (Figure 2a(iii,iv)). The entire thickness of the applied thermal epoxy was approximately 2 mm, and the electrode sites of the neural electrode and the metal electrode of the FPCB for electrical connection were covered with a stretchable polymer film (Parafilm, Bemis Company, Inc., Neenah, WI, USA), and a parylene layer of thickness of 5 µm was deposited using a commercial parylene deposition instrument (Figure 2a(v)). After parylene deposition, the stretchable polymer film was removed and the mechanical rigidity was reinforced on FPCB using a heat-shrink tube to complete the manufacturing of the prototype as shown in Figure 2a(vi),b.

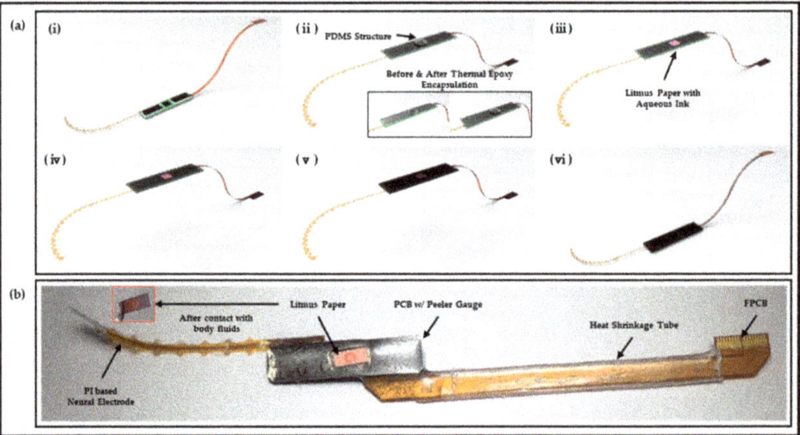

Figure 2. Packaging sample preparation. (**a**) Fabrication process of neural interface system with polymer-based packaging and (**b**) result of sample preparation.

2.2. Electrophysiological Experiment Setups

In vitro and in vivo experiments were conducted to verify whether there was a sufficiently applicable performance of the neural interface to which the polymer-based packaging was applied or not. In order to verify the signal acquisition performances of the neural interfaces to be used before the relevant experiment, the impedance at each frequency was measured using electrochemical impedance spectroscopy. The impedances of a total of two neural interfaces were measured, and each measurement result exhibited an average 29.08 kOhm and 32.63 kOhm at 1 kHz. This is a sufficient impedance value suitable for acquiring a neural signal, and the standard deviation of each average impedance was simultaneously analyzed to be 5.02 kOhm and 8.17 kOhm; thus, it was also verified that it had stability. In addition, the neural interface obtained from the measurement result of Figure 3a was used for in vitro experiment, and the neural interface with the impedance of Figure 3b was used for in vivo experiment, respectively. As shown in Figure 3c, the in vitro experiments were configured in a state in which the anode/cathode electrodes, which were connected to the electric pulse generator (HSE-HA stimulator CS for isolated cells, Hugo Sachs Elektronik Harvard Apparatus GmbH, Baden-Württemberg, Germany), and the neural interface and reference electrode, which were connected to the neural signal acquisition equipment (RHS stim/recording system, Intan Technologies, Los Angeles, CA, USA), were immersed in a PBS solution. This experiment was a process to verify the performance and capability of the neural interface that acquired the artificial signal generated through an electric pulse generator, and the stimulation parameter was constructed based on the criteria of the general neural stimulation parameter [30–32]. As shown in Figure 3d, the stimulus parameters were composed of two types (50 µA-100 µs-10 Hz and 200 µA-100 µs-100 Hz). In addition, the corresponding stimulation signal generated by the electrical pulse generator was acquired and recorded in real time from a total of 32 channel electrodes using a neural signal acquisition device.

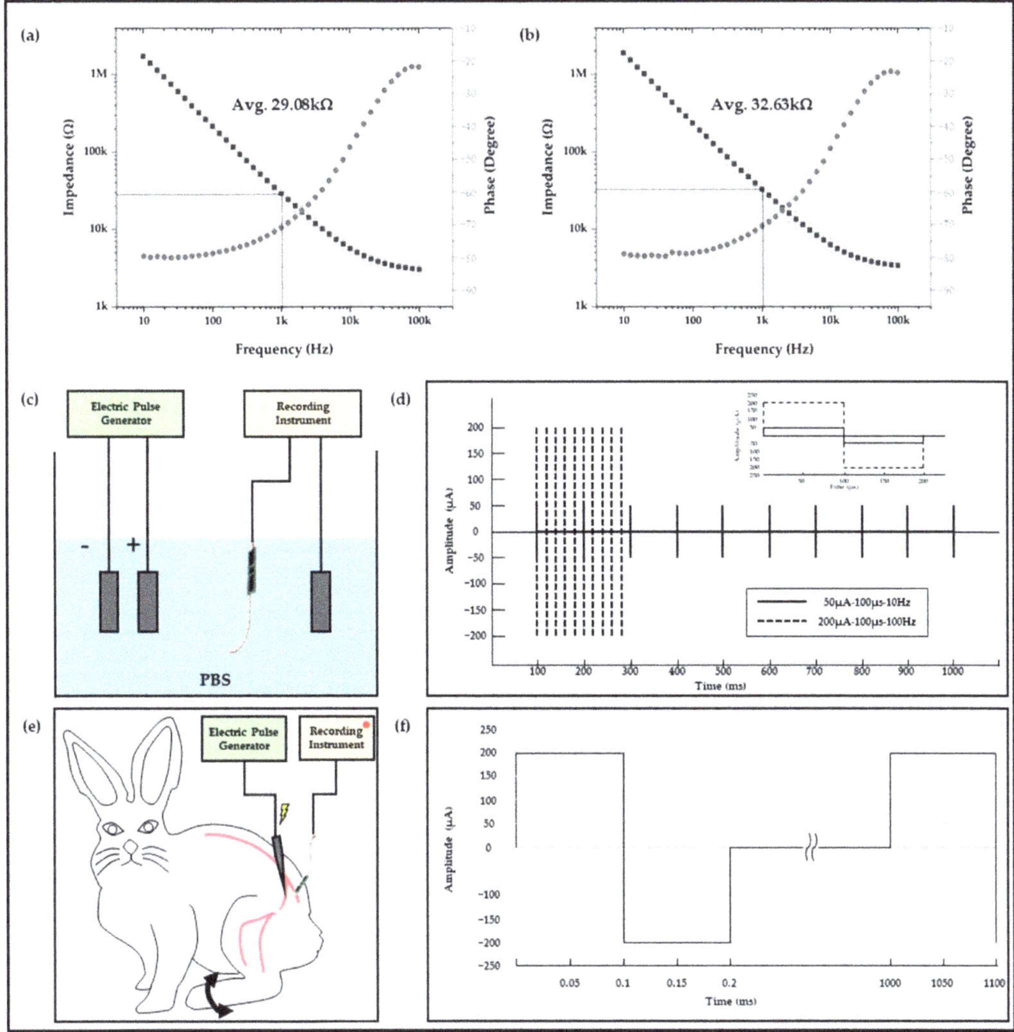

Figure 3. The results of impedance measurements of (**a**) neural electrode for in vitro experiment and (**b**) neural electrode for in vivo experiment. Electrophysiological experiment setup and stimulation parameters. (**c**) In vitro experiment setup in PBS, (**d**) a couple of stimulation parameters, (**e**) in vivo experiment setup with New Zealand white rabbit, and (**f**) in vivo stimulation parameters.

2.3. In Vivo Experiment Setups

In addition to verification through the in vitro experiment, the in vivo experiments using the New Zealand male white rabbit was configured as shown in Figure 3e. The stimulation probe was connected to the electric pulse generator to apply stimulation, and the neural electrode applied to the neural interface was implanted into the sciatic nerve of the rabbit for signal acquisition. Implantation of the neural electrode was performed using a surgical method established using rats in previous study [25], and a 12-week-old New Zealand male white rabbit was anesthetized through laparoscopic injection of a Zoletil (Virbac, Carros, France)–Rompun (BAYER, Leverkusen, Germany) with composition ratio of 3:1. After anesthetizing, the neural electrode was inserted into the sciatic nerve, which was exposed through a muscle incision. The verification experiment was conducted under breathing anesthesia, and the stimulation signal interval was set to 1000 ms so that the movement of the leg of the rabbit by the sciatic nerve stimulation could be visually classified. Figure 3e,f show the configuration and results of the in vivo experiment, for which a system capable of acquiring and recording stimulation signals in real time using neural signal acquisition equipment was constructed. All procedures for animal testing were performed through the Institutional Animal Care and Use Committee (IACUC) guidelines of the Korea Institute of Science and Technology (KIST).

2.4. Test Sample Preparation for Polyimide (FPCB)-Epoxy Peel-Off Test

The polymer-based packaging for neural interface in this study was mainly configured using thermal epoxy encapsulation. After thermal epoxy encapsulation, thin parylene film, which is widely used as a coating material to prevent immune reactions and achieve biocompatibility, was deposited. [33–35] In this configuration, thermal epoxy was applied as a major material for polymer-based packaging, and the adhesion between thermal epoxy and the material forming the neural interface is directly related to the packaging's life and performance. Because the interface is bound to form between thermal epoxy and polyimide, which is the main component of the FPCB applied for electrical connection with the neural interface and external system, this interface in packaging is the most vulnerable site. Accordingly, samples for experiments were manufactured in the order shown in Figure 4 to verify the adhesion between the thermal epoxy and polyimide depending on the curing conditions of the thermal epoxy. The experiment was constructed in a form that could apply tear force and shear force by attaching a specific area of thermal epoxy and FPCB. First, a sample with a specific area of thermal epoxy and FPCB was manufactured as shown in Figure 4a. A structure capable of containing a certain amount of thermal epoxy and a #-shaped structure for preventing the separation of the thermal epoxy from the sample structure during the adhesion measurement experiment were printed by using a commercial 3D printer (CUBICON Single Plus-320C, Cubicon, Seongnam, Korea) and were manually assembled (Figure 4a(i,ii)). Thereafter, thermal epoxy was poured into the structure and cured at room temperature for 2 h as per recommended curing condition of the manufacturer (Figure 4a(iv)), and the thermal epoxy surface was covered except for a pre-determined area with a commercial detachable tape (Scotch® Magic™ tape, 3M, Saint Paul, MN, USA; Figure 4a(v)). The exposed surface area of the thermal epoxy is established by 1 mm offset from the outer line of the FPCB to the inside. This offset was set to prevent the applied thermal epoxy from being attached to the outer line or edge of the FPCB and acting as an obstacle to adhesion measurement of the interface. To specify a volume of the second thermal epoxy through the thickness (about 60 μm) of commercial detachable tape, the identical thermal epoxy was spread to the exposed surface and the squeezing method was carried out in stencil printing using a peeler gauge with a thickness of 100 μm (Figure 4a(v)) [36]. In the final step shown in Figure 4a(vi), the second thermal epoxy curing was performed with an FPCB placed thereon and a weight of 3 kg placed on top for conformal attachment. Though different products from the thermal epoxy were used in this research, the curing condition was established based on research results that when the thermal epoxy, for which studies recommend a room-temperature curing condition, is cured at a higher temperature, its curing rigidity can be increased [37]. The thermal epoxy of the sample applied in the experiment was cured at 45 °C/2h and 65 °C/2 h, respectively, which were higher than room temperature.

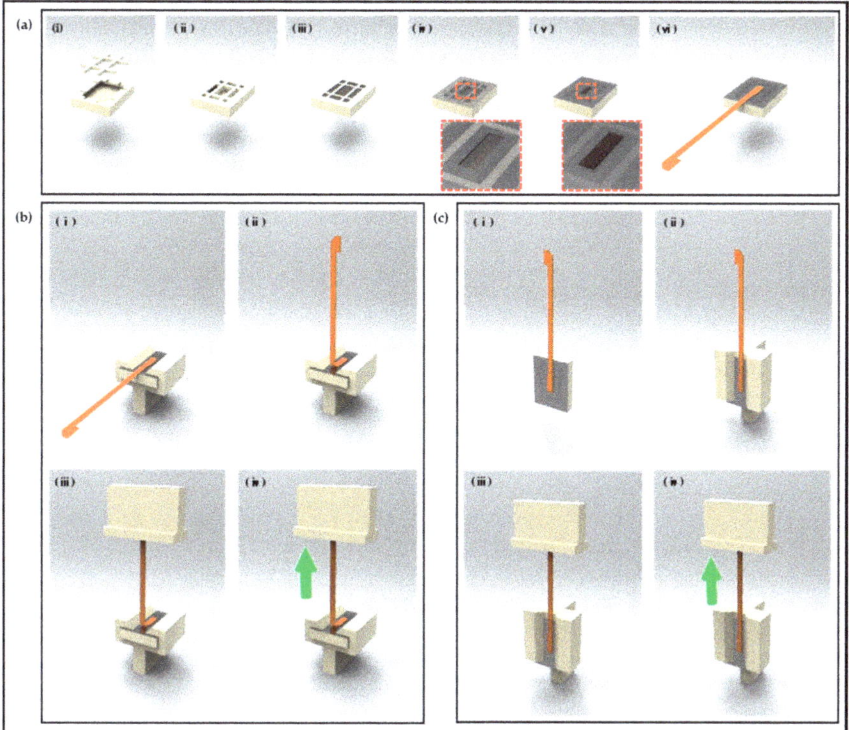

Figure 4. Procedures of adhesion evaluation test sample preparation. (**a**) Process of sample preparation with uniform adhered area between epoxy and FPBC; (**a-i–a-ii**) preparation and assembly of 3D-printed sample holder; (**a-iii**) filling and curing epoxy into vacancy of sample holder; (**a-iv**) covering epoxy surface with detachable tape excluding uniform area; (**a-v**) additional epoxy filling and squeezing epoxy into uncovered area; and (**a-vi**) applying FPCB to uncured epoxy and curing. (**b**) Tear-test procedures. (**c**) Shear-test procedures.

2.5. Peel-Off Test Setup for Polyimide (FPCB) Epoxy

Additional structures capable of applying tear force and shear force for the measurement of adhesion between thermal epoxy and polyimide were configured as shown in Figure 4b,c. A structure for fixing the FPCB thermal epoxy-attached sample to the chuck of the tensile experimental equipment and a structure that can be grasped the end of the FPCB by another chuck of the tensile experimental equipment were output using a 3D printer. The structure for fixing the FPCB thermal epoxy-attached sample was configured to have a difference of 90 degree angle to enable tear force and shear force applications, respectively. In the experiment for evaluating the adhesion between thermal epoxy and polyimide (FPCB) using the prepared sample, commercial tensile experimental equipment (Shimadzu EZ-S machine, Shimadzu, Kyoto, Japan) was used and tensile stroke was applied to one end of the FPCB under conditions of 1 mm/min; the generated force was measured in mN units.

2.6. Acceleration Test Setup

The reliabilities of acceleration experiments using temperature applications higher than the used temperature are generally derived using the Arrhenius equation as shown below in Equation (1).

$$A_T = \frac{\lambda_{T1}}{\lambda_{T2}} = \exp\left[\left(\frac{-E_a}{k}\right)\left(\frac{1}{T_1} - \frac{1}{T_2}\right)\right] \tag{1}$$

where E_a is the activation energy (eV), k is Boltzmann's constant as 8.62×10^{-5} eV/K, T_1 is the absolute temperature of application condition, T_2 is the absolute temperature of experiment condition, λ_{T1} is the observed failure rate at application condition, and λ_{T2} is the observed failure rate at experiment condition. Although the life of polymer-based packaging according to the acceleration experiment can be predicted based on the Arrhenius equation, an acceleration experiment was constructed on the basis that about 100 days of lifetime verification with 300 h of aging can be performed on an environment 30 °C higher than the usage temperature based on Donaldson's report [38]. Three samples (a total of 24 samples, 12 45 °C cured samples and 12 65 °C cured samples) were each taken out at 50, 100, 200, and 300 h of aging while immersed in a 65 °C PBS solution, and then tear force was applied to evaluate the changes in interfacial adhesion between the thermal epoxy and FPCB.

2.7. In Vivo Chronic Implantation

To verify the practical applicability of fabricated neural interfaces to which polymer-based packaging was applied, in vivo experiments were constructed. Neural electrodes were inserted into rats' sciatic nerves, and sites, including PCB and FPCB, were placed in the rats' hind leg muscles and subcutaneously. Insertion surgeries were performed on a total of two individual rats, and the electrodes were extracted at the times of 8 months and 18 months after implantation to check for changes in the conditions of the overall polymer-based packaging, especially applied litmus papers. At the time of extractions, it was difficult to separate the neural electrodes from the sciatic nerves, so the neural electrodes were cut and extracted.

3. Results

3.1. In Vitro and In Vivo Characterizations of Polymer-Based Neural Interface Packaging Prototype

In the PBS solution-based in vitro experiment, two artificial stimulation signals (50 μA-100 μs-10 Hz and 200 μA-100 μs-100 Hz) were applied, and the results acquired using 32 electrode channels formed in the neural electrode and commercial neural signal acquisition equipment are shown in Figure 5a. For each artificial stimulation signal, it can be evaluated that signals similar to stimulation signals were acquired from all 32 channels. Similarly, for in the in vivo experiment using a New Zealand male white rabbit, it was confirmed that stimulation signals of 200 μA-100 μs-1 Hz could be obtained from all 32 channels (Figure 5a). In addition, the relevant procedures and results of the in vivo experiment can be confirmed through Video S1, and the contraction of the leg muscles of the rabbit with a certain cycle by a stimulation signal can be clearly verified. Through in vitro and in vivo experiments, it was confirmed that the neural interface to which the polymer-based packaging was applied operated as an initial purpose. In addition, a quantitative analysis of the signals acquired through the experiments was carried out, and the results are summarized in Table 1. The maximum, minimum, average, and standard deviation values were derived from the positive and negative regions of the acquisition signals for a total of three stimulation conditions. Also, the average and standard deviation values of the entire signal amplitude were also calculated. The difference between the maximum and minimum values in each region was up to 342.0 μV under 200 μA-100 μs-10 Hz stimulation conditions, but the standard deviation of the region was 89.7 μV. Even under other stimulation conditions, the standard deviation in the positive and negative regions was from a minimum of 26.5 μV to a maximum of 90.8 μV. In a process of acquiring a signal using a neural electrode, the deviation may sufficiently occur depending on system parameters, such as electrode impedance, cross-talk between electrodes, and unknown recording conditions, and the result of impedance measurement research is reported to be different over time in the same system configuration [39]. In comparison, although the maximum/minimum value of the total acquisition signal differs, it cannot be determined that the non-uniform signal was obtained on a single electrode itself because the average and standard deviation of the total signal size specified in Table 1 are not abnormally large.

Accordingly, it is determined that every acquired signal amplitude is a result of different distances from the stimulation signal-generation position, the difference in impedance for each electrode, and the cross-talk between electrodes. As a result, it can be evaluated that a sufficient neural signal acquisition performance and post-processing, such as clustering for acquired neural signal analyses, have sufficiently possible signal levels.

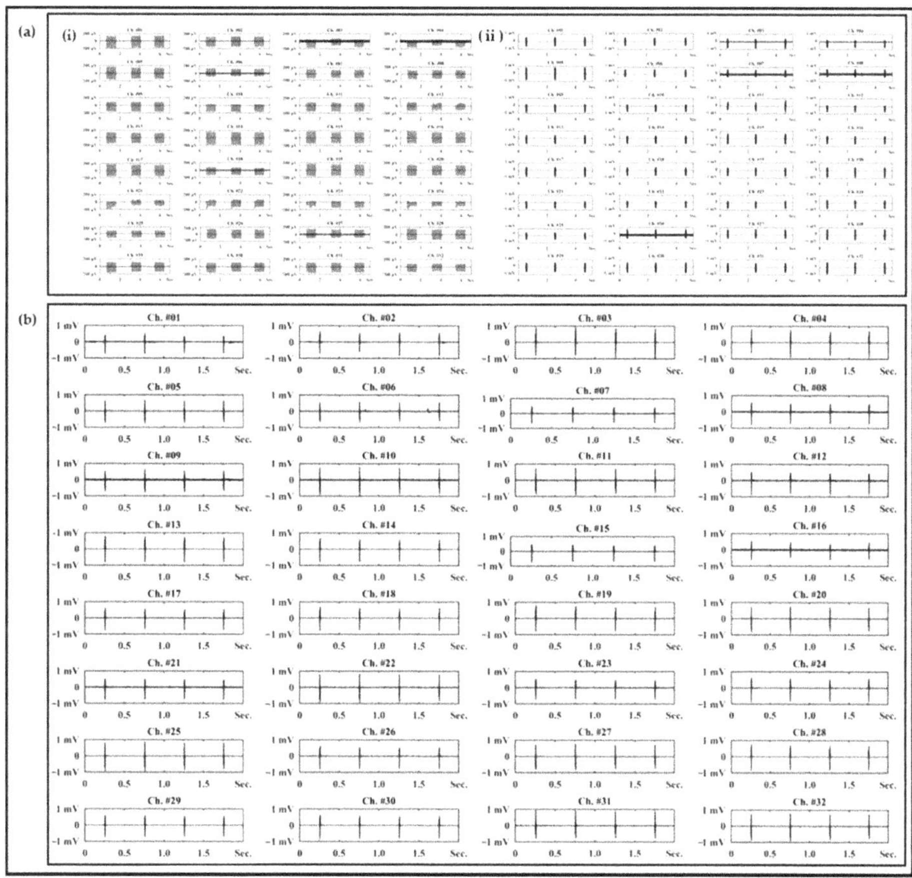

Figure 5. Results of in vitro and acute in vivo experiments. (**a**) Recording results in PBS in vitro experiment with different biphasic stimulation parameters (50 μA-100 μs-10 Hz and 200 μA-100 μs-100 Hz) and (**b**) in vivo recording results in sciatic nerve of New Zealand white rabbit with biphasic stimulation parameter (200 μA-100 μs-1 Hz).

Table 1. Comparison of the characteristics of signals acquired through in vitro and in vivo experiments.

Samples	Stimulation Parameter	Signal Amplitude (μV)									
		Positive Max.	Positive Min.	Negative Max.	Negative Min.	Positive Avg.	Positive Std. Dev.	Negative Avg.	Negative Std. Dev.	Total Amplitude Avg.	Total Amplitude Std. Dev.
In vitro sample	50 μA-100 μs-10 Hz	183.3	15.3	−63.9	−189.8	128.3	50.8	−165.5	26.5	293.8	54.0
	200 μA-100 μs-10 Hz	896.5	554.5	−592.0	−897.6	641.7	89.7	−816.2	90.8	1457.6	123.0
In vivo sample	200 μA-100 μs-1 Hz	884.8	650.8	−811.8	−899.6	797.9	80.8	−848.2	33.9	1646.0	107.7

3.2. Adhesion Evaluation of Peel-Off Test

The results of the tear force and shear force experiments conducted for the evaluation of the interface adhesion between the thermal epoxy and FPCB are as shown in Figure 6. When shear force is applied to a sample manufactured under a curing condition of 65 °C/2 h, the FPCB is not separated from the thermal epoxy; rather, it fractures at the middle point of the FPCB. As shown in the graph of Figure 6a, the measurement results of the shear forces are similar or higher to the tensile strength of the FPCB itself. It means that the sheer force experiment is not suitable for the evaluation of the interfacial adhesion between the thermal epoxy and the FPCB (polyimide) because the FPCB can be deformed or fractured before detaching it. Conversely, in the tear force experiment, it was verified that the FPCB was separated from the thermal epoxy based on 3.486 N and 8.358 N, respectively, in the samples prepared under curing conditions of 45 °C/2 h and 65 °C/2 h. As a result, in the situation with the same physically adhered interface conditions, it was evaluated that the interfacial adhesion differed more than twice in accordance with the curing conditions of the thermosetting epoxy.

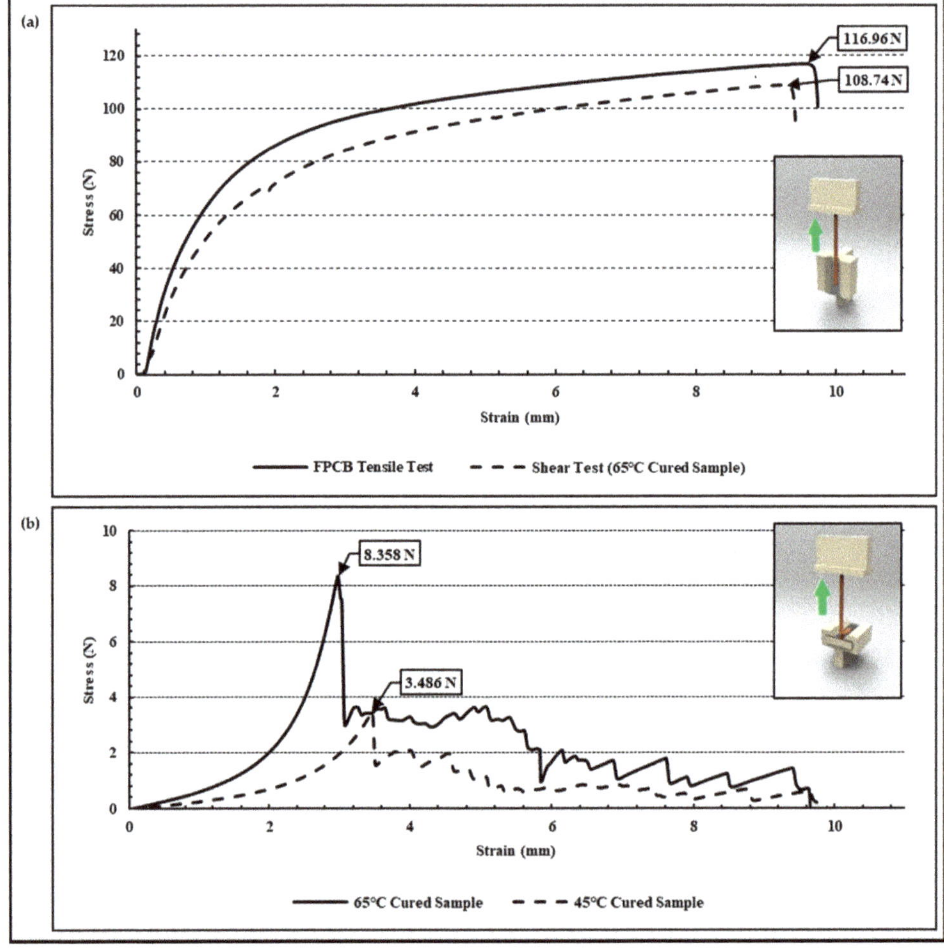

Figure 6. Results of Adhesion test. (**a**) Shear-force induced samples and (**b**) tear-force induced samples.

3.3. Lifetime and Characteristics of Acceleration Test

In an acceleration experiment conducted to confirm the changes in interfacial adhesion according to the curing conditions, a tear force application experiment was conducted on the samples, which were exposed to a 65 °C PBS solution for 50, 100, 200, and 300 h aging times. The interfacial adhesion characteristics of the samples, which were prepared under curing conditions of 65 °C/2 h, are expressed for each aging time period in Figure 7a. It can be confirmed that the interfacial adhesion increases rather than the results of the initial tear force experiment until 50 h, 100 h, and 200 h elapse. This means that although the room-temperature-curable thermal epoxy used in the experiment was cured under a higher temperature at 65 °C/2 h curing conditions, a full curing was not achieved. In this regard, it can be determined through the results of Lapique's research [37] that show a difference in the degree of full curing depending not only on the curing temperature but also on the curing time for the room-temperature-curable thermal epoxy. Although it can be seen that the interfacial adhesion decreased to about 32% when 300 h had elapsed compared to when 200 h had elapsed, based on the situation where the 300 h elapsed point was maintained at a higher level than the initial interface adhesion, it can be predicted that the interface between the thermal epoxy and FPCB maintains its performance. The acceleration experiment results for the samples manufactured under 45 °C/2 h and 65 °C/2 h curing conditions were summarized in Figure 7b, and the samples with 4 5 °C/2 h curing conditions showed a tendency to increase interfacial adhesion until 50 h but decreased from 100 h and were lower than the initial value at 200 h. Assuming that the body's temperature was 35 °C, it can be predicted that the lifetime of the sample manufactured under the 45 °C/2 h curing condition does not exceed 2 months and that the sample with the 65 °C/2 h curing condition has a lifespan of 100 days or more.

3.4. In Vivo Chronic Implantation

As shown in Figure 8, the polymer-based packaged neural interfaces, which were inserted into the sciatic nerves of, fixed to the muscles of, and fixed subcutaneously to rats, were extracted at the times of 8 months (Figure 8b) and 18 months (Figure 8c) after their insertions to confirm the discoloration and circle-shaped aqueous ink smudging of the litmus papers, respectively. Since the discoloration of the litmus papers and the diffusion of the aqueous ink were not confirmed, it was verified that bodily fluids were not flowing into the interface between the thermal epoxy and the FPCB interface formed in the polymer-based packaging; additionally, the polymer-based packaging was not deformed or damaged even from the hind limb movements of the rats. It was additionally confirmed that the rats were in healthy conditions, which could be estimated based on their behavior, the average amounts of daily intake 18 months after the implantations, and the fact that inflammation caused by the immune response had not formed around the implanted neural interfaces due to the thin parylene film coating for biocompatibility.

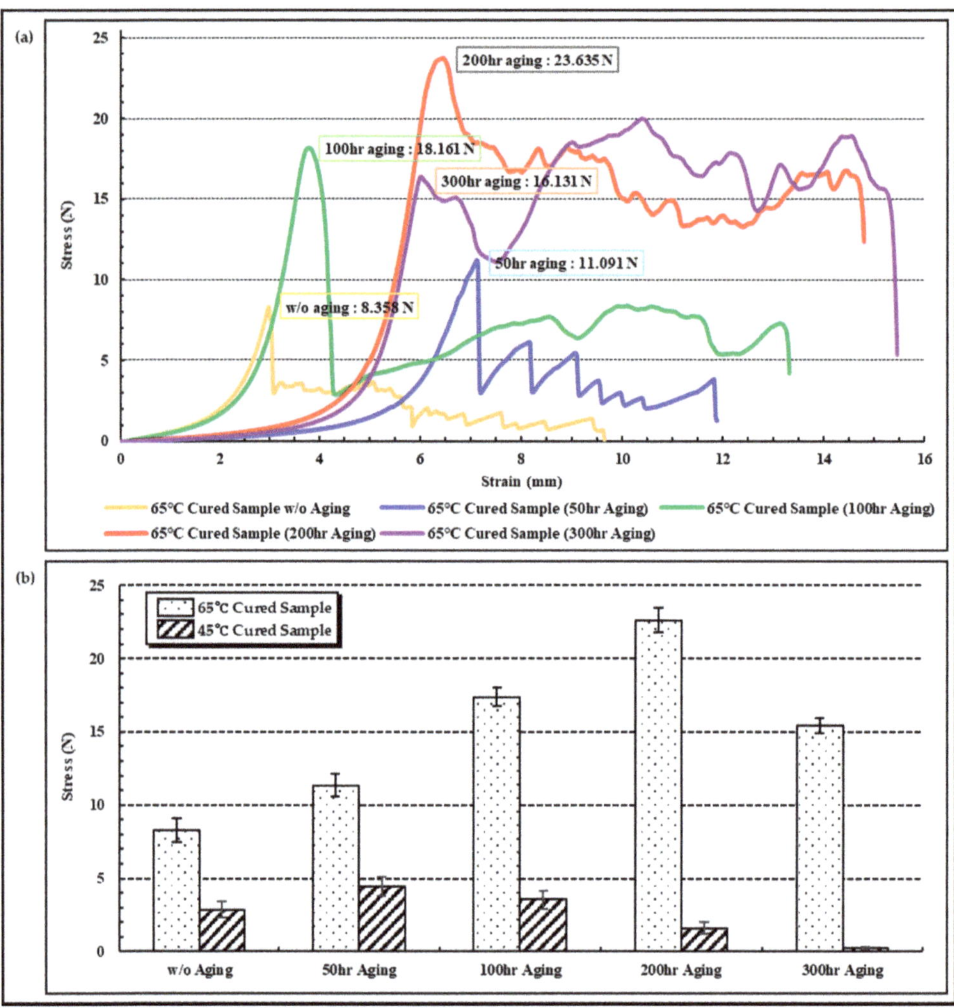

Figure 7. Results of acceleration aging test in 65 °C PBS solution at 50, 100, 200, 300 h. (**a**) Results of tear test of 65 °C cured sample after each aging time and (**b**) maximum tear force comparison between 45 °C and 65 °C cured samples.

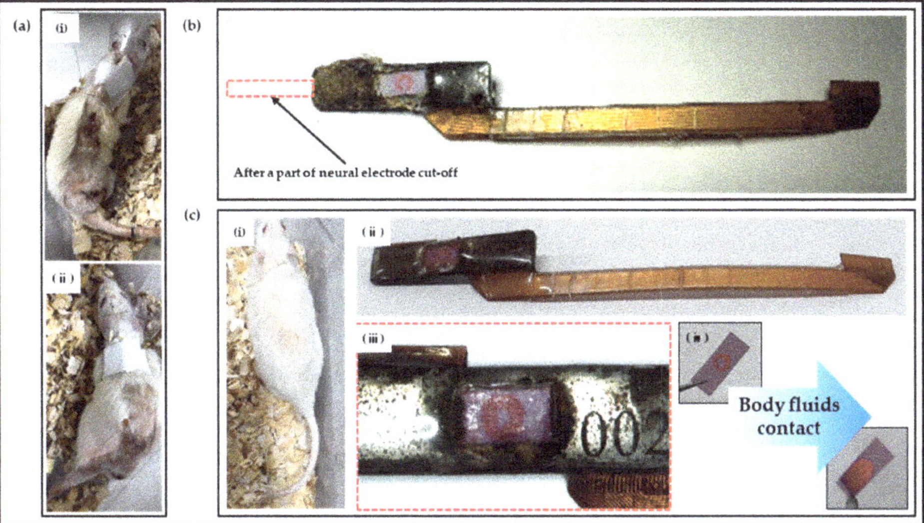

Figure 8. Results of in vivo experiments. (**a**) Two rats after packaged neural electrode implantations, (**b**) photograph of neural interface extracted after 8 months of implantation, (**c-i**) a rat of 18 months after implantation (its hair grew back in the surgical area and maintained a healthy state), (**c-ii**) photograph of neural interface extracted after 18 months of implantation, (**c-iii**) no discoloration of litmus paper and diffusion of aqueous ink, and (**c-iv**) an example of discoloration of litmus paper and diffusion of aqueous ink with bodily fluid contact.

4. Conclusions

To achieve the ultimate purpose of the neural interface, a polymer-based packaging approach to chronic implantations that was capable of biological monitoring and applications of various stimulation signals was proposed. The possibility of chronic implantation was deduced through mechanical property evaluations, acceleration experiments, and in vitro and in vivo experiments. In addition, it was confirmed that when applying commercial thermal epoxy to polymer-based packaging, it is necessary to establish a full curing condition for each, not a curing condition proposed by a product manufacturer. Due to the realistic limitation that the existing neural interface does not operate normally over 18 months, it was necessary to verify the packaging performance by examining for a discoloration of litmus paper and a spreading of aqueous ink. However, it was found that the life of the proposed polymer-based packaging was more than 18 months as bodily fluids did not penetrate the interface between the thermal epoxy and the polyimide through as confirmed by examining changes in litmus paper. The packaging technology verified through this study can be used as a novel polymer-based packaging method for the integration of neural electrodes and pre-amplifiers for implementing neural interfaces with various purposes, such as neural signal monitoring or neural stimulation. The packaging method is consequently expected to improve the possibility of chronic implantations of neural interfaces. Furthermore, analyses of interfacial properties and the optimization of application conditions between all the materials used in the packaging including the proposed packaging combination have to accompany for reliable polymer-based packaging accomplishment.

Supplementary Materials: The following supporting information can be downloaded at: https://www.mdpi.com/article/10.3390/mi13040516/s1. Video S1: Stimulation & Recording Acute in-vivo Experiment.

Author Contributions: Conceptualization, S.S. and J.K.; methodology, H.P. and Y.-J.K.; validation, H.P. and S.O.; visualization, H.P.; investigation, H.P. and W.C.; resources, H.P.; data curation, H.P.; writing—original draft preparation, H.P.; writing—review and editing, S.S., Y.-J.K., and J.K.; visualization, H.P.; supervision, S.S., Y.-J.K., and J.K.; project administration, J.K.; funding acquisition, J.K. All authors have read and agreed to the published version of the manuscript.

Funding: This research was supported by the Korea Medical Device Development Fund grant funded by the Korean government (the Ministry of Science and ICT, the Ministry of Trade, Industry and Energy, the Ministry of Health and Welfare, and the Ministry of Food and Drug Safety, project number 1711138281, KMDF_PR_20200901_0145-2021) and the Korea Institute of Science and Technology institutional program (2E31642).

Institutional Review Board Statement: The animal study protocol was approved by the Institutional Review Board of Korean Institute of Science and Technology (KIST) (protocol code KIST-2020-210 approved on 30 June 2020).

Conflicts of Interest: The authors declare no conflict of interest.

References

1. Patschger, A.; Hopf, A.; Güpner, M.; Bliedtner, J. Laser Material Processing of Medical Titanium: Titanium boxes for hermetic transport of medical implants. *Laser Tech. J.* **2016**, *13*, 24–27. [CrossRef]
2. Yang, H.; Wu, T.; Zhao, S.; Xiong, S.; Peng, B.; Humayun, M.S. Chronically implantable package based on alumina ceramics and titanium with high-density feedthroughs for medical implants. In Proceedings of the 2018 40th Annual International Conference of the IEEE Engineering in Medicine and Biology Society (EMBC), Honolulu, HI, USA, 17–21 July 2018; pp. 3382–3385.
3. Jiang, G.; Zhou, D.D. Technology advances and challenges in hermetic packaging for implantable medical devices. In *Implantable Neural Prostheses 2*; Springer: New York, NY, USA, 2009; pp. 27–61.
4. Dewald, H.A.; Lukyanenko, P.; Lambrecht, J.M.; Anderson, J.R.; Tyler, D.J.; Kirsch, R.F.; Williams, M.R. Stable, three degree-of-freedom myoelectric prosthetic control via chronic bipolar intramuscular electrodes: A case study. *J. Neuroeng. Rehabil.* **2019**, *16*, 147. [CrossRef] [PubMed]
5. Pope, J.E.; Carlson, J.D.; Rosenberg, W.S.; Slavin, K.V.; Deer, T.R. Peripheral nerve stimulation for pain in extremities: An update. *Stimul. Peripher. Nerv. Syst.* **2016**, *29*, 139–157.
6. Tan, D.W.; Schiefer, M.A.; Keith, M.W.; Anderson, J.R.; Tyler, J.; Tyler, D.J. A neural interface provides long-term stable natural touch perception. *Sci. Transl. Med.* **2014**, *6*, ra138–ra257. [CrossRef]
7. Oddo, C.M.; Raspopovic, S.; Artoni, F.; Mazzoni, A.; Spigler, G.; Petrini, F.; Micera, S. Intraneural stimulation elicits discrimination of textural features by artificial fingertip in intact and amputee humans. *eLife* **2016**, *5*, e09148. [CrossRef]
8. Koopman, F.A.; Schuurman, P.R.; Vervoordeldonk, M.J.; Tak, P.P. Vagus nerve stimulation: A new bioelectronics approach to treat rheumatoid arthritis? *Best Pract. Res. Clin. Rheumatol.* **2014**, *28*, 625–635. [CrossRef]
9. Haidar, A.; Legault, L.; Messier, V.; Mitre, T.M.; Leroux, C.; Rabasa-Lhoret, R. Comparison of dual-hormone artificial pancreas, single-hormone artificial pancreas, and conventional insulin pump therapy for glycaemic control in patients with type 1 diabetes: An open-label randomised controlled crossover trial. *Lancet Diabetes Endocrinol.* **2015**, *3*, 17–26. [CrossRef]
10. Güemes, A.; Georgiou, P. Review of the role of the nervous system in glucose homoeostasis and future perspectives towards the management of diabetes. *Bioelectron. Med.* **2018**, *4*, 9. [CrossRef]
11. Sung, C.; Jeon, W.; Nam, K.S.; Kim, Y.; Butt, H.; Park, S. Multimaterial and multifunctional neural interfaces: From surface-type and implantable electrodes to fiber-based devices. *J. Mater. Chem. B* **2020**, *8*, 6624–6666. [CrossRef]
12. Coker, R.A.; Zellmer, E.R.; Moran, D.W. Micro-channel sieve electrode for concurrent bidirectional peripheral nerve interface. Part B: Stimulation. *J. Neural Eng.* **2019**, *16*, 026002. [CrossRef]
13. Christensen, M.B.; Pearce, S.M.; Ledbetter, N.M.; Warren, D.J.; Clark, G.A.; Tresco, P.A. The foreign body response to the Utah Slant Electrode Array in the cat sciatic nerve. *Acta Biomater.* **2014**, *10*, 4650–4660. [CrossRef] [PubMed]
14. Kundu, A.; Harreby, K.R.; Yoshida, K.; Boretius, T.; Stieglitz, T.; Jensen, W. Stimulation selectivity of the "thin-film longitudinal intrafascicular electrode"(tfLIFE) and the "transverse intrafascicular multi-channel electrode"(TIME) in the large nerve animal model. *IEEE Trans. Neural Syst. Rehabil. Eng.* **2013**, *22*, 400–410. [CrossRef] [PubMed]
15. Sharma, A.; Rieth, L.; Tathireddy, P.; Harrison, R.; Oppermann, H.; Klein, M.; Solzbacher, F. Long term in vitro functional stability and recording longevity of fully integrated wireless neural interfaces based on the Utah Slant Electrode Array. *J. Neural Eng.* **2011**, *8*, 045004. [CrossRef] [PubMed]
16. Shon, A.; Chu, J.U.; Jung, J.; Kim, H.; Youn, I. An implantable wireless neural interface system for simultaneous recording and stimulation of peripheral nerve with a single cuff electrode. *Sensors* **2018**, *18*, 1. [CrossRef] [PubMed]

17. Kang, Y.N.; Chou, N.; Jang, J.W.; Byun, D.; Kang, H.; Moon, D.J.; Kim, S. An intrafascicular neural interface with enhanced interconnection for recording of peripheral nerve signals. *IEEE Trans. Neural Syst. Rehabil. Eng.* **2019**, *27*, 1312–1319. [CrossRef] [PubMed]
18. Pena, A.E.; Kuntaegowdanahalli, S.S.; Abbas, J.J.; Patrick, J.; Horch, K.W.; Jung, R. Mechanical fatigue resistance of an implantable branched lead system for a distributed set of longitudinal intrafascicular electrodes. *J. Neural Eng.* **2017**, *14*, 066014. [CrossRef]
19. Skok, T.; Tabakow, P.; Chmielak, K. Methods of integrating the human nervous system with electronic circuits. *Adv. Clin. Exp. Med.* **2019**, *28*, 1125–1135. [CrossRef]
20. Rijnbeek, E.H.; Eleveld, N.; Olthuis, W. Update on peripheral nerve electrodes for closed-loop neuroprosthetics. *Front. Neurosci.* **2018**, *12*, 350. [CrossRef]
21. George, J.A.; Page, D.M.; Davis, T.S.; Duncan, C.C.; Hutchinson, D.T.; Rieth, L.W.; Clark, G.A. Long-term performance of Utah slanted electrode arrays and intramuscular electromyographic leads implanted chronically in human arm nerves and muscles. *J. Neural Eng.* **2020**, *17*, 056042. [CrossRef]
22. Kaijankoski, H.; Nissen, M.; Ikäheimo, T.M.; von Und Zu Fraunberg, M.; Airaksinen, O.; Huttunen, J. Effect of spinal cord stimulation on early disability pension in 198 failed back surgery syndrome patients: Case-control study. *Neurosurgery* **2019**, *84*, 1225–1232. [CrossRef]
23. Seok, S. Polymer-Based Biocompatible Packaging for Implantable Devices: Packaging Method, Materials, and Reliability Simulation. *Micromachines* **2021**, *12*, 1020. [CrossRef] [PubMed]
24. Jeong, J.; Bae, S.H.; Seo, J.M.; Chung, H.; Kim, S.J. Long-term evaluation of a liquid crystal polymer (LCP)-based retinal prosthesis. *J. Neural Eng.* **2016**, *13*, 025004. [CrossRef]
25. Kim, O.; Choi, W.; Jung, W.; Jung, S.; Park, H.; Jeong, J.; Chu, J.U.; Park, J.W.; Kim, J. Spirally Arrayed Electrode for Spatially Selective and Minimally Displacive Peripheral Nerve Interface. *J. Microelectromechanical Syst.* **2020**, *29*, 514–521. [CrossRef]
26. Martin, S.; Bhushan, B. Transparent, wear-resistant, superhydrophobic and superoleophobic poly (dimethylsiloxane)(PDMS) surfaces. *J. Colloid Interface Sci.* **2017**, *488*, 118–126. [CrossRef] [PubMed]
27. Wang, Y.; Ling, Y.; Zhou, S.; Chen, Y.; Liang, M.; Zou, H. Enhanced mechanical and adhesive properties of PDMS based on novel PDMS-epoxy IPN structure. *J. Polym. Res.* **2021**, *28*, 171. [CrossRef]
28. Wang, Q.; Xiong, L.; Liang, H.; Chen, L.; Huang, S. Synthesis of a novel polysiloxane containing phosphorus, and boron and its effect on flame retardancy, mechanical, and thermal properties of epoxy resin. *Polym. Compos.* **2018**, *39*, 807–814. [CrossRef]
29. Amerian, M.; Amerian, M.; Sameti, M.; Seyedjafari, E. Improvement of PDMS surface biocompatibility is limited by the duration of oxygen plasma treatment. *J. Biomed. Mater. Res. Part A* **2019**, *107*, 2806–2813. [CrossRef]
30. Merrill, D.R.; Bikson, M.; Jefferys, J.G. Electrical stimulation of excitable tissue: Design of efficacious and safe protocols. *J. Neurosci. Methods* **2005**, *141*, 171–198. [CrossRef]
31. Newbold, C.; Richardson, R.; Millard, R.; Seligman, P.; Cowan, R.; Shepherd, R. Electrical stimulation causes rapid changes in electrode impedance of cell-covered electrodes. *J. Neural Eng.* **2011**, *8*, 036029. [CrossRef]
32. Günter, C.; Delbeke, J.; Ortiz-Catalan, M. Safety of long-term electrical peripheral nerve stimulation: Review of the state of the art. *J. Neuroeng. Rehabil.* **2019**, *16*, 13. [CrossRef]
33. Li, J.; Kang, L.; Yu, Y.; Long, Y.; Jeffery, J.J.; Cai, W.; Wang, X. Study of long-term biocompatibility and bio-safety of implantable nanogenerators. *Nano Energy* **2018**, *51*, 728–735. [CrossRef] [PubMed]
34. Kuo, W.C.; Wu, T.C.; Wu, C.F.; Wang, W.C. Bioperformance analysis of parylene C coating for implanted nickel titanium alloy. *Mater. Today Commun.* **2021**, *27*, 102306. [CrossRef]
35. Iacovacci, V.; Naselli, I.; Salgarella, A.R.; Clemente, F.; Ricotti, L.; Cipriani, C. Stability and in vivo safety of gold, titanium nitride and parylene C coatings on NdFeB magnets implanted in muscles towards a new generation of myokinetic prosthetic limbs. *RSC Adv.* **2021**, *11*, 6766–6775. [CrossRef]
36. Rusdi, M.S.; Abdullah, M.Z.; Aziz, M.S.A.; Abdullah, M.K.; Chellvarajoo, S.; Husin, A.; Rethinasamy, P.; Veerasamy, S. Multiphase flow in solder paste stencil printing process using CFD approach. *J. Adv. Res. Fluid Mech. Therm. Sci.* **2018**, *46*, 147–152.
37. Lapique, F.; Redford, K. Curing effects on viscosity and mechanical properties of a commercial epoxy resin adhesive. *Int. J. Adhes. Adhes.* **2002**, *22*, 337–346. [CrossRef]
38. Donaldson, P.E.K.; Sayer, E. A technology for implantable hermetic packages. Part 2: An implementation. *Med. Biol. Eng. Comput.* **1981**, *19*, 403–405. [CrossRef]
39. Perge, J.A.; Homer, M.L.; Malik, W.Q.; Cash, S.; Eskandar, E.; Friehs, G.; Donoghue, J.; Hochberg, L.R. Intra-day signal instabilities affect decoding performance in an intracortical neural interface system. *J. Neural Eng.* **2013**, *10*, 036004. [CrossRef]

micromachines

MDPI

Article

Characteristics of Cracking Failure in Microbump Joints for 3D Chip-on-Chip Interconnections under Drop Impact

Zhen Liu [1,2], Mingang Fang [1], Lei Shi [3], Yu Gu [4], Zhuo Chen [1,*] and Whenhui Zhu [1,*]

[1] The School of Mechanical and Electrical Engineering, Central South University, Changsha 410083, China; zhenliu@csu.edu.cn (Z.L.); mingangfang@csu.edu.cn (M.F.)
[2] Hunan Vocational College of Science and Technology, Changsha 410004, China
[3] China Electronic Technology Corporation (CETC) 48 Research Institute, Changsha 410083, China; iheartlan23@163.com
[4] Qingdao Electronics School, Qingdao 266014, China; guyu_qddz@aliyun.com
[*] Correspondence: zhuochen@csu.edu.cn (Z.C.); zhuwenhui@csu.edu.cn (W.Z.)

Abstract: With the rapid development of microelectronics packaging and integration, the failure risk of micro-solder joints in packaging structure caused by impact load has been increasingly concerning. However, the failure mechanism and reliability performance of a Cu-pillar-based microbump joint can use little of the existing research on board-level solder joints as reference, due to the downscaling and joint structure evolution. In this study, to investigate the cracking behavior of microbump joints targeted at chip-on-chip (CoC) stacked interconnections, the CoC test samples were subjected to repeated drop tests to reveal the crack morphology. It was found that the crack causing the microbump failure first initiated at the interface between the intermetallic compound (IMC) layer and the solder, propagated along the interface for a certain length, and then deflected into the solder matrix. To further explore the crack propagation mechanism, stress intensity factor (SIF) of the crack tip at the interface between IMC and solder was calculated by contour integral method, and the effects of solder thickness and crack length were also quantitatively analyzed and combined with the crack deflection criterion. By combining the SIF with the fracture toughness of the solder–Ni interface and the solder matrix, a criterion for crack deflecting from the original propagating path was established, which can be used for prediction of critical crack length and deflection angle for the initiation of crack deflection. Finally, the relationship between solder thickness and critical deflection length and deflection angle of main crack was verified by a board level drop test, and the influence of grain structure in solder matrix on actual failure lifetime was briefly discussed.

Keywords: crack propagation; microbump; deflection angle; stress intensity factor (SIF)

Citation: Liu, Z.; Fang, M.; Shi, L.; Gu, Y.; Chen, Z.; Zhu, W. Characteristics of Cracking Failure in Microbump Joints for 3D Chip-on-Chip Interconnections under Drop Impact. *Micromachines* 2022, 13, 281. https://doi.org/10.3390/mi13020281

Academic Editor: Seonho Seok

Received: 7 January 2022
Accepted: 5 February 2022
Published: 10 February 2022

Publisher's Note: MDPI stays neutral with regard to jurisdictional claims in published maps and institutional affiliations.

1. Introduction

Three-dimensional (3D) integration of silicon dies or wafers has received considerable attention in the past decade, due to its advantages of higher I/O density, lower RC delay, capability of heterogeneous integration, and footprint shrinking. Microbumps containing solder alloys have been deployed for establishing electrical and mechanical connection between vertically stacked chips [1,2]. Although similar in principle to the well-developed flip-chip technology, the interconnections using microbumps are still subjected to process adaptations. Therefore, solder joint reliability plays a vital role in the quality of electronic products.

Among all reliability issues, drop impact reliability of a solder joint, in particular, is of great importance and has attracted many researchers. For ball grid array (BGA) level solder joints typically 200–500 μm in size, the main failure mode during drop impact loading is manifested as cracking along the interface of solder bump and the intermetallic compounds (IMCs) formed by soldering [3,4], and the joint at the outermost corner is found as the most critical, which fails along the solder–pad interface [5,6]. F. X. Che et al. found that the constitutive model of solder used in the input-G simulation has a major impact on the stress and strain in a solder joint and on the hardening effect of bulk solder under a high strain rate during drop impact, which prevents the drop impact energy from dissipating through the bulk solder and accounts for the interface cracking [7]. However, downsizing of the interconnection joint size entails the reconsideration of a failure mechanism and characteristics of the micro-interconnections, as the joints in a chip-on-chip stacking scenario could use little of the previous studies at a larger scale as direct reference. Therefore, recent research of drop reliability also focuses on the 3D die-stacking structure. This includes the study by Chen et al. who determined that the critical position under the board-level drop impact is the corner of bottom layer of copper via [8], and the reliability improvement with a thinner IMC layer was revealed by Hsien-Chie Cheng et al. [9,10]. They also found that the interconnects under the drop test would exhibit a cohesive fracture inside the solder, which is different from the BGA cases studied by Suh [6,11]. M. O. Alam studied the parameters of stress intensity factors (SIF, K_I and K_{II}) around predefined cracks in the IMC layer of a solder butt joint by using linear elastic fracture mechanics (LEFM) and found that the SIF values increase sharply when the placement of the crack approaches near the interface. In summary, the reliability of microbumps for 3D integration under drop impact draws increasing concerns in interfacial fracture mechanics, as the cracking is strongly affected by the interfacial mechanical mismatch, and the propagation path will be complexly determined by both the interface feature and the solder matrix. Some research works have involved the path selection of the crack near the interface [12], but there is no description of the dynamic process of crack propagation near the interface.

In this study, we first observed the crack failure of a microbump joint in a chip-on-chip (CoC) test vehicle under drop test conditions and found that the crack formed at the edge of the soldering interface, propagated along the interface for a certain length, and deflected into the solder layer, eventually causing failure of the joint. To elucidate this phenomenon, a finite-element model was constructed to investigate the crack propagation behavior, based on basic fracture mechanics theories. The stress intensity factor of the crack tip at the interface between the IMC and solder is calculated by the contour integral method, and the propagation path of the solder joint interface crack is studied by using the criterion of energy release rate versus the fracture toughness in both the original and the deflected propagation path. Experimental tests for the joints of different solder thicknesses were carried out and compared with the numerical calculations to validate the model. Finally, the experimental observations revealed how the grain structure of the solder layer may affect the actual cracking path and drop lifetime.

2. Setup for Drop Experiment

Figure 1 shows the schematic diagram of the drop experiment. A Chip-on-chip test vehicle was used, which consisted of a $6 \times 6 \times 0.5$ mm top chip, and a $12 \times 12 \times 0.5$ mm bottom chip. Both the top chip and bottom chip had a microbump array fabricated on the surface through a standard lithography–etching–electroplating bumping process. Each microbump consisted of a Cu pillar, a Ni barrier layer, and a SnAg cap. The two chips were bonded through a flip-chip thermo compression process by an Athlete CB-600 flip-chip bonder with an alignment accuracy of ± 1 μm. The temperatures of the bonding head and bottom suction tool were set at 340 and 100 °C to obtain a peak temperature of 260 °C at the soldering interface, and the bonding pressure was 0.06 N per bump. Target temperature and pressure were applied for 30 s. The 5 μm Cu traces on both chips linked each bump to form two daisy chains, each comprising 24 pair of bumps.

A JEDEC-compliant Salon Teknopajia drop tester executed the drop experiments. The CoC module was firmly assembled on the center of test board where the impact-induced distortion is highest. The dimension of the test board also complies with the JEDEC standard, although only the 1-chip arrangement was used. A daisy chain in the module was electrically connected to a high speed data acquisition circuit to allow for transient resistance recording in real-time during drop test. The test board was then fastened onto the base plate by four screws. For each drop, the base plate was raised to the height specified in JEDEC standard and dropped on the strike surface with the acceleration G measured to follow the curve shown in Figure 2. For the observation of microstructure evolution, CoC modules after certain numbers of drops were cross-sectioned and examined under a field-emission scanning electron microscope (FE-SEM) working at the backscattered electron imaging mode.

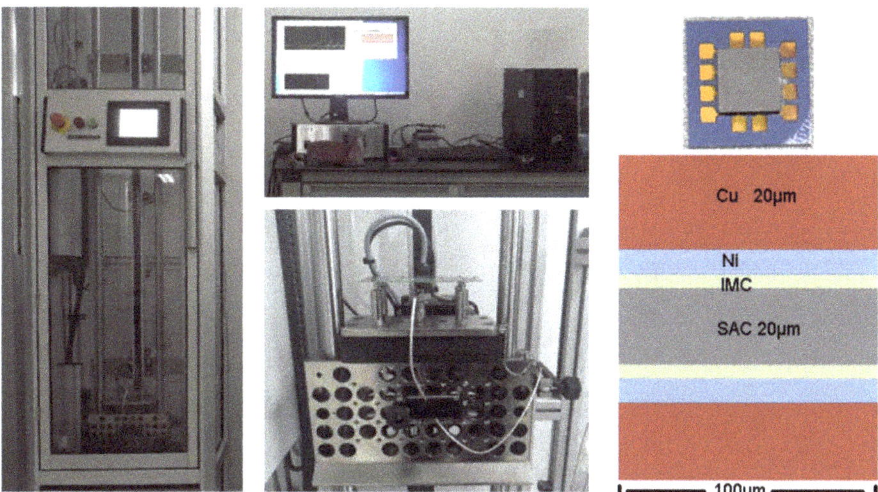

Figure 1. The schematic diagram of drop test and the cross-section images of the unit of the daisy chain.

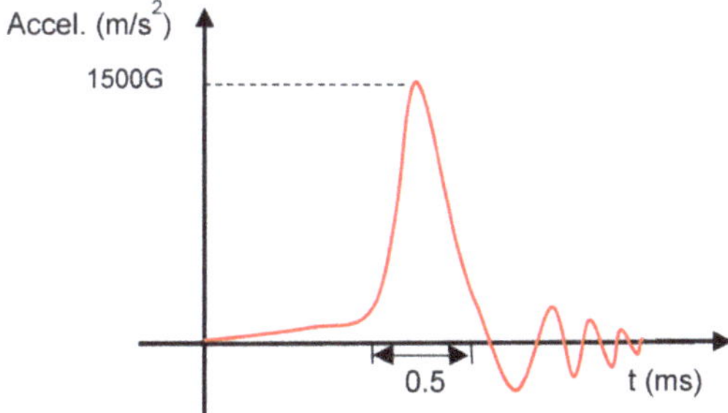

Figure 2. Impact acceleration of test results.

3. Set up for Simulation and Experiment

A finite-element (FE) code that employs transient dynamics was applied to investigate the mechanical response of the bump joint structure in a mechanical simulation for the drop test related above. The material properties in the model are all linear elastic models, as shown in Table 1. Von Mises stress distribution in the whole model at the moment of highest impact acceleration is shown in Figure 3a. According to the literature, in the board-level drop test of BGA, the failure of solder balls was mainly due to peel stress [13]. Here, the reliability of the microbumps is likewise focused by simulating the stress built in the joints between top and bottom chips. For the outer corner joints, which were subjected to highest impact stress, the maximum peeling stress is shown in Figure 3b. In the top-side, IMC was 75.8 MPa, while in the bottom, IMC of the same joints was 91.4 MPa. Therefore, the applied load was set from 10 to 90 MPa in the following FE model for the analysis of interfacial cracking behavior.

Table 1. Material properties of the main parts modeled as linearly elastic [14].

Part	Density (g/cm³)	Elastic Modulus (MPa)	Poisson Ratio
SAC305	0.00736	81,000	0.347
IMC	0.00855	114,000	0.318
Ni	0.0089	199,000	0.312

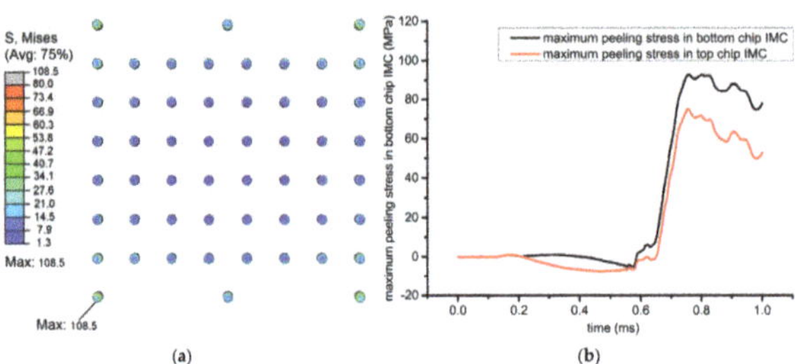

(a)   (b)

Figure 3. (**a**) Von Mises stress distribution of solder joint; (**b**) maximum peeling stress curves in a microbump joint during the impact load.

Because the solder joint has a cylindrical symmetry, the model for the calculation of the stress intensity factor at the crack tip is a two-dimensional model based on plane strain (Figure 1), which has an Sn-3.0Ag-0.5Cu solder(SAC305)–IMC–Ni sandwich configuration with dimensions of 100×20 μm, 100×1 μm, and 100×2 μm, respectively. A zero-thickness crack is preset at the interface between the IMC and solder layers, and the crack length is variable. The method of presetting the zero-thickness crack is the common point method. The surface morphology of IMC is ignored, and the interface between IMC and the solder is assumed to be flat. With an IMC thickness of only 1 μm, the possible void formation around the IMC layer was ignored, and the Ni–IMC interface was considered as ideal. The bottom of the copper pad is a fixed end, and a static-type tensile load is uniformly applied on the upper surface of the solder.

The interaction integral method is used to solve the stress intensity factor at the crack tip. Because the crack in the model is on the interface between the IMC and the solder, the elasticity of the material on each side is different; thus, discontinuity appears on the interface. To ensure the calculation accuracy, the integral path of the contour is processed in sections. The mesh of the model adopts the region division method, and the smaller mesh size is used at the crack tip to ensure the solution accuracy, as shown in Figure 4. Affected by the thickness of the IMC, the mesh quality of the grid in the crack tip decreases sharply from the first to the fourth layer. Therefore, the average stress intensity factor calculated by taking the four integral contours at the innermost layer in the crack tip is utilized as the stress intensity factor around the crack tip.

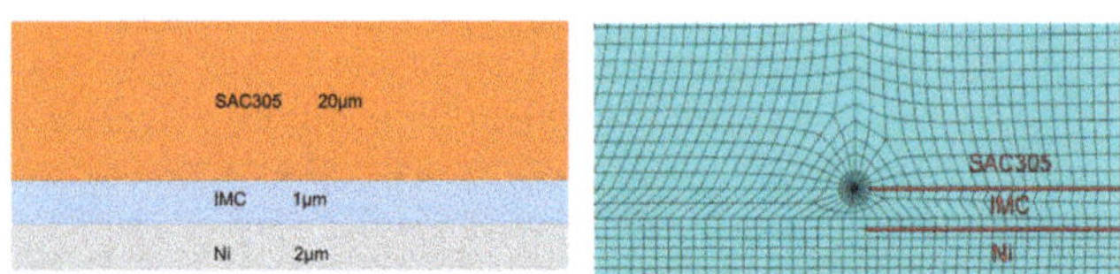

Figure 4. Model for calculating the stress intensity factor under different crack lengths.

4. Results and Discussion

4.1. Failure Mode and Mechanism of Microbumps

In order to determine the failure characteristics of microbump interconnections under drop impact, first, the recording of the transient resistance of the daisy chain was plotted against the drop counts, as shown in Figure 5. The resistance change contains three distinct stages. Stage I denotes the period in which resistance value R remained unchanged; this stage typically lasts for the first 60 drops. Then, in several tens of following drops, denoted as stage II, fluctuation of R is detected, with the peak value not exceeding 120% of the original value. Later, R experiences a period of drastic fluctuation that it increases to far more than the initial value, and the daisy chain becomes completely open in less than 80 drop counts. In order to further explore the crack propagation mechanism, the drop samples were sliced and analyzed at different stages of circuit damage during continuous drop test. Figure 6a is a cross-sectional SEM of the sample without a drop test, and it can be seen from the figure that the IMC interface formed under the hot pressing bonding conditions used in the experiment is of good quality. As shown in Figure 6b,c, after the first 50 drops, a micro-crack was visible at the end of the IMC–solder interface of the bottom chip side. After the circuit was completely disconnected, a through crack could be observed. It can be concluded that the solder joint accelerated failure after crack propagation and deflection. Therefore, the resistance change pattern can be used to estimate the extent to which the structural damage of a critical microbump has progressed. It can also be seen that the joint degradation accelerated after the crack deflection since a significant spurt of resistance corresponds to the rapid shrinking of the residual joint area in this stage.

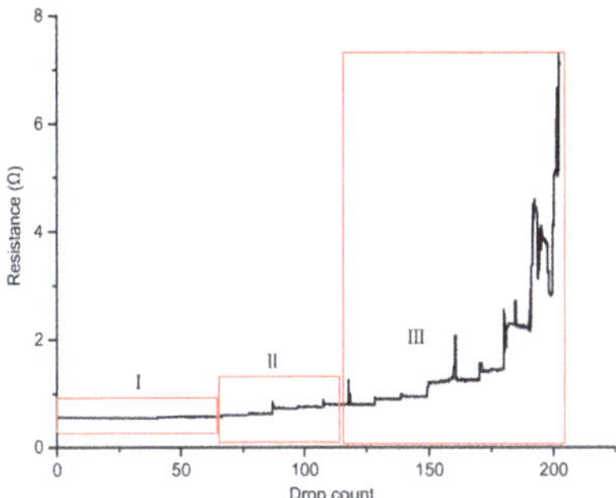

Figure 5. Typical resistance curves of daisy chains under drop test.

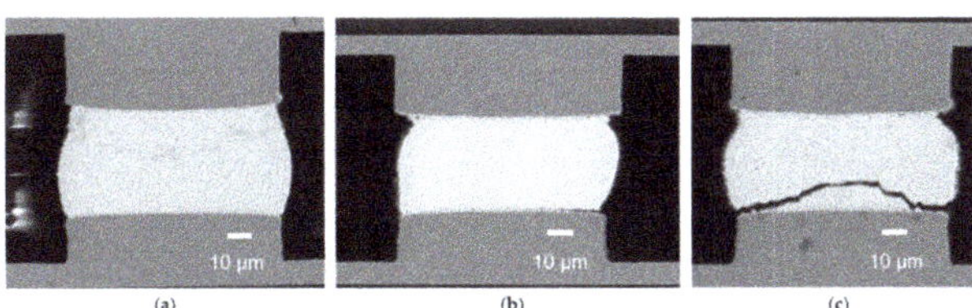

Figure 6. Cross-sectional SEM of the microbumps at different stages of drop impact. (**a**) Cross-sectional SEM of the sample without drop test; (**b**) after 50 drops; (**c**) after the circuit is completely disconnected.

4.2. Stress Intensity Factor Analysis of Solder–IMC Interface Crack under Quasi-Static Load

4.2.1. Relationship between Stress Intensity Factor at Interface Crack Tip and Crack Length

Figure 7 shows the von Mises equivalent stress distribution at the crack tip when the load is 10 MPa and the crack length is 10 and 20 µm. It can be seen that the equivalent stress of the crack tip increases as the crack length increases, and the high equivalent stress appears both on the solder and IMC. However, it does not mean that failure or crack propagation will definitely occur in these locations. Under tensile load, the crack between the upper layer and the substrate initiate from the free edge of the actual specimen, especially where defects such as cracking or void brought by the bonding process existed. The initial crack first expands along the interface to a certain depth and then propagates along the interface or is deflected to the solder matrix, which depends on the energy release rate of the two propagation paths. Therefore, the energy release rate will be used to judge whether the crack is initiated and propagated, and the stress intensity factor will be used to determine the crack tip propagation path.

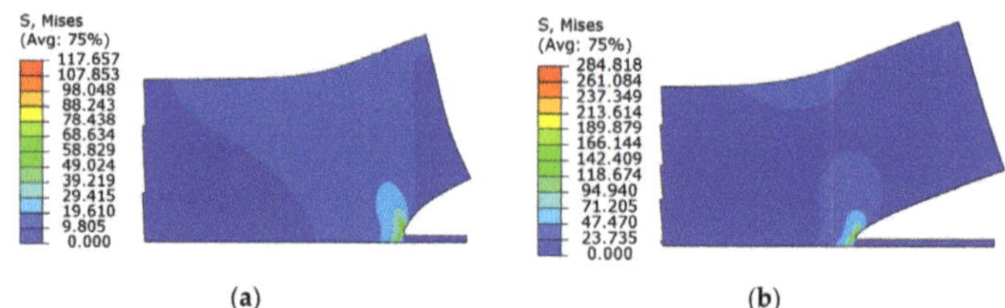

Figure 7. The Von Mises equivalent stress distribution in the crack tip with the crack length of: (**a**) 10 μm; (**b**) 20 μm.

The relationship between the stress intensity factor at the crack tip of the IMC–solder interface calculated by the interaction integral and the crack depth is shown in Figure 8. It can be seen that the stress intensity factors of K_I and K_{II} of the interfacial crack tip increase with an increase in the crack length under the same load, and the K_{II} will increase quickly due to the elastic deformation of the solder, which leads to the increasing tendency of the type II cracking mode and possibly the crack deflection as well.

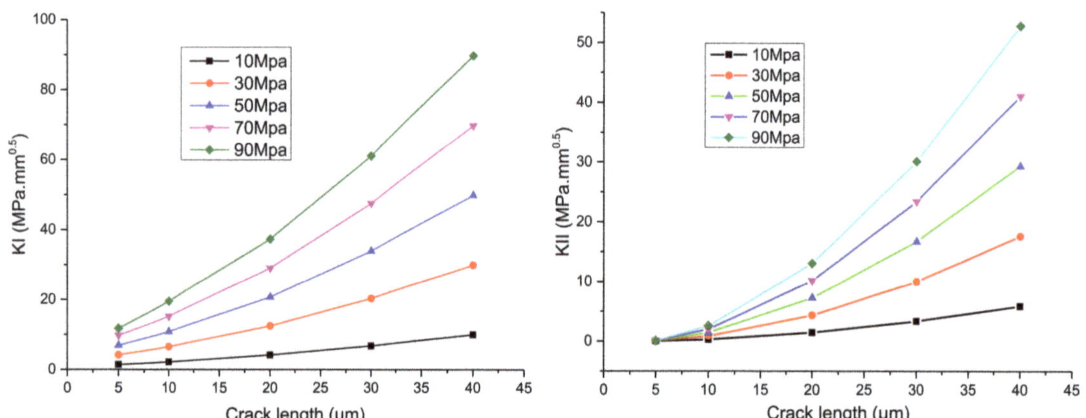

Figure 8. The relationship between the stress intensity factor of the interface crack between the IMC and the solder and crack depth.

Polynomial fitting is performed for the stress intensity factor at the crack tip with different crack lengths in the figure, and the fitting expression is as follows:

$$K = \begin{cases} 3.7 \times 10^3 a^2 + 1.7 \times 10^2 a + 0.013 & \sigma = 10\text{MPa} \\ 18.6 \times 10^3 a^2 + 8.5 \times 10^2 a + 0.65 & \sigma = 50\text{MPa} \\ 33.3 \times 10^3 a^2 + 15.4 \times 10^2 a + 1.2 & \sigma = 90\text{MPa} \end{cases} \tag{1}$$

$$K = \begin{cases} 5.44 \times 10^3 a^2 - 1.27a - 0.054 & \sigma = 10\text{MPa} \\ 27.2 \times 10^3 a^2 - 5.68a - 0.265 & \sigma = 50\text{MPa} \\ 48.9 \times 10^3 a^2 - 9.77a - 0.48 & \sigma = 90\text{MPa} \end{cases} \qquad (2)$$

where σ is the peel stress loaded on the upper surface of the solder and a is the crack length. Comparing the stress intensity factors of K_I and K_II under three loads, it can be seen that K_I and K_II are proportional to the load, because the material model used in the simulation is a linear elastic model. Therefore, the expressions of K_I and K_II can be rewritten as follows:

$$K = 3.7 \times 10^2 a^2 \sigma + 17a\sigma + 0.013\sigma$$
$$K = 5.44 \times 10^2 a^2 \sigma - 0.127a\sigma - 0.0054\sigma \qquad (3)$$

4.2.2. Influence of Solder Thickness on Stress Intensity Factor of Interface Crack

In the existing research on solder joints of several hundred micrometers, due to the much lower elastic modulus of solder versus the rest part of a joint, the solder volume plays an important role in the mechanical properties of the microbumps. If the thickness of the solder layer is too small, the mechanical properties of the microbumps will be adversely affected. The solder thickness in a microbump-based die stacking 3D integration structure is greatly reduced compared to the flip chip interconnection, which necessitates the research on the dependence of SIF on the solder thickness quantitatively. Figure 9 compares the SIF evolution with a progressing crack under different solder thicknesses from 15 to 30 μm. It can be seen that both K_I and K_II increase with a decrease in solder thickness. This phenomenon is plainly explained by the stress distribution around the crack tip, as shown in Figure 10. The elastic mismatch between IMC and the solder causes stress concentration around the crack tip, which is better alleviated with a thicker solder layer, as can be judged from the more uniform distribution of stress across the cross section of analysis. Therefore, switching from the spherical solder bumps to the Cu pillar-based microbump joints is believed to pose additional failure risk under the drop impact condition.

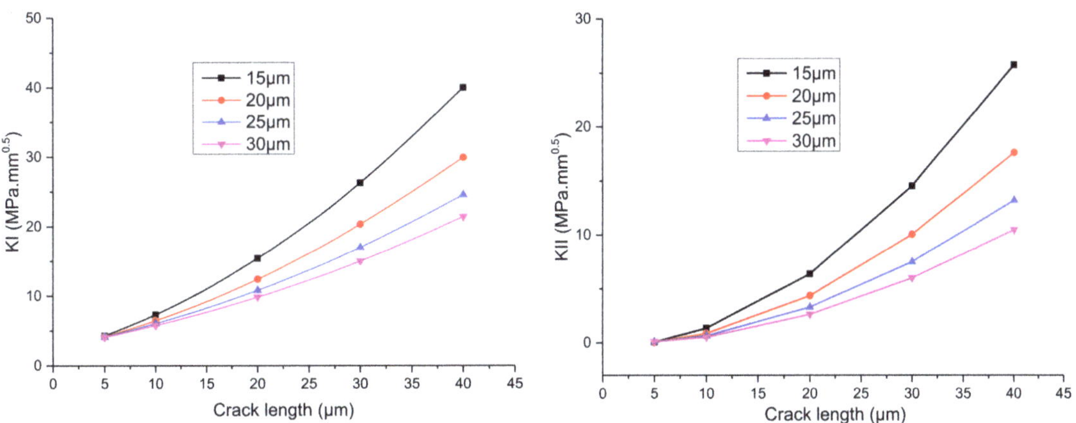

Figure 9. The relationship between the stress intensity factor K_I of the interfacial crack tip and solder thickness under different solder thicknesses.

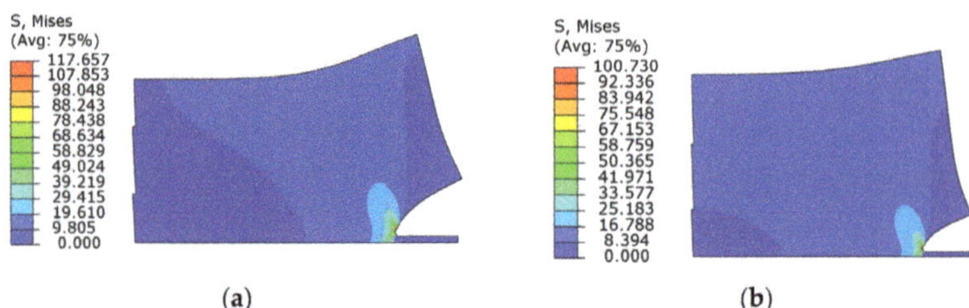

Figure 10. The von Mises equivalent stress distribution in the crack tip with different solder thicknesses: (**a**) 20 μm; (**b**) 30 μm.

4.3. Investigation on Crack Growth Behavior

The analyses above have revealed the increase in the stress intensity factors K_I and K_{II} with increasing crack length. Further investigation of the crack propagation behavior, especially the propagation path, needs the quantitative analyses on the crack tip energy release rates J_1 and J_2. Hu [15] found the propagation behavior of a semi-infinite plane crack at the interface of a two-phase material in 1989 and revealed that theoretically the crack deviated from the original main crack propagation path by a minimum length. They further deduced the relationship between the stress intensity factor after crack deflection and along the original path. The maximum energy release rate can be used to determine the crack deflection angle. The criterion of deflection of quasi-static interface crack propagation behavior is as follows:

$$\frac{G_S}{G} > \frac{\Gamma}{\Gamma_i} \tag{4}$$

Among them: $G_s = J = \sqrt{J_1^2 + J_2^2}$, $G = \frac{K^2}{E^*}$, Γ is the fracture toughness of the solder, and Γ_i is the fracture toughness of Ni_3Sn_4 IMC. In this paper, the maximum fracture toughness of solder SAC305 is set to be 295 N/m, which is measured by Loo [16]. To be able to directly compare the fracture toughness values of the Ni-Sn–IMC interface from the various existing research, the fracture toughness is converted into a critical stress intensity factor. For the Ni_3Sn_4 layer, a critical stress intensity factor of 4.22 ± 0.45 MPa $m^{1/2}$ measured by Ghosh [14] was adopted, which equals 165.5 N/m; thus, we obtain $\frac{\Gamma}{\Gamma_i} = 1.78$. It can be seen from the expression of the crack tip energy release rate that when the material is of linear elastic property, the ratio $\frac{G_S}{G}$ is irrelevant to load. For the convenience of calculation, the ERR is calculated with the uniaxial load of 50 MPa, and the IMC and solder thicknesses are set as 1 and 20 μm, respectively. The energy release rate at the interface crack tip under different crack lengths is calculated as follows:

For the homogeneous two-material interface:

$$J_1 = \frac{K\overline{K}}{E^* \cosh^2(\pi\varepsilon)} \tag{5}$$

$$J_2 = -\frac{\mathrm{Re}[Kr^{i\varepsilon}]\mathrm{Im}[Kr^{i\varepsilon}]}{\pi\varepsilon \cosh^2(\pi\varepsilon)} \times \left[\frac{1-\nu_1}{4\mu_1}(1 - e^{-2\pi\varepsilon}) + \frac{1-\nu_2}{4\mu_2}(e^{2\pi\varepsilon} - 1)\right] \tag{6}$$

where ε is the oscillatory index

$$\varepsilon = \frac{1}{2\pi} \ln\left(\frac{1-\beta}{1+\beta}\right) \tag{7}$$

β is the second Dundurs' constant

$$\beta = \frac{\mu_1(\kappa_2 - 1) - \mu_2(\kappa_1 - 1)}{\mu_1(\kappa_2 + 1) + \mu_2(\kappa_1 + 1)} \tag{8}$$

and κ is Kolosov's constant

$$\kappa = \begin{cases} \frac{3-v_p}{1+v_p} & \text{plane stress} \\ 3 - 4v_p & \text{plane strain} \end{cases} \tag{9}$$

where

$$\frac{1}{E^*} = \left[\frac{1-v_1}{4\mu_1} + \frac{1-v_2}{4\mu_2} \right] \tag{10}$$

The results of the relevant parameters of the dual-material SAC305–IMC interface in the above formula are shown in Table 2. The trend of $\frac{G_s}{G}$ with crack length is calculated, as shown in Figure 11.

Table 2. Parameters of two-material SAC305–IMC interface.

Parameters	E^*	ϵ	β	μ_1	v_1	μ_2	v_2
Values	6.06×10^{10}	-0.028	0.088	3.59×10^{10}	0.314	1.388×10^{10}	0.347

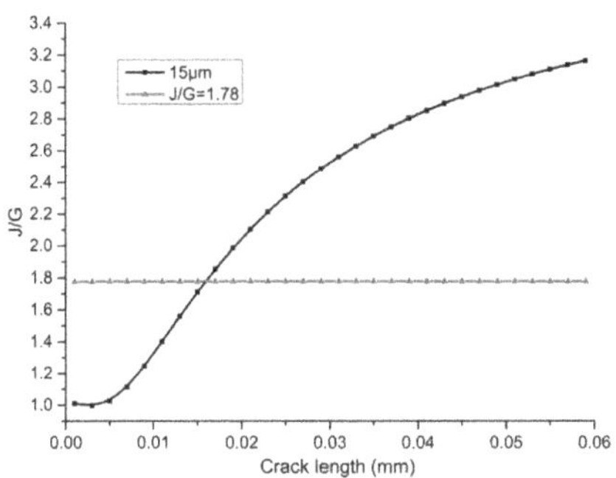

Figure 11. The variation trend of $\frac{G_s}{G}$ with crack length.

Hu found that the interface cracks start from the free edge of the sample, propagate at one to two times the thickness of the film along the interface, and then deflect into the matrix, expanding to a depth of four to five times the thickness of the film and finally parallel to the interface. From Figure 11, it can be seen that the ratio of the crack tip ERR after deflection to that propagating along the interface increases with the increase in the main crack length. When the main crack expands to a length of about 16 µm, the ratio will be greater than the ratio of the fracture toughness of the solder matrix to the fracture toughness of the interface. At this time, the crack will deviate from the original interface path and deflect into the matrix. The deflection angle is calculated by $\omega = \arctan\left|\frac{J_2}{J_1}\right|$, and we can find $w = 42°$. It can also be seen from the above figure that if the ratio of the fracture toughness of the solder matrix to the interface fracture toughness is greater than the ratio between two paths, then the crack will always expand along the interface without deflecting to the solder matrix.

Figure 12 compares the influence of solder thickness on the interfacial crack growth behavior. It can be seen from the figure that when the solder thickness decreases, the critical main crack length for crack deflection will decrease. When the solder thickness is 15, 20, 25 and 30 um, the critical crack deflection length is 16, 23, 27 and 29 um, respectively, due to the reason related in Section 4.2.2, i.e., the decrease in solder cushioning causes an increase in stress concentration in the solder matrix, thus increasing the advantage of deflected cracking path.

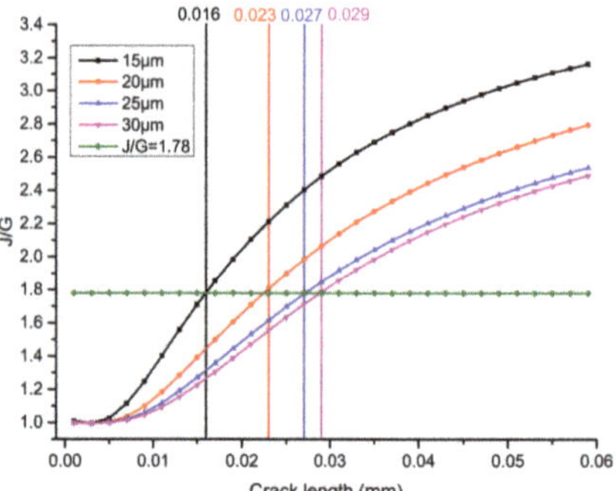

Figure 12. The influence of different solder thicknesses on interfacial crack growth behavior.

Figure 13 shows the variation of arctan |J2/J1|, or in other words, the virtual crack deflection angle, whether or not deflection actually takes place. With the main crack length under different solder thickness conditions, the angle increases rapidly at first, and then closes to a constant value. The crack deflection angle trend is consistent with the research of HH YU et al. on the interfacial cracking behavior of chromium films on silica substrates [12]. The asymptotic value of the crack deflection angle is about 42°.

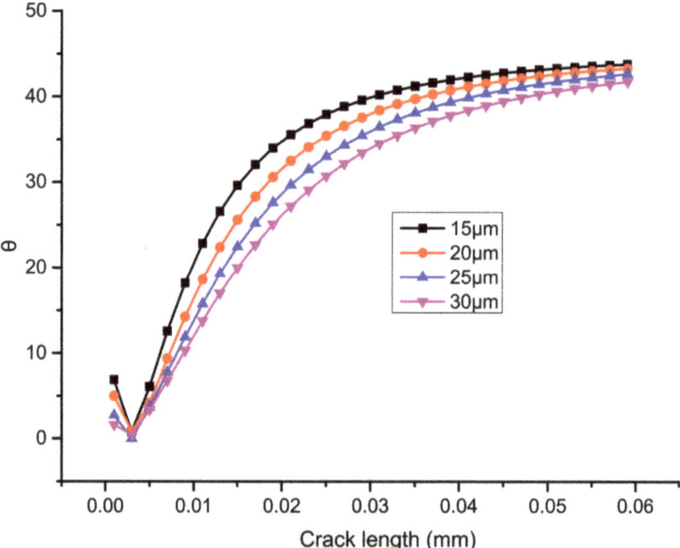

Figure 13. Relationship between the crack deflection angle and the main crack length under different solder thicknesses.

5. Experimental Validation and Discussion

The cross section of the solder joint in the case of drop failure with different solder thickness is shown in Figure 14. According to the SEM analysis, when the solder thickness is 20 μm, the crack length of 8 μm deflects, and the deflection angle is 32.8°. When the solder thickness is 30 μm, the crack length of 28 μm deflects, and the deflection angle is 37°. When the solder height is 37 μm, the crack length of 32 μm deflects, and the deflection angle is 30.4°. The measured deflection angle of the interfacial crack is 30° to 40°, which is larger than the asymptotic value of deflection angle at the moment of deflection initiation, calculated by numerical simulation. This is owed to the microbump not only being subjected to normal stress, but it is also subjected to a shear force parallel to the interface during the drop experiment, while the load used in the numerical calculation is only the normal stress. In practice, when the crack propagates to a certain length, the portion of type II cracking produced by the shear stress cannot be negligible. The change of initial deflection angle versus the solder thickness is in good agreement with the numerical calculation based on ERR and fracture toughness. Therefore, in general, the numerical methods adopted in this paper can be used as an effective way to predict the cracking behavior in an actual microbump joint.

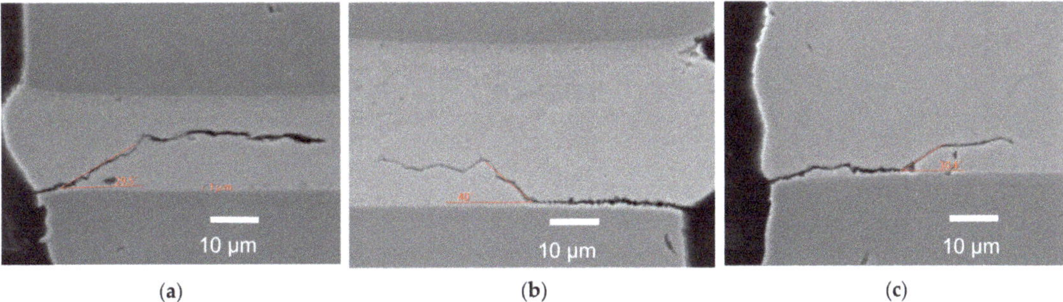

(a)   (b)   (c)

Figure 14. Cross section of the solder joint with different solder thicknesses, listed as: (**a**) 20 μm; (**b**) 30 μm; (**c**) 37 μm.

The actual crack propagation behavior is affected by many factors, such as interfacial defects or inhomogeneity of microstructure. There is a clear competition between interfacial propagation and solder matrix propagation; for example, it was found in the test vehicles of inferior interfacial strength, e.g., the bonding was carried out at lower than optimal temperatures, and the crack would not deflect due to the increased value. In addition, the competition of the crack path in a well-bonded test vehicle is often observed as minute crack branching, as shown in Figure 15. These small-scale branched cracks often terminated within 1 μm. As the fracture progresses, the deflected path gradually gains favor.

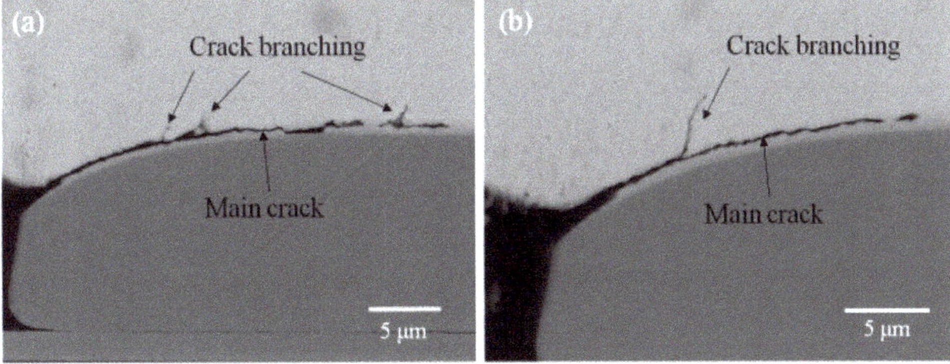

Figure 15. Crack branching during the initial stage of the drop test: (**a**) The minute crack branching increase as the fracture progresses; (**b**) The first crack branching.

The explanation for the crack branching is that the grain boundary is the low strength region and alternative crack propagation path. Here, a test vehicle with Cu–SnAg–Cu microbump structure was used to enhance the interfacial reaction, and the joint was ion-milled cross-sectionally before SEM observation to exhibit grain contrast, as shown in

Figure 16. The second phase was identified as Cu_6Sn_5 IMC. IMC particles can be seen clearly in the junction of Sn grains, which is formed by Cu atoms diffusing along the grain boundary and precipitating in the junction in the form of Cu_6Sn_5 during the solidification process. These Cu_6Sn_5 particles play a significant role in the arresting and deflection of cracks. As can be seen in Figure 16a,b, crack tips meet the second phase and stop propagating. A higher driving force is required to either propagate around the second phase by deflection, or to continue through the second phase, the latter being less probable from an energetic point of view. Therefore, once arrested by the boundary junction, cracks would further proceed along the boundary of the IMC particle and Sn grain, while the preferred direction of all possible ones is related to the deflection angle, finally forming fracture patterns that differ from one sample to another in shape. The Ag_3Sn IMC grains were believed to not have a significant impact on the crack propagation path since they were present in the form of a primary eutectic component located inside each Sn grain [17,18]. It has been previously reported that under thermal cycling or coupled thermomechanical–electrical load, the fatigue crack preferred an intergranular propagation path [19,20], in which case, the reconstructed grain structure and recrystallization might contribute to the weakening of grain boundary strength. This inclination seems to apply well to the highly dynamic and purely mechanical drop impact scenario. We can also reasonably suspect that if the interfacial IMC grows to a certain thickness that leaves visible voids due to the volume shrinkage effect, the bonding interface will be much weakened in that the crack will only propagate along the voided interface.

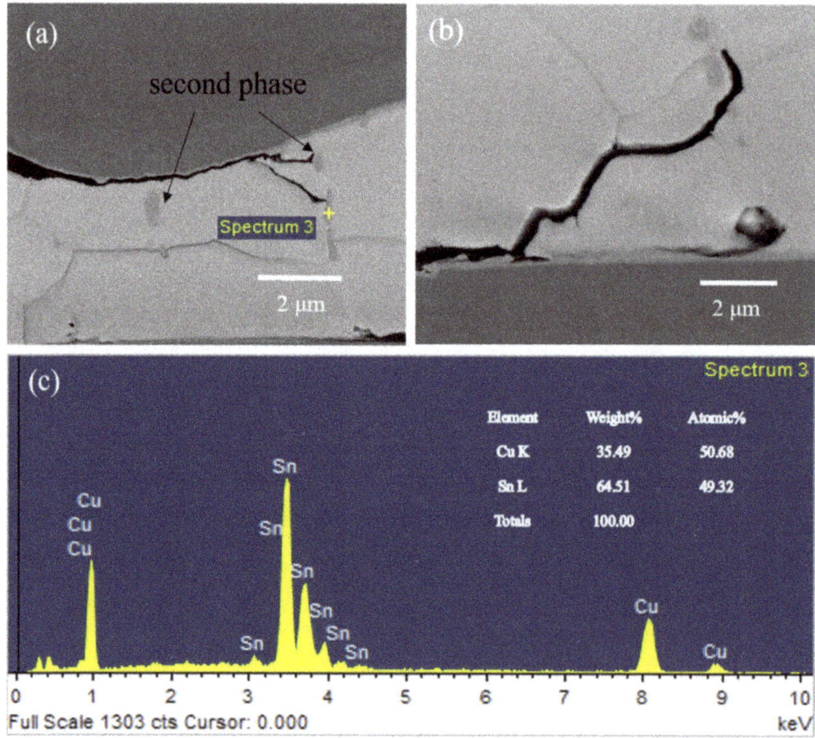

Figure 16. Influence of second-phase particles on crack propagation path: (**a**) crack tips meet the second phase and stop propagating; (**b**) The crack propagate around the second phase by deflection; (**c**) EDX analysis diagram.

Combining the results in Figure 5, it can also be further deduced that the stage III of resistance change plays a significant role in determining the joint lifetime under drop impact, and one possible way to enhance the durability is to eliminate the grain boundaries; thus, the deflected path would cost higher energy than in a joint of the multi-grain solder layer. The research of controlling the grain number of the solder layer in a microbump joint is currently ongoing among various researchers [21,22].

6. Conclusions

In this paper, we report for the first time the cracking failure characteristics in microbump joints for chip-on-chip stacked interconnections. Experimental tests were carried out using a JEDEC standard test board to reveal the joint resistance change and the crack morphology. To elucidate the crack deflection during the joint degradation process, a local finite-element model was established to calculate the stress intensity factor at the crack tip, and the numerical results were further incorporated into a fracture mechanics model to obtain the crack deflection criteria. The main conclusions are summarized below:

(1) The main failure mode of microbump interconnections for 3D CoC packaging is that cracks were first initiated at the edge of the IMC–solder interface. After propagating along the interface for a distance, they deflected into the solder matrix, eventually penetrating the entire joint. The electrical resistance change is closely linked to the cracking progress.

(2) Stress intensity factor of a zero-thickness crack tip at the interface of the solder and IMC is calculated under quasi-static load by the method of interaction integral method. Both K_I and K_{II} increase with the increase in the crack length under the same load, and reducing the solder thickness causes higher SIF due to less alleviated mechanical mismatch.

(3) The crack propagation path is studied using a criterion based on energy release rate and fracture toughness. The calculation results show that the cracks on the interface between the solder and IMC will deflect into the solder matrix after extending to a certain depth along the interface. The deflection angle for crack initiation converges to 40° with the increase in crack length. The critical length of the main crack for crack deflection increases with the increase in solder thickness, which is experimentally confirmed by an actual drop test on samples with different solder heights.

(4) The crack propagation path in actual drop test samples was influenced by factors, including the actual strength of the bonding interface and the grain structure of the solder layer. Grain boundaries are the favored path for the deflected cracks.

Author Contributions: Conceptualization, W.Z. and Z.C.; methodology, Z.L.; software, L.S. and Z.L.; validation, M.F. and Z.L.; investigation, Z.C.; writing—original draft preparation, Z.L.; writing—review and editing, Z.C., Y.G. and Z.L.; supervision, W.Z.; project administration, W.Z.; funding acquisition, W.Z. All authors have read and agreed to the published version of the manuscript.

Funding: This research was funded by "National Natural Science Foundation of China, grant number U20A6004" and "Natural Science Foundation of Hunan Province, grant number 2021JJ40734", and "State Key Laboratory of High Performance Complex Manufacturing, grant number ZZYJKT2020-08", and "Natural science research project of Hunan Provincial Department of Education, grant number 19B228".

Institutional Review Board Statement: The study did not require ethical approval.

Data Availability Statement: Data sharing not applicable.

Conflicts of Interest: The authors declare no conflict of interest.

References

1. Lin, J.; Chiou, W.; Yang, K.; Chang, H.; Lin, Y.; Liao, E.; Hung, J.; Tsai, P.; Shih, Y.; Wu, T.; et al. High density 3D integration using CMOS foundry technologies for 28 nm node and beyond. In Proceedings of the 2010 International Electron Devices Meeting, San Francisco, CA, USA, 6–8 December 2010; pp. 2.1.1–2.1.4. [CrossRef]
2. Yu, A.; Lau, J.H.; Ho, S.W.; Kumar, A.; Hnin, W.Y.; Yu, D.-Q.; Jong, M.C.; Kripesh, V.; Pinjala, D.; Kwong, D.-L. Study of 15 μm pitch solder microbumps for 3D IC integration. In Proceedings of the 2009 59th Electronic Components and Technology Conference, San Diego, CA, USA, 26–29 May 2009; pp. 6–10. [CrossRef]

3. Suh, D.; Kim, D.W.; Liu, P.; Kim, H.; Weninger, J.A.; Kumar, C.M.; Prasad, A.; Grimsley, B.W.; Tejada, H.B. Effects of Ag content on fracture resistance of Sn–Ag–Cu lead-free solders under high-strain rate conditions. *Mater. Sci. Eng. A* **2007**, *460–461*, 595–603. [CrossRef]
4. Kim, D.; Suh, D.; Millard, T.; Kim, H.; Kumar, C.; Zhu, M.; Xu, Y. Evaluation of High Compliant Low Ag Solder Alloys on OSP as a Drop Solution for the 2nd Level Pb-Free Interconnection. In Proceedings of the 57th Electronic Components and Technology Conference, Sparks, NV, USA, 29 May–1 June 2007; pp. 1614–1619. [CrossRef]
5. Tee, T.Y.; Ng, H.S.; Lim, C.T.; Pek, E.; Zhong, Z. Impact life prediction modeling of TFBGA packages under board level drop test. *Microelectron. Reliab.* **2004**, *44*, 1131–1142. [CrossRef]
6. Tee, T.Y.; Ng, H.S.; Lim, C.T.; Pek, E.; Zhong, Z. Board level drop test and simulation of TFBGA packages for telecommunication applications. In Proceedings of the 53rd Electronic Components and Technology Conference, New Orleans, LA, USA, 27–30 May 2003; pp. 121–129. [CrossRef]
7. Pang, J.H.L.; Che, F. Drop impact analysis of Sn-Ag-Cu solder joints using dynamic high-strain rate plastic strain as the impact damage driving force. In Proceedings of the 56th Electronic Components and Technology Conference, San Diego, CA, USA, 30 May–2 June 2006; p. 6. [CrossRef]
8. Chen, Z.; Wang, X.; Liu, Y.; Liu, S. Drop test simulation of 3D stacked-die packaging with Input-G finite element method. In Proceedings of the 2010 11th International Conference on Electronic Packaging Technology & High Density Packaging, Xi'an, China, 16–19 August 2010; pp. 742–746. [CrossRef]
9. Cheng, H.-C.; Cheng, H.-K.; Lu, S.-T.; Juang, J.-Y.; Chen, W.-H. Drop Impact Reliability Analysis of 3-D Chip-on-Chip Packaging: Numerical Modeling and Experimental Validation. *IEEE Trans. Device Mater. Reliab.* **2014**, *14*, 499–511. [CrossRef]
10. Cheng, T.-H.; Cheng, H.-C.; Chen, W.-H.; Huang, H.-Y.; Chang, T.-C. Reliability characterization of 2.5D multi-chip module on board under drop impact. In Proceedings of the 2014 International Conference on Electronics Packaging (ICEP), Toyama, Japan, 23–25 April 2014; pp. 215–218. [CrossRef]
11. Tee, T.Y.; Luan, J.-E.; Pek, E.; Lim, C.T.; Zhong, Z. Novel numerical and experimental analysis of dynamic responses under board level drop test. In Proceedings of the 5th International Conference on Thermal and Mechanical Simulation and Experiments in Microelectronics and Microsystems, 2004. EuroSimE 2004, Brussels, Belgium, 10–12 May 2004; pp. 133–140. [CrossRef]
12. Yu, H.-H.; He, M.; Hutchinson, J. Edge effects in thin film delamination. *Acta Mater.* **2001**, *49*, 93–107. [CrossRef]
13. Chong, D.Y.; Che, F.; Pang, J.H.; Ng, K.; Tan, J.Y.; Low, P.T. Drop impact reliability testing for lead-free and lead-based soldered IC packages. *Microelectron. Reliab.* **2006**, *46*, 1160–1171. [CrossRef]
14. Ghosh, G. Elastic properties, hardness, and indentation fracture toughness of intermetallics relevant to electronic packaging. *J. Mater. Res.* **2004**, *19*, 1439–1454. [CrossRef]
15. Ming-Yuan, H.; Hutchinson, J.W. Crack deflection at an interface between dissimilar elastic materials. *Int. J. Solids Struct.* **1989**, *25*, 1053–1067. [CrossRef]
16. Loo, S.Z.Y.; Lee, P.C.; Lim, Z.X.; Yantara, N.; Tee, T.Y.; Tan, C.M.; Chen, Z. Interface fracture toughness assessment of solder joints using double cantilever beam test. *Int. J. Mod. Phys. B* **2010**, *24*, 164–174. [CrossRef]
17. Ding, Y.; Wang, C.; Li, M. Scanning electron microscope in-situ investigation of fracture behavior in 96.5Sn3.5Ag lead-free solder. *J. Electron. Mater.* **2005**, *34*, 1324–1335. [CrossRef]
18. Kanchanomai, C.; Miyashita, Y.; Mutoh, Y. Low-cycle fatigue behavior and mechanisms of a lead-free solder 96.5Sn/3.5Ag. *J. Electron. Mater.* **2002**, *31*, 142–151. [CrossRef]
19. Chen, H.; Mueller, M.; Mattila, T.T.; Li, J.; Liu, X.; Wolter, K.-J.; Paulasto-Kröckel, M. Localized recrystallization and cracking of lead-free solder interconnections under thermal cycling. *J. Mater. Res.* **2011**, *26*, 2103–2116. [CrossRef]
20. Le, V.-N.; Benabou, L.; Tao, Q.-B.; Etgens, V. Modeling of intergranular thermal fatigue cracking of a lead-free solder joint in a power electronic module. *Int. J. Solids Struct.* **2017**, *106–107*, 1–12. [CrossRef]
21. Chang, Z.Y.; Zhao, N.; Wu, C.M.L. Effects of cooling rate and joint size on Sn grain features in Cu/Sn–3.5Ag/Cu solder joints. *Materialia* **2020**, *14*, 100929. [CrossRef]
22. Darbandi, P.; Bieler, T.R.; Pourboghrat, F.; Lee, T.-K. The Effect of Cooling Rate on Grain Orientation and Misorientation Microstructure of SAC105 Solder Joints Before and After Impact Drop Tests. *J. Electron. Mater.* **2014**, *43*, 2521–2529. [CrossRef]

micromachines

MDPI

Article

Investigation of Integrated Reactive Multilayer Systems for Bonding in Microsystem Technology

El-Mostafa Bourim *, Il-Suk Kang and Hee Yeoun Kim *

National NanoFab Center, Department of Nanostructure Technology, KAIST, 291 Daehak-ro, Yuseong-gu, Daejeon 34141, Korea; iskang@nnfc.re.kr
* Correspondence: e.bourim@gmail.com (E.-M.B.); hyeounkim@nnfc.re.kr (H.Y.K.); Tel.: +82-10-2585-9828 (E.-M.B.)

Abstract: For the integration of a reactive multilayer system (iRMS) with a high exothermic reaction enthalpy as a heat source on silicon wafers for low-temperature bonding in the 3D integration and packaging of microsystems, two main conflicting issues should be overcome: heat accumulation arising from the layer interface pre-intermixing, which causes spontaneous self-ignition during the deposition of the system layers, and conductive heat loss through the substrate, which leads to reaction propagation quenching. In this work, using electron beam evaporation, we investigated the growth of a high exothermic metallic Pd/Al reactive multilayer system (RMS) on different Si-wafer substrates with different thermal conduction, specifically a bare Si-wafer, a RuO_x or PdO_x layer buffering Si-wafer, and a SiO_2-coated Si-wafer. With the exception of the bare silicon wafer, the RMS grown on all other coated wafers underwent systematic spontaneous self-ignition surging during the deposition process once it reached a thickness of around 1 µm. This issue was surmounted by investigating a solution based on tuning the output energy by stacking alternating sections of metallic reactive multilayer Pd/Al and Ni/Al systems that have a high and medium enthalpy of exothermic reactions, respectively. This heterostructure with a bilayer thickness of 100 nm was successfully grown on a SiO_2-coated Si-wafer to a total thickness of 3 µm without any spontaneous upsurge of self-ignition; it could be electrically ignited at room temperature, enabling a self-sustained propagating exothermic reaction along the reactive patterned track without undergoing quenching. The results of this study will promote the growth of reactive multilayer systems by electron beam evaporation processing and their potential integration as local heat sources on Si-wafer substrates for bonding applications in microelectronics and microsystems technology.

Keywords: multilayer reactive bonding; integrated nanostructure-multilayer reactive system; spontaneous self-ignition; self-propagating exothermic reaction; Pd/Al reactive multilayer system; Ni/Al reactive multilayer system; low-temperature MEMS packaging

Citation: Bourim, E.-M.; Kang, I.-S.; Kim, H.Y. Investigation of Integrated Reactive Multilayer Systems for Bonding in Microsystem Technology. *Micromachines* **2021**, *12*, 1272. https://doi.org/10.3390/mi12101272

Academic Editor: Seonho Seok

Received: 31 August 2021
Accepted: 9 October 2021
Published: 19 October 2021

1. Introduction

The adoption of integrated reactive multilayer systems (iRMS) as a bonding technique in microelectronics and micromechanical systems (MEMS) has recently started to gain more attention and traction in microsystems technology [1–6]. The reactive bonding at micrometer scales using bond frames with dimensions of a few micrometers makes reactive bond interface engineering using traditional freestanding reactive multilayer foils not practically feasible (i.e., due to the difficulty of foil handling, patterning, and positioning). Therefore, integration processing by the deposition/patterning or the patterning/deposition of reactive multilayer film systems directly on silicon wafers or other substrate components presents an interesting research challenge.

Reactive bonding uses a highly reactive nanoscale multilayer system as a self-heating source between joining substrates. The heat generation after an external initiation is created by a self-propagating exothermic reaction of the integrated RMS [4,5,7,8]. The integrated reactive bond induces local heat to the bonding interface; such limited heat/temperature quenches locally through the substrate material. This allows temperature-sensitive microdevices located outside the interface and materials with different coefficients of thermal expansion (CTE) to be bonded without thermal damage.

Generally, the reactive multilayer system typically consists of several alternating layers (up to hundreds) of two or more different reactant films combined as metal/metal, metal/oxide or metal/metalloid [9,10]. The bonding thermal energy source results from the exothermic reaction by the interdiffusion of adjacent material layers [8,11,12]. The required bonding energy level is chosen based on the negative enthalpy of the formation of material combinations [8,10]. It is also necessary to consider that the integrated reactive multilayer systems in bonding should be composed of highly reactive materials, which can release higher amounts of exothermic energy. This is important in order to assure a self-sustained propagating exothermic reaction and compensate for the heat absorbed in the bonding interfaces [2,8,13,14], such as the conductive heat losses through the hosting substrate or bonded component partners [14,15]. However, the use of highly reactive reactants has given rise to different issues. The most disadvantageous is the spontaneous self-ignition of RMS during thin reactant layer deposition. With existing standard sputtering equipment without an active substrate cooling system, it is not possible to deposit any number of individual layers; after just a few layer superpositions, these piled layers self-ignite and react before the coating process has reached the final requested total layer number [2,8].

Other issues include the reaction initiation effect during RMS handling, premature intermixing at the interfaces of stacked RMS reactant layers [12,16–18], and RMS ignition in a strong explosive exothermic reaction leading to the vaporization or ejection of the reaction product [19–22]. It should also be noted that the partly self-reacted RMS and pre-intermixed reactant layer interfaces are considered to be among the main factors preventing the reliable initiation of self-propagating exothermic reactions at room temperature in integrated RMSs [12,16–18]. These partially consumed reactants reportedly reduce the potential heat energy required for the reaction ignition and propagation [12,16–18].

The goal of this investigation is to integrate highly reactive reactant film layers directly on Si-wafer substrates using a conventional electron beam deposition system that is not equipped with an active substrate cooling setup. To avoid the delamination of deposited layers and reduce internal mechanical stresses within the reactive multilayer system, the total thickness of the integrated reactive multilayer system (iRMS) should not exceed 5 μm [2,7,8]. Furthermore, to overcome the limitation of the standard photoresist lift-off patterning technique used in the preparation of iRMS pattern samples, the maximum total thickness of the deposited RMS was limited to 3 μm. Hence, a smaller iRMS thickness contributes to cost-efficiency by using smaller amounts of noble metals, which are commonly used as reactants in RMSs that supply high thermal energy.

In this work we investigated the integration of Pd/Al RMS on a silicon substrate coated with different thermal barrier layers. The RMS deposition was performed on a bare silicon wafer, a RuO_x-buffered Si-wafer, a PdO_x-buffered Si-wafer, and a SiO_2-coated Si-wafer. It has been shown that the Pd/Al iRMS can, in principle, be used for reactive bonding. However, as Pd/Al iRMS is highly exothermically reactive, spontaneous self-ignition and propagating reactions during deposition could happen frequently when the thickness of the deposited RMS reaches a critical value. It was also observed that for a substrate that has a low thermal conductivity, the high-confined released heat and its associated high temperature in the ignited iRMS led to reaction propagation in an explosive-like manner with a partial evaporation/ejection of the reaction product. To overcome these issues, we attempted to mitigate the reactants' reactivity effect by alternately stacking a highly exothermic heat-releasing RMS with a relatively lower or medium RMS. The alternating combination of multi-sections of a pure metallic stack Pd/Al RMS with a high enthalpy of mixing with a pure metallic stack Ni/Al RMS that had a moderate enthalpy of mixing, deposited together on a SiO_2-coated Si-wafer, demonstrated the successful growth of a full, intact iRMS with a thickness of 3 μm with no spontaneous self-ignition surging. It was also confirmed by thermal measurements that this combination, by alternately stacking different RMSs with different exothermic heat enthalpy, is an efficient way of modulating the reaction heat release. Moreover, these grown multi-section Pd/Al-Ni/Al iRMS samples exhibited ignition, which led to a self-sustained propagating reaction that was feasible by a simple triggering with an electrical DC pulse at room temperature.

2. Materials and Experimental Techniques

The reactive multilayer thin-film systems were deposited using an electron-beam evaporation system (KVE & T-C500200, Korea Vacuum Tech, Ltd., Gimpo-si, Korea). The substrate holder of this system was not equipped with an active substrate cooling accessory. The substrates were prepared in accordance with the standard RCA cleaning method. Depositions were performed at room temperature and at a base pressure of 5×10^{-8} Torr. Generally, such deposition conditions promote nano-grain formation in a columnar-like structure that is basically controlled by the reduced atomic mobility of the deposited species on the substrate, and self-shadowing during film growth [23,24]. The substrates on which the iRMS was deposited were 4″ (100) bare Si-wafers, Si-wafers coated with thermally grown SiO_2 (1 μm thick), and Si-wafers buffered either with $SiN_x/RuO_x/Cr$ or $SiN_x/PdO_x/Cr$ stacks with respective layer thicknesses of 50 nm/60 nm/10 nm. The iRMS systems were deposited by an alternating e-beam evaporation of the reactant layers (Pd and Al for Pd/Al RMS and Ni and Al for Ni/Al RMS), from high purity targets (Al 99.9995%, Pd 99.9995%, and nickel 99.9995%). For the buffering (Ru, Pd) and adhesive (Ti, Cr) layers with respective thicknesses of 60 and 10 nm, the deposition was from targets with respective specific purities of 99.95%, 99.9995% and 99.9995%, 99.9995%. The thermal oxidation of Ru- and Pd-buffering Cr/Si-wafers was performed in a quartz tube furnace with flowing oxygen at 350 °C for 30 min (MTI Korea). After oxidation, the elemental distribution in the thickness direction was achieved by Auger electron spectroscopy (AES) using the depth profile technique (VG Scientific MicroLab 350). The SiN_x diffusion barrier layer, capping RuO_x and PdO_x-buffered Si-wafers, was grown by a low-pressure chemical vapor deposition method at 200 °C (LPCVD; E-1200, Centrotherm).

The thickness ratio of the bilayer reactants was determined in relation to a stoichiometric ratio of 1:1, corresponding to the maximum heat release from both Pd/Al and Ni/Al RMS. The Pd/Al-iRMS is a high-energy system and the Ni/Al-iRMS is a medium-energy system. The standard formation enthalpies for the stoichiometric ratio of 1:1 used here in this work for both systems are, respectively, -90 and -59 kJ/mol [25,26]. The different investigated iRMSs were grown with bilayer periods (δ) of either 50 nm, 100 nm or 200 nm, and total heights of either 1 μm, 2 μm, 2.4 μm or 3 μm.

The RMS films were integrated by a lift-off technique. The different tailored RMS host geometries on the photoresist-coated Si-wafer were a photo-lithographically transferred replica of motifs patterned on a chromium glass photomask using a contact aligner (EVG 640, EV Group, Austria) for ultraviolet light exposure up to 170 mJ/cm^2. The patterned geometry of iRMS traces emerged by dissolving the photoresist with acetone and, thus, systematically releasing the metal deposited on it. All iRMS samples were patterned mainly in shapes of small, squared pads connected either to rectangular frames or to serpentine paths, simple long stripes, and some simple large rectangular areas (see Figure 1). Such patterning would be practical for measuring, under a high-speed recording camera, the front speed of the propagating reaction as well as confirming the reaction's ability to propagate through different complicated bonding paths. The square pad in the iRMS patterns served as a starting local area for the ignition triggering.

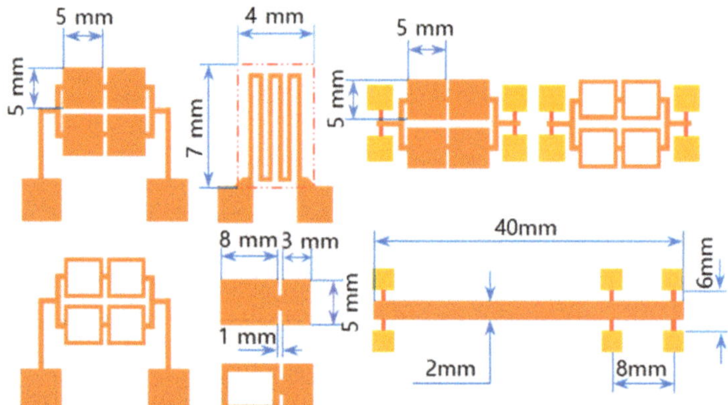

Figure 1. Different integrated reactive multilayer systems (iRMS) pattern shapes used for the observation of reaction propagation behavior (the initiation of the iRMS ignition using a Joule heating effect can be carried out by the application of the DC bias current either directly on the reactive pads or through gold contacts (yellow pads) connected with a titanium line (red stripes)).

Microstructure, composition, and elemental distribution analyses of the as-deposited iRMS films and the reaction products were performed on the lamellae of selected cross sections by using a field emission transmission electron microscope (FE-TEM, JEM-2100F HR, JEOL, Japan) and a Cs-corrected scanning transmission electron microscope (STEM, JEM-ARM200F, JEOL, Japan) equipped with an energy dispersive X-ray spectrometer (EDS) for elemental mapping. Transmission electron microscope (TEM) image scans were taken with an emissive gun operating at an acceleration voltage of 200 kV, and EDS analyses were performed with an acceleration voltage of 15 kV and a step size of ~0.25 μm. The lamellae-like samples were prepared by means of a dual-beam focused ion beam apparatus (FIB, Helios NanoLab, FEI, Netherlands) equipped with an omniprobe lift-out system.

The reaction propagation speed was determined by recording the reaction propagation path in the patterned iRMS with a high-speed camera (i-SPEED 221, iX Cameras, Frames at Max. Res.: 600 FPS, European Union). The initiation of the self-propagating reaction front was made by setting a DC power supply to 10 V/max. 1 A, which was applied between two tungsten needle probes with sharp tips.

The reaction heat of the as-prepared reactive multilayer systems was assessed by using a differential scanning calorimeter (DSC) (NETZSCH, DSC 404F1, Germany). The samples were freestanding, reactive multilayer nano-foil strips previously integrated by depositing an RMS film on 250 nm sacrificial Cu-layer-coated Si-wafers. The nano-foil strips were released from wafers by selective Cu-layer etching in an acid immersion solution of ammonium persulfate (20% $(NH_4)_2S_2O_8 + H_2O$). DSC measurements were carried out on samples of ~5 mg placed in an alumina crucible and heated in a temperature ranging from room temperature (RT) to 800 °C at a constant rate of 40 °C/min in an atmosphere of N_2 and Ar gases flowing at rates of 50 and 20 mL/min, respectively.

In cases where the cross section preparation of intact TEM specimens by FIB cutting was not possible, the morphological surface characterization of iRMS was carried out using a digital optical microscope (KEYENCE VHX-6000) and the crystalline structure and phase composition were examined by an X-ray diffractometer (XRD, SmartLab, Rigaku Corporation, Japan) operated in Bragg-Brentano θ-2θ geometry mode with a CuKα radiation source (k = 1.5405 Å) at 40 kV and 40 mA. The scans were performed with a 2θ step size of 0.02° in the 2θ range from 20° to 90°.

3. Results and Discussion

Before presenting the results of this work, we need to give information about our targeted experiments that allowed us to obtain the following experimental results. Our calculation based on thermal transfers (not presented here) showed that in order to obtain ignition at room temperature with a sustained propagating reaction in Pd/Al iRMS with a bilayer period of 100 nm grown directly on a Si-wafer, a reactive multilayer stack with a total thickness over 5 μm is needed. However, this is technically contradictory to what is mentioned in the introduction (the limitations of mechanical stress and photoresist patterning). Furthermore, the calculated thickness leading to ignition with reaction propagation at room temperature for a Pd/Al RMS with a bilayer period of 100 nm grown on a SiO_2-coated Si-wafer was determined to be around 1.6 μm and higher. However, the high thermal insulating SiO_2 layer did not help to reach the aimed iRMS thickness due to heat accumulation, which led to spontaneous self-ignition and a propagating reaction during deposition. Therefore, to remedy such issues by assuring heat sinking during multilayer stack deposition and heat damming during the self-sustained propagating reaction, a technical solution based on building an instant thermal barrier was carried out by inserting thin metal oxide buffer layers of either RuO_x or PdO_x between the Si-wafer and the iRMS film. These buffering layers, with a thermal conductivity that was much higher than that of SiO_2 during RMS deposition (for comparison, the thermal conductivity value of wafer-covering layers and the equivalent thermal conductivity of their superposition are given in Table 1), would easily dissipate the accumulated heat in the multilayer stack down to the Si-substrate sink and would, thereby, avoid self-ignition. In contrast, in an effective RMS ignition test, for which a sustained propagating reaction is expected, the high temperature attained as well as the dissipated heat would diffuse oxygen down from the oxide buffer layer and simultaneously oxidize the Si-wafer surface progressively along the propagation path. This results in the formation of an instant local thermal barrier interface, thereby avoiding self-propagating reaction quenching.

Table 1. Thermal conductivity values of the covering layers used in the preparation of Si-wafers before the integrated reactive multilayer systems (iRMS) deposition.

Wafer and Covering Layers	Wafer and Covering Layer Thicknesses	Thermal Conductivity (W/m.K)	Equivalent Thermal Conductivity (W/m.K)
Silicon (wafer)	500 μm (bare wafer)	148 [27]	
SiO_2	1 μm	1.4 [27]	
SiN_x/RuO_x/Cr	50 nm/60 nm/10 nm	25/50/91.3 [28–30]	36.2
SiN_x/PdO_x/Cr	50 nm/60 nm/10 nm	25/37.5 $\star$/91.3	32.3
Photoresist	3 μm	0.19 [31]	

$\star$ As the PdO_x thermal conductivity value has not been measured and was also not available in the literature, and since the palladium oxidation at low temperature is basically not complete, we estimated its thermal conductivity to be half of the pure metallic palladium 75 W/m.K [30].

Further, as supplementary information, Pd/Al iRMS grown on bare Si-wafers, which theoretically requires a thickness greater than 5 μm to be ignitable at RT, as well as on SiO_2-coated Si-wafers, which practically undergoes a systematic spontaneous self-ignition during the deposition process, was prepared and investigated for comparison.

3.1. Pd/Al iRMS Grown on a Bare Si-Wafer and a RuO_x-Buffered Si-Wafer (iRMS Total Thickness ~ 1 μm)

3.1.1. Characterization of the as-Deposited Pd/Al-iRMS

Figure 2 shows the microstructural analyses by TEM and the corresponding EDS elemental mappings of a Pd/Al-iRMS consisting of 20 bilayers with a period of 50 nm and a stoichiometric ratio of Pd:Al = 1:1. The iRMS was patterned and deposited simultaneously on two different substrate surfaces: a bare Si-wafer (Figure 2a) and a RuO_x-buffered Si-wafer (Figure 2b). For the TEM cross-sectional analysis of the Pd/Al-iRMS deposited directly on the bare Si-wafer, a successful multilayer stack having a layered microstructure with a sharp separation of single layers was observed. The EDS analysis, in turn, confirmed the presence of alternating metallic layers of Pd and Al (Figure 2a). However, for the Pd/Al-iRMS deposited on the RuO_x-buffered Si-wafer, the TEM cross section demonstrates that a spontaneous self-ignition occurred and was instantly accompanied by an explosive detachment of the reacted product. Thus, Figure 2b shows the layers deposited immediately afterwards when the previous deposited layers had been reacted and detached by self-ignition. The EDS analysis, in turn, confirmed the stability of the used buffering $Ti/SiN_x/RuO_x/Cr$ stack layers and the remaining alternately deposited metallic layers of Pd and Al (Figure 2b).

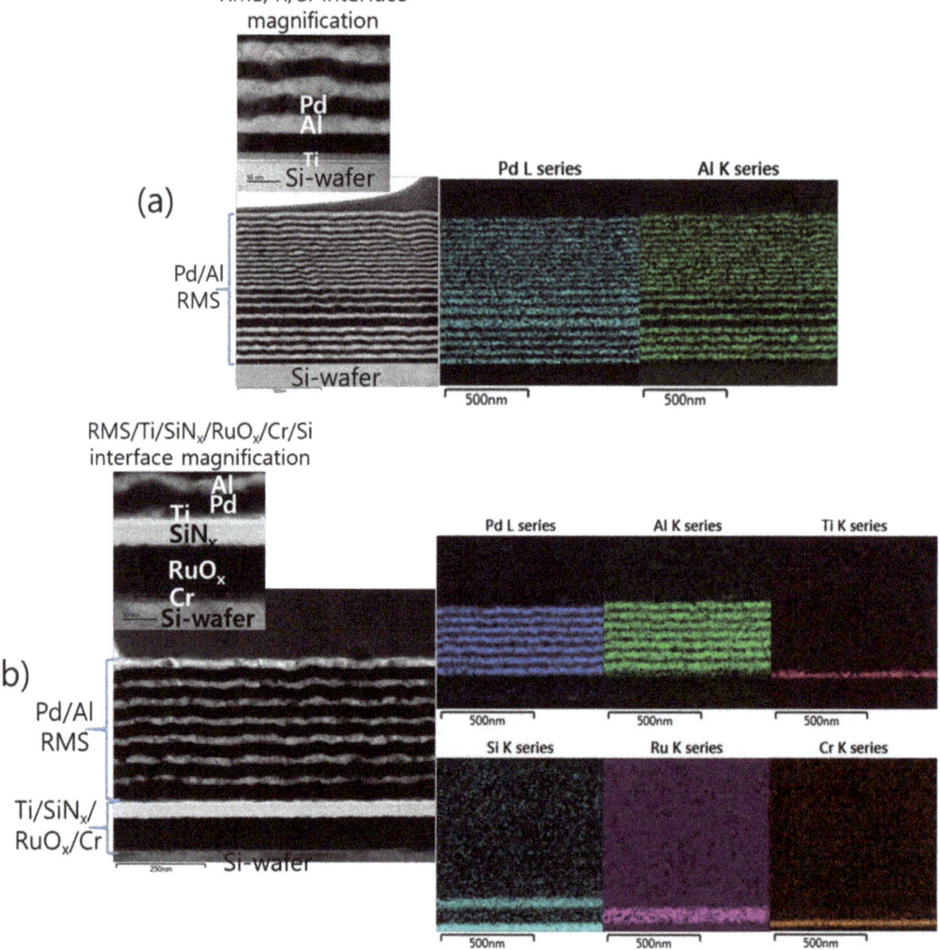

Figure 2. Transmission electron microscope (TEM) cross sections (**left**) and corresponding energy dispersive X-ray spectrometer (EDS) elemental mappings (**right**) of as-deposited Pd/Al RMS on the Ti/Si-wafer (**a**), and on the Ti/SiN$_x$/RuO$_x$/Cr/Si-wafer (**b**).

3.1.2. Characterization of the Reacted Pd/Al-iRMS

Figure 3 shows the reaction propagation across a patterned Pd/Al-iRMS of square frames grown on a Si-wafer. Since the total multilayer stack thickness (~1 µm) was lower than the minimal value (~5 µm) calculated for a self-maintained propagating reaction, preheating of the substrate at 100 °C for 5 min was indeed needed for the Pd/Al-iRMS ignition. The three snapshots presented in Figure 3, respectively, illustrate the ignition step initiated by a DC pulse of 10 V/max. 1 A (Figure 3a), the reaction propagation step manifested in an explosive yellow bright glow (Figure 3b), and the final morphology of the reaction product after the propagating reaction occurred (Figure 3c). The same explosive reaction propagation behavior was observed for the Pd/Al-iRMS grown on the RuO_x-buffered Si-wafer; however, higher preheating at 150 °C for 5 min was needed to initiate the reaction propagation, since the successfully deposited iRMS thickness was smaller.

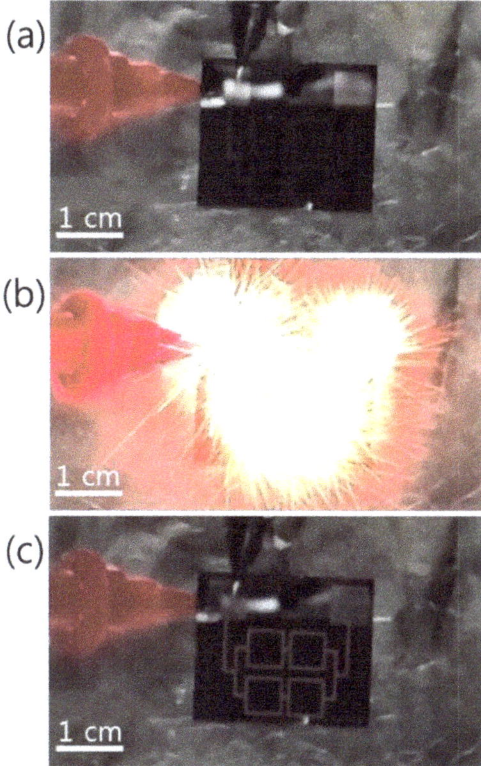

Figure 3. Reaction propagation across a patterned Pd/Al-iRMS of squared frames grown on a Si-wafer. (**a**) Initiation step; (**b**) Maintained propagating reaction step; (**c**) Reaction product morphology after the reaction step.

Close snapshots taken with a high-speed camera, used to determine the front speed of the propagating reaction, showed that the reaction behavior of Pd/Al-iRMS on the Si-wafer manifested an explosive combustion process with small quantities of ejected product particles (Figure 4a). Meanwhile, for the Pd/Al-iRMS on the RuO_x-buffered Si-wafer, a fiercer explosive combustion process with larger quantities of ejected product particles, leading to an almost full detachment of the patterned structure, was observed (Figure 4b). The propagation speeds of the reaction front, determined by the reactions in the Pd/Al-iRMS grown on the bare Si-wafer and the RuO_x-buffered wafer, were 38 and 50 m/s, respectively.

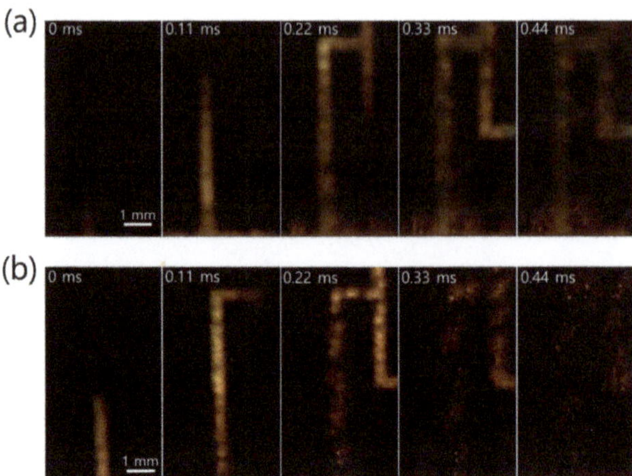

Figure 4. High-speed snapshots of the explosive reaction propagation in Pd/Al-iRMS, from left to right. (**a**) Pd/Al-iRMS grown on a bare Si-wafer; (**b**) Pd/Al-iRMS grown on a RuO_x-buffered Si-wafer.

The explosive reaction process with a high ejection or evaporation of the reaction product could be related to the attainment of a high local temperature. In fact, the small bilayer period of 50 nm that generates fast reaction propagation leads to a very high temperature level, as the produced heat does not have enough time to dissipate through the substrate. It should also be mentioned that a Pd/Al-iRMS with a smaller bilayer period has a higher self-initiation risk than systems with a larger bilayer period. Therefore, to overcome the bilayer size issues, only the iRMS with a bilayer period higher than 50 nm was investigated in the following study. This approach, according to the literature, should be seen as beneficial since reactive multilayer systems with bilayer periods between 100 nm and 200 nm are almost free of residual stress. RuO_x buffering is also suspected to provide an additional heat supply. In fact, RuO_x, at a high ambient temperature, could undergo more exothermic oxidation transformations such as RuO_3, RuO_4 etc. [32–34], which highly increase the temperature in the ignited iRMS and, thus, lead to the evaporation and ejection of the reaction product. Hence, in the subsequent experiments, Si-wafer buffering is prepared by replacing the RuO_x with a thin PdO_x buffer layer. This one possesses a high chemical stability in high temperatures [34,35].

3.2. Pd/Al-iRMS Grown on SiO_2-Coated and PdO_x-Buffered Si-Wafers (iRMS Total Thickness ~2 to 2.4 µm)

3.2.1. Microstructural Characterization of the as-Deposited Pd/Al-iRMS

Figure 5a,b show microstructural analyses by TEM and corresponding EDS elemental mappings of Pd/Al-iRMS structures grown on SiO_2-coated and PdO_x-buffered Si-wafers with a multilayer structure design composed of N = 24 bilayers with δ = 100 nm and N = 12 bilayers with δ = 200 nm, respectively. For each bilayer period, the iRMS frames were patterned and deposited simultaneously on both SiO_2-coated Si-wafer and PdO_x-buffered Si-wafer. In Figure 5a,b for the Pd/Al-iRMS with δ = 100 nm, it can be seen that spontaneous self-ignition occurred simultaneously on both the prepared Si-wafer-types, after about a 1.1 µm thickness of multilayer stack deposition corresponding to around 11 deposited bilayers. The intact reactive multilayers, grown next on the intermixed part, show for both prepared Si-wafer-types a neat, layered microstructure with a sharp separation between single layers with no significant premature intermixing. The thicknesses of the deposited reactant layers approached the expected nominal values corresponding to a 1:1 atomic ratio. EDS analyses also, by a chemical composition probing of the grown microstructure, confirmed the presence of an Al_xPd_y reaction product, which resulted from spontaneous self-ignition, and an additional stack on it of alternating reactant layers with Al and Pd compositions (Figure 5a,b).

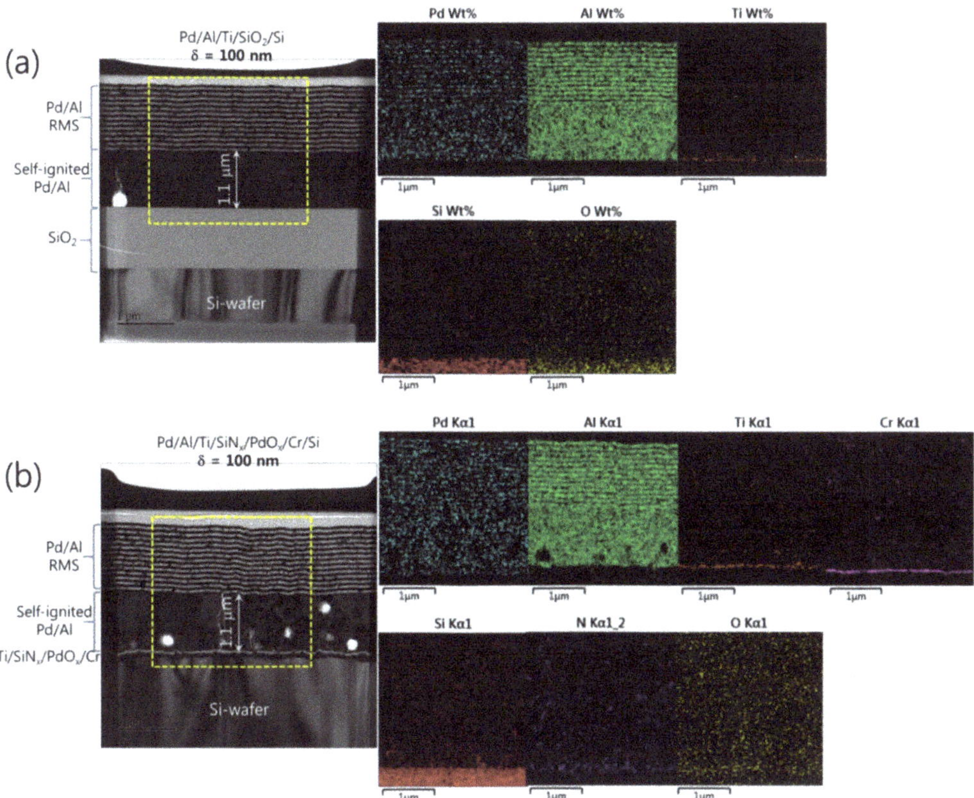

Figure 5. TEM cross sections (**left**) and corresponding EDS elemental mappings (**right**) of Pd/Al-iRMS with a bilayer thickness of δ = 100 nm and a bilayer number of N = 24. (**a**) Pd/Al-iRMS grown on a SiO_2-coated Si-wafer; (**b**) Pd/Al-iRMS grown on a PdO_x-buffered Si-wafer.

In Figure 6 for the Pd/Al-iRMS with δ = 200 nm and a total bilayer number of N = 12, different results were obtained. It can be seen that the spontaneous self-ignition and the resulting intermixing product occurred only for the Pd/Al-iRMS grown on the SiO_2-coated Si-wafer (Figure 6a), whereas the Pd/Al-iRMS grown on the PdO_x-buffered Si-wafer showed continuous stacking of individual (Al, Pd) reactant layers, confirming the absence of any self-ignition surging (Figure 6b). The stacking faults observed in the Pd/Al reactant layers are likely linked to the rough morphology of the surface of the PdO_x buffer layer. The reacted product on the SiO_2-coated Si-wafer formed when the Pd/Al reactants layer stack reached a thickness of about 1.5 µm, after which around five Pd/Al bilayers were further deposited. The delay of the spontaneous self-ignition in this iRMS with a 200 nm Pd/Al bilayer grown on a SiO_2-coated Si-wafer, compared to a 100 nm Pd/Al bilayer grown on either SiO_2 or PdO_x layers covering Si-wafers, could be due to fewer layer interfaces (potential sites of premature intermixing), which resulted in less barrier hindrance to heat conduction and, hence, less heat accumulation in the iRMS, thereby avoiding precocious start of spontaneous ignition. Again, EDS analyses of the Pd/Al-iRMS with a 200 nm bilayer period confirmed, for the iRMS on a SiO_2-coated Si-wafer, the presence of the intermetallic Al_xPd_y reaction product composition with an additional deposited Pd/Al layered structure over it (Figure 6a). On the other hand, for the iRMS grown on the PdO_x-buffered Si-wafer, the analysis probed a full layered structure of a continuous alternating reactant stack of Pd and Al compositions (Figure 6b).

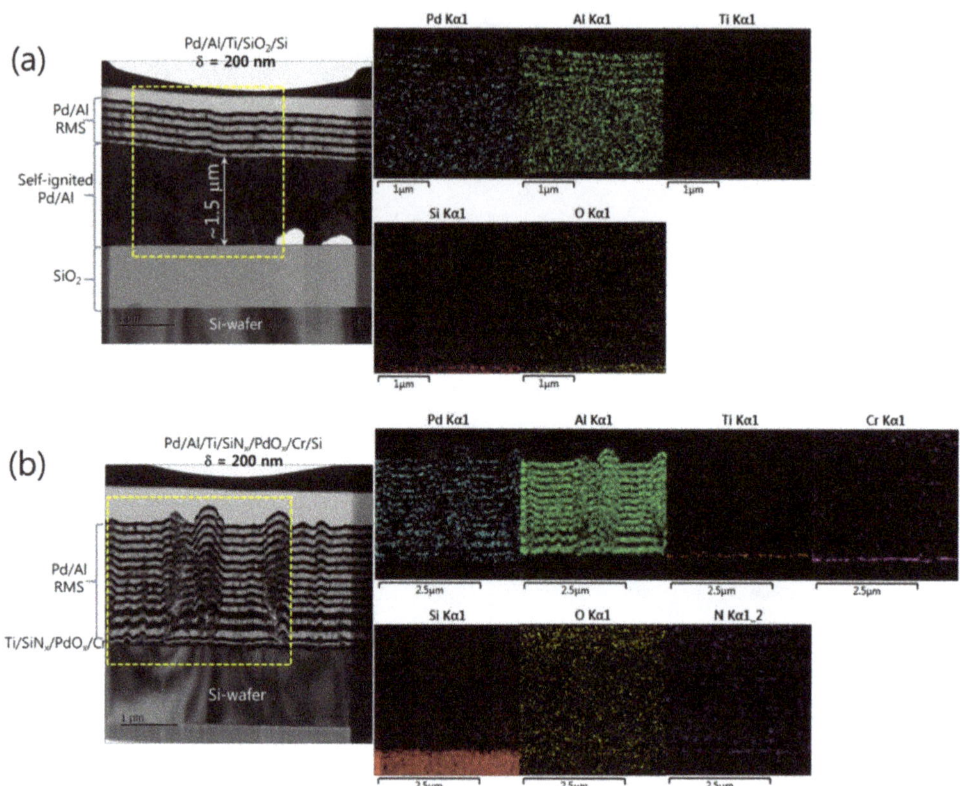

Figure 6. TEM cross sections (**left**) and corresponding EDS elemental mappings (**right**) of Pd/Al-iRMS with a bilayer thickness of δ = 200 nm and a bilayer number of N = 12. (**a**) Pd/Al-iRMS grown on a SiO_2-coated Si-wafer; (**b**) Pd/Al-iRMS grown on a PdOx-buffered Si-wafer.

3.2.2. Microstructural Characterization of the Reacted Pd/Al-iRMS

The following presents the microstructural analyses of the reaction product produced by electrically igniting the Pd/Al iRMS. The self-propagating exothermic reaction was observed only after preheating each of the four investigated samples. For samples that underwent spontaneous self-ignition during deposition (i.e., the Pd/Al-iRMS grown on SiO_2-coated Si-wafers with δ = 100 nm and 200 nm; and the Pd/Al-iRMS grown on a PdO_x-buffered Si-wafer with δ = 100 nm), ignition triggering happened at a preheating temperature of ~150 °C for 10 min. Meanwhile, for the sample with entire layer stack deposition (Pd/Al-iRMS grown on a PdO_x-buffering Si-wafer with δ = 200 nm), ignition triggering happened at a preheating temperature of ~100 °C for 10 min. For the convenience of presentation, Figure 7 depicts the microstructure of the reaction product of the samples of Pd/Al-iRMS with δ = 100 nm grown on a SiO_2-coated Si-wafer and Pd/Al-iRMS with δ = 200 nm grown on a PdO_x-buffered Si-wafer, after the self-propagating exothermic reaction passed through the reactive Pd/Al multilayer system. The individual Al and Pd layers are completely intermixed and, thus, no layered structure can be seen (TEM images, Figure 7a,b). EDS analyses also clearly confirm that all the Al and Pd reactant elements are completely intermixed, and no individual layer can be detected (EDS mapping, Figure 7a,b). Hence, since the ignition of the Pd/Al multilayer stack system had a stoichiometric ratio of Pd:Al = 1:1, it can systematically be demonstrated that the formed reaction product phase consisted of a homogeneous AlPd intermetallic compound.

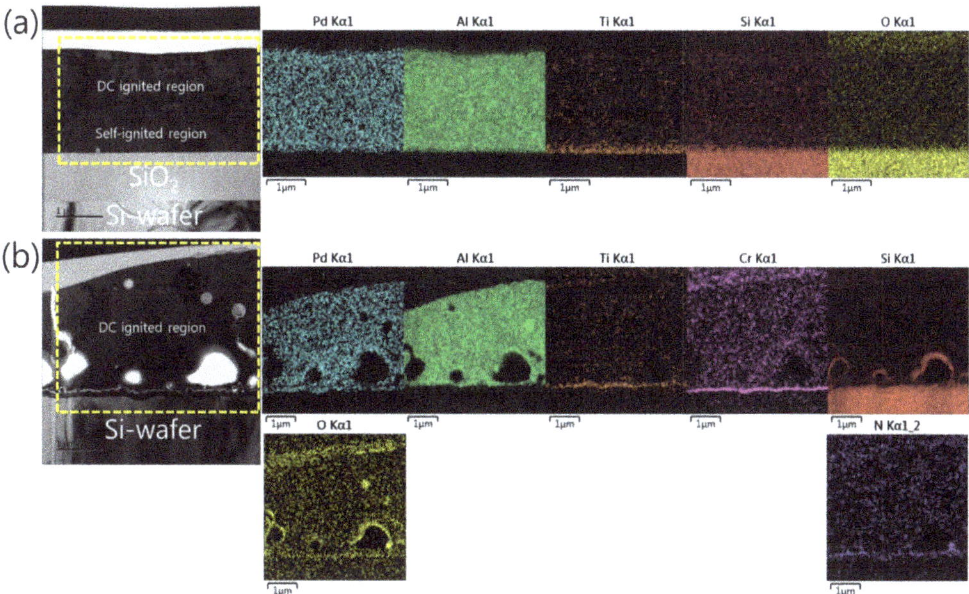

Figure 7. TEM cross sections (**left**) and corresponding EDS elemental mappings (**right**) of the Pd/Al-iRMS reaction product microstructure after the exothermic propagation reaction. (**a**) Pd/Al-iRMS with a bilayer thickness of δ = 100 nm grown on a SiO$_2$-coated Si-wafer; (**b**) Pd/Al-iRMS with a bilayer thickness of δ = 200 nm grown on a PdO$_x$-buffered Si-wafer.

Note that the difference in temperatures of the preheating process, which is required to ignite and maintain the self-propagating reaction process, indicates that in the iRMS samples that underwent spontaneous self-ignition during multilayer deposition, the preformed AlPd intermetallic reaction product located under the additional upper Pd/Al-iRMS stack could sink more heat into the Si-wafer substrate. Thus, for the upper iRMS to be ignited, it required more energy and a higher temperature.

Furthermore, the analysis of the interface structure between the iRMS and Si-wafer substrate showed that no remarkable change occurred before and after the ignition of the samples. For the Pd/Al-iRMS grown on the SiO$_2$-coated Si-wafer substrate, the interface structure configuration was found to be similar before and after the reaction propagation. For the Pd/Al-iRMS grown on the PdO$_x$-buffered Si-wafer substrate, the iRMS/(Ti/SiN$_x$/PdO$_x$/Cr)/Si-wafer interface of the as-deposited, and then ignited, iRMS is presented by the cross-sectional TEM images in Figure 8a,b. It is shown that, as a part of the PdO$_x$ buffer layer diffused down through the adhesive Cr layer, the result was that the adhesive Cr layer was totally pushed a few nanometers up and, thus, the PdO$_x$ established direct contact with the Si-wafer substrate. Note that the interface structure configuration was found to be practically similar for both situations, before and after the reaction ignition.

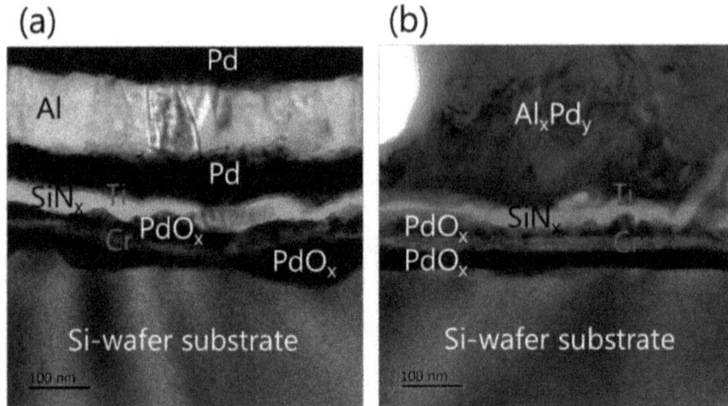

Figure 8. TEM images showing the interface structure configuration of the iRMS grown on a PdO$_x$-buffered Si-wafer. (**a**) As-deposited RMS; (**b**) After reaction ignition.

Such an interface structure configuration was also confirmed by STEM and EDS analyses, as shown in Figure 9. In fact, Pd diffusion happened, in the preparation of the buffered Si-wafer, during the 60 nm thick Pd film thermal oxidation. This behavior was elucidated by an AES analysis of the as-deposited, room temperature Pd film on a Cr-coated Si-wafer substrate and its thermal oxidation at 350 °C for 30 min (Figure 10). For the as-deposited Pd film at room temperature, a neat separation between Pd and Cr layers and the Si-wafer was probed (Figure 10a). However, for the thermally oxidized Pd film, the deep diffusion of the Pd element through the Cr layer and inward into the Si-wafer with oxidation limited at the Cr layer was probed (Figure 10b). It is also worth mentioning that the operation of the instantaneous oxidation of the Si-substrate surface during the iRMS ignition remains effective. However, given that the amount of oxygen stored in the PdO$_x$ buffer layer was not very high, the diffusion and distribution of the oxygen into the Si-substrate surface could not be detected clearly by the probing EDS system used in this study.

3.2.3. Thermal Characterization of Pd/Al-iRMS

Figure 11 shows DSC curves of freestanding nano-foils of Pd/Al-RMS with bilayer thicknesses of 100 and 200 nm. These RMS nano-foils were released from the SiO$_2$/Si-wafer coated with a Cu layer of 250 nm by immersion in an acidic solution bath. Irrespective of the bilayer period, two larger exothermic peaks were produced with their start and end temperatures having nearly the same values. The first peak began at around 240 °C and the second peak started at around 425 °C. Furthermore, it can be seen that the first peak height increased with increasing bilayer period thickness, a behavior commonly related to the stored energy loss by the higher intermixed interface volume in the reactive system having a lower bilayer period. The heat released was calculated using the principle of integrating the heat flow with respect to time. The total heat of the reaction was determined to be 797.3 J/g for the system with a 100 nm bilayer and 778.6 J/g for the system with a 200 nm bilayer. These detected reaction enthalpy values are significantly lower than the expected theoretical value of 1260 J/g [8,36]. Therefore, the relatively large difference between the experimentally measured reaction heat and the theoretically predicted value could be mainly attributed to the formation of the reaction product by the spontaneous self-reaction that happened in the prior deposited layers by self-ignition into the e-beam evaporator chamber during the deposition process.

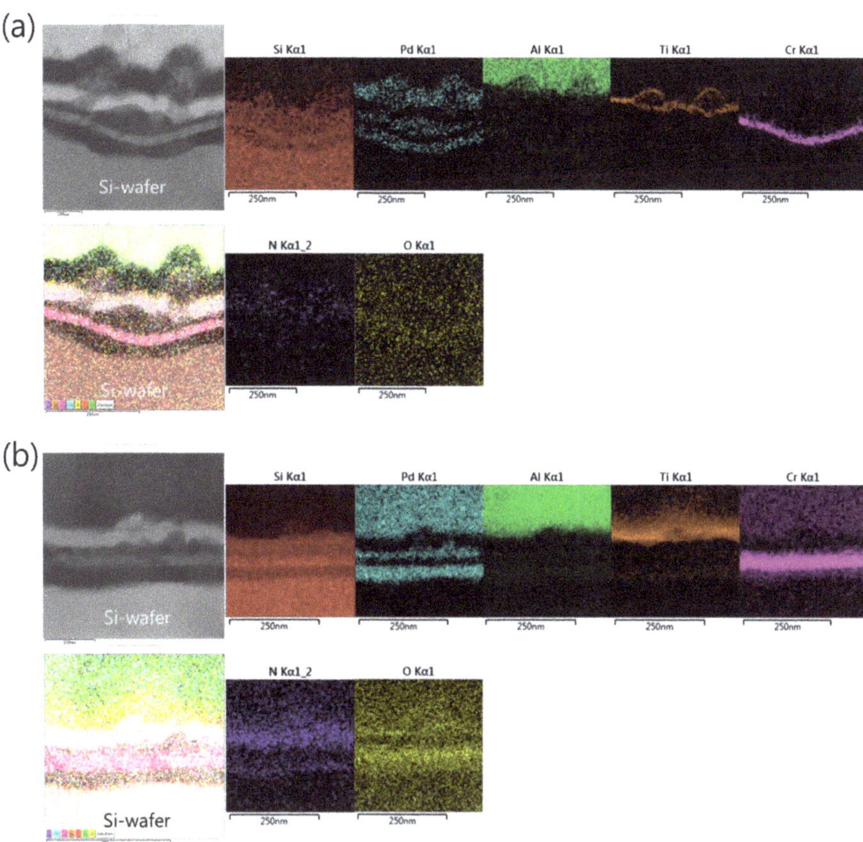

Figure 9. Scanning transmission electron microscope (STEM) images (upper left) and corresponding EDS elemental mappings of the interface structure configuration of the iRMS grown on a PdO_x-buffered Si-wafer. (**a**) As-deposited RMS; (**b**) After reaction ignition.

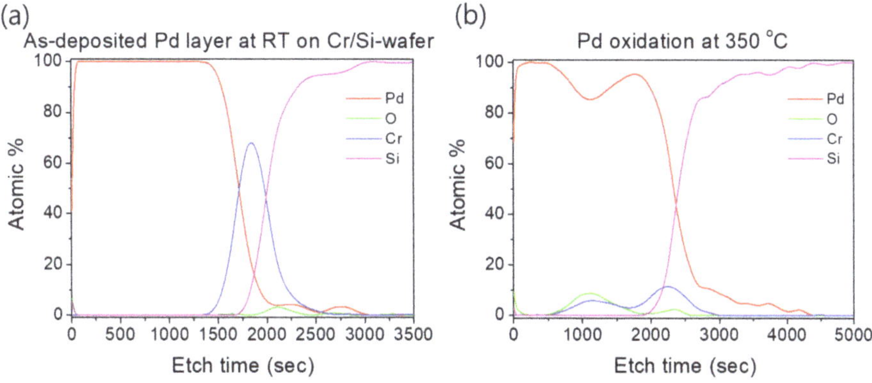

Figure 10. AES profiles showing the effect of temperature on Pd and Cr distribution during the thermal oxidation process. (**a**) As-deposited Pd film on a Cr-coated Si-wafer; (**b**) Pd film oxidation at 350 °C for 30 min.

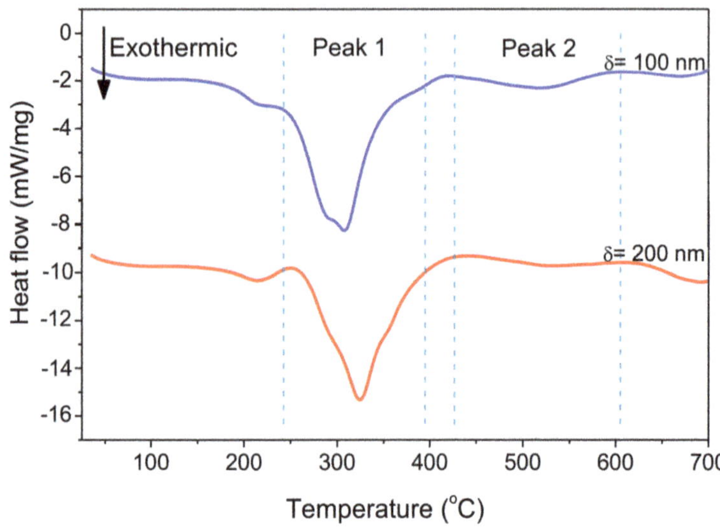

Figure 11. DSC curves of Pd/Al-iRMS with different bilayer thickness.

The spontaneous self-ignition occurred in the prepared Pd/Al-RMS freestanding nano-foils despite the intercalation of a highly thermal-conductive layer of Cu material (which can sink the accumulated heat in the reactant layers and, thus, avoid the reaction ignition), was confirmed by the observed formation of the reaction product Al_xPd_y in the microstructural characterization depicted in Figure 12. This figure presents the TEM microstructure and the corresponding EDS analysis obtained from a Pd/Al-iRMS with a bilayer period of 100 nm and an expected total thickness of 30 bilayers grown on a SiO_2/Si-wafer coated with a thin Cu layer of 250 nm. The cross-sectional images clearly depict the presence of three surface sections, respectively, corresponding to the lower copper layer, the reaction product formed by spontaneous self-ignition, and the deposited intact Pd/Al iRMS layers at the top. The thickness of the Al_xPd_y reaction product stated that the self-ignition was initiated after around 1 µm thickness of the earlier deposited Pd/Al stack layers. This scenario was also similar to the spontaneous self-ignition that had taken place, at around 1 µm of layer stack deposition, in the Pd/Al-iRMS with a bilayer period of 100 and 200 nm grown directly on a SiO_2-coated Si-wafer.

From the deposition experiments described above, the investigated Pd/Al-iRMS with a bilayer period of 100 nm showed spontaneous self-ignition accompanied systematically by a sustained exothermic propagation reaction for all stacked Pd/Al reactive multilayers with thicknesses exceeding 1 µm, deposited either directly on a SiO_2/Si-wafer (iRMS on a high-thermal insulating substrate), metal/SiO_2/Si-wafer or metal oxide/Si-wafer (iRMS on a moderate thermal insulating substrate). Such behavior can apparently be related to the Pd/Al system's high exothermic enthalpy of formation ($\Delta H_{Al-Pd} \approx 90$ kJ/mol-atom), a value that is substantially higher than the 30 kJ/mol-atom required for generating self-sustained propagating reactions in thin reactive multilayer films [10]. Accordingly, small nanometric inter-diffusions at the Pd/Al reactants interfaces, even for a system with a small number of bilayers, could release enough energy output for spontaneous self-ignition triggering and to simultaneously sustain continuous exothermic propagation reactions on the silicon substrate. Therefore, the main issue to be addressed regarding the Pd/Al-iRMS is the low excitation energies causing instabilities, which lead to an enhanced risk of spontaneous reaction ignition as well as pre-intermixing activation at the reactant interfaces during the deposition processes.

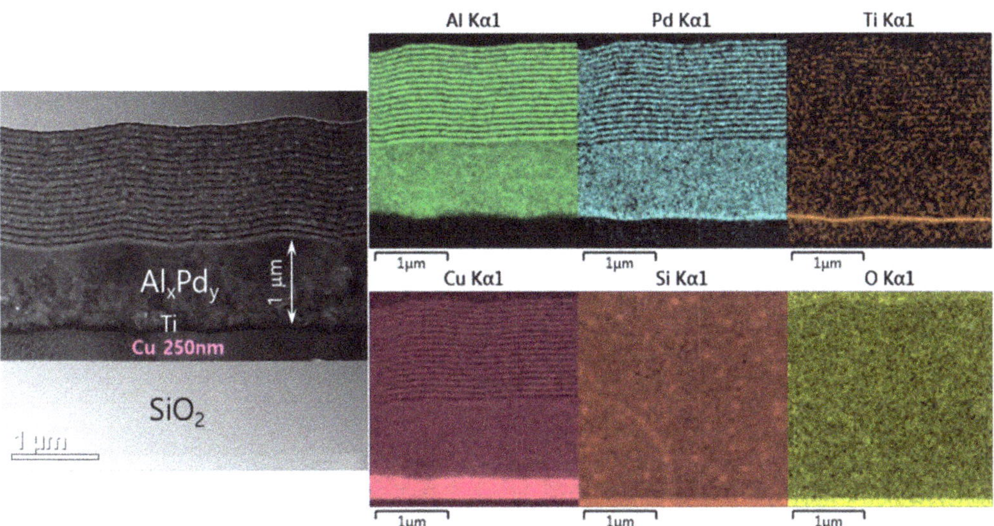

Figure 12. TEM cross section microstructure (**left**) and corresponding EDS elemental mappings (**right**) of Pd/Al-iRMS with a bilayer thickness of δ = 100 nm grown on a 250 nm Cu film-coated SiO_2/Si-wafer.

Since, in this work, the photoresist was adopted as the material for the lift-off patterning of the integrated RMS structures, further information worth noting is related to the interaction of the RMS and the photoresist. Because of the very low thermal conductivity of the photoresist, the few bilayer number depositions of the reactive Pd/Al system directly on the photoresist would undergo spontaneous self-ignition and a continuous propagation reaction spreading over the whole four-inch Si-wafer surface. In fact, the high reactive Pd/Al system, the mixing enthalpy and the low thermal conduction through the photoresist, caused the spontaneous reaction to occur in a strongly explosive manner, simultaneously accompanied by the detachment/ejection of the reacted product, while leaving the photoresist film still tightly adhered to the wafer with an appearance of some dendrite-like branch structures outlined on its surface (Figure 13a). This explosive exothermic reaction with the detachment/ejection of the reacted system could take place serval times once the deposited multilayer reactants reach a critical thickness (Figure 13a′). The explosive reaction could also occur during the last step of depositing the last upper reactant layer, which lets only the iRMS motifs lodged into the litho-patterned photoresist trenches appear on the wafer surface. These embedded iRMS did not undergo ignition due to their direct contact with the substrate having slightly higher thermal conductivity compared to the one of the photoresist (Figure 13a). Images in Figure 13b,b′ present the patterned wafer surface morphology after the photoresist lift-off.

To overcome these encountered problems, which mostly are related to the high enthalpy of mixing/formation leading to an explosive reaction and the high inter-diffusion rate leading to spontaneous self-ignition, the modulation of the stored chemical energy and the rate of heat release would induce a substantial change in the characteristics of the reaction's behavior. Accordingly, a reactivity tuning approach based on alternately stacking two different reactive multilayer systems, respectively, with different enthalpies of the mixing and formation of intermetallic compounds as well as different inter-diffusion rates will be investigated in the next section.

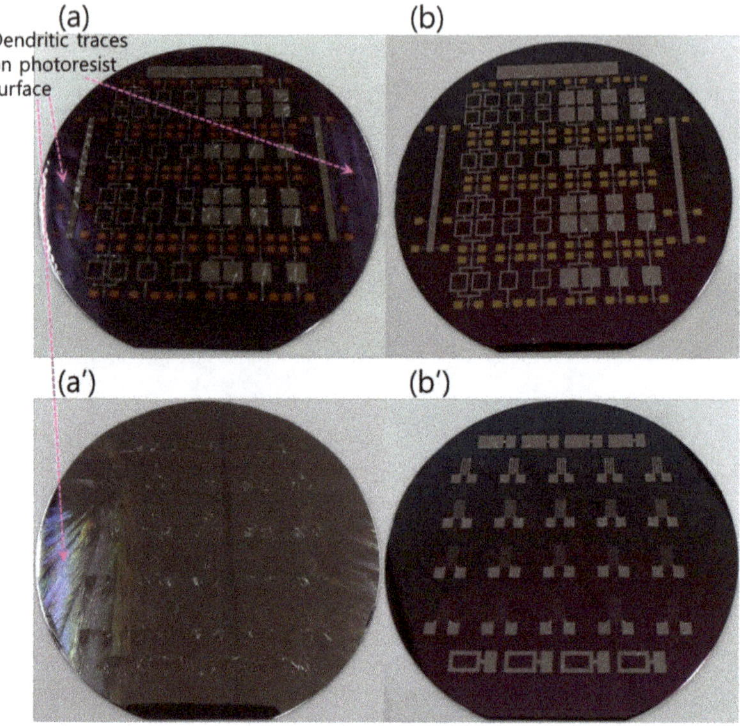

Figure 13. Dendritic traces on a photoresist film surface after spontaneous self-ignition. (**a**,**a'**) Wafer surfaces before photoresist lift-off; (**b**,**b'**) Wafer surfaces after photoresist lift-off.

3.3. Pd/Al-Ni/Al Multi-Section (MS)-iRMS Grown on a SiO$_2$-Coated Si-Wafer (MS-iRMS Expected Total Thickness ~3 µm)

As the experiments outlined above demonstrate, for an RMS deposited on a substrate that has a low thermal conductivity, the high exothermic enthalpy of reactants mixing, and the resulting high temperature systematically lead either to spontaneous self-ignition during reactant deposition or to an explosive reaction with the detachment/ejection of the reacted product. Therefore, the modulation of the released heat and its subsequent temperature by stacking two reactive material systems with high and medium energy release (designed as hRMS and mRMS), respectively, is considered to be a potential solution. In this context, we investigated a sandwiched hRMS-mRMS reactive structure (see Figure 14), wherein the alternating hRMS sections correspond to the already used Pd/Al-RMS that had a higher enthalpy of mixing, and the alternating mRMS sections corresponded to the Ni/Al-RMS that had a medium enthalpy of mixing. The reactive Ni/Al system was selected for the relatively small difference between the CTE value of nickel (12 mm/mm.°C) and that of palladium (11.7 mm/mm.°C) in the reactive Pd/Al system. Indeed, as the temperature is increased, a CTE mismatch will induce compressive stress in a multilayer system that has a higher CTE value, and this tends to inhibit atomic diffusion into compressed system layers [37]. Hence, such effect would increase the activation energy of ignition, delay the intermixing process, and also reduce the propagation of the reaction rate [38]. For the dimensions of the sections in the reactive system structure, the thickness of the Pd/Al-hRMS section (S$_{hRMS}$) was selected to be 0.5 µm to avoid spontaneous self-ignition, which was previously demonstrated to take place in Pd/Al-RMS on a SiO$_2$-coated Si-substrate once the sputter deposition exceeded 1 µm. Moreover, since the CTE of Ni/Al-mRMS is slightly higher than that of Pd/Al-hRMS, the thickness of the Ni/Al-mRMS section (S_{mRMS}) was selected to be smaller or equal to 0.2 µm, and, thus, its resulting volume would allow low tensile stress in the high exothermic Pd/Al system sections of the major volume fraction in the structure and would, consequently, avoid any noticeable change in the behavior of its properties.

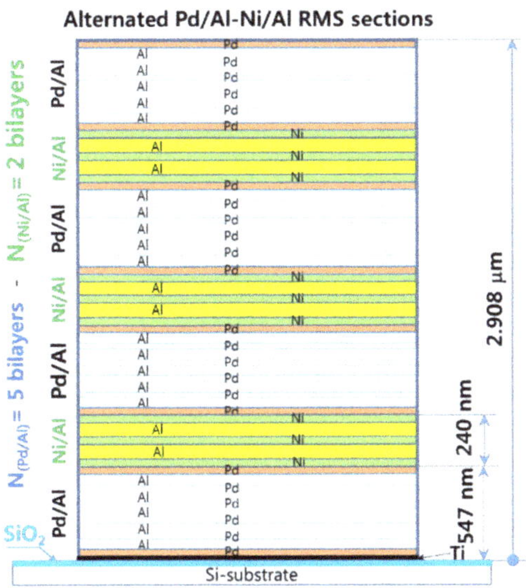

Figure 14. A cross-sectional schematic of the Pd/Al-Ni/Al MS-RMS design showing alternating sections of reactive materials with the corresponding section thicknesses. Design parameters: Pd(47 nm)/Al(53 nm) 5 bilayers, Ni(40 nm)/Al(60 nm) 2 bilayers, Pd/Al-section thickness S_{hRMS} = 500 nm (4 sections) and Ni/Al-section thickness S_{mRMS} = 200 nm (3 sections).

Furthermore, as Pd and Ni reactant components have a very negligible enthalpy of formation and mixing and could form a nearly ideal solution [39,40], therefore, to minimize further heat accumulation, which reduces the risk of spontaneous reactions and more premature interdiffusion during the deposition of hRMS-mRMS reactive sections, each Pd/Al-hRMS section deposition started and ended with the Pd layer and each Ni/Al-mRMS section deposition started and ended with the Ni layer. This structural arrangement ensures that Pd and Ni layers are always in contact at the interfaces between both hRMS and mRMS sections.

3.3.1. Microstructural Characterization of the as-Deposited Pd/Al-Ni/Al MS-iRMS

Figure 15 presents TEM and EDS microstructural analyses of a Pd/Al-Ni/Al reactive multi-section system grown on SiO$_2$-coated Si-wafer. The TEM cross section shows a total grown Pd/Al-Ni/Al system composed of four Pd/Al hRMS sections alternating with three Ni/Al mRMS sections. Both under-systems, hRMS and mRMS, have the same bilayer period δ of 100 nm and the same molar ratio of 1:1. The TEM and EDS images show a neat typical multilayered structure where the alternating single layers are clearly identified in each section composing the whole reactive multi-section system. The clear interfaces between the reactant layers in each section reveal that neither pre-intermixing reaction nor spontaneous self-ignition and reaction propagation occurred during the sputtering process. The thicknesses of the single deposited layers of each section approach the expected nominal values corresponding to a 1:1 atomic ratio: 47 and 53 nm for the bilayer Pd/Al and 40 and 60 nm for the bilayer Ni/Al.

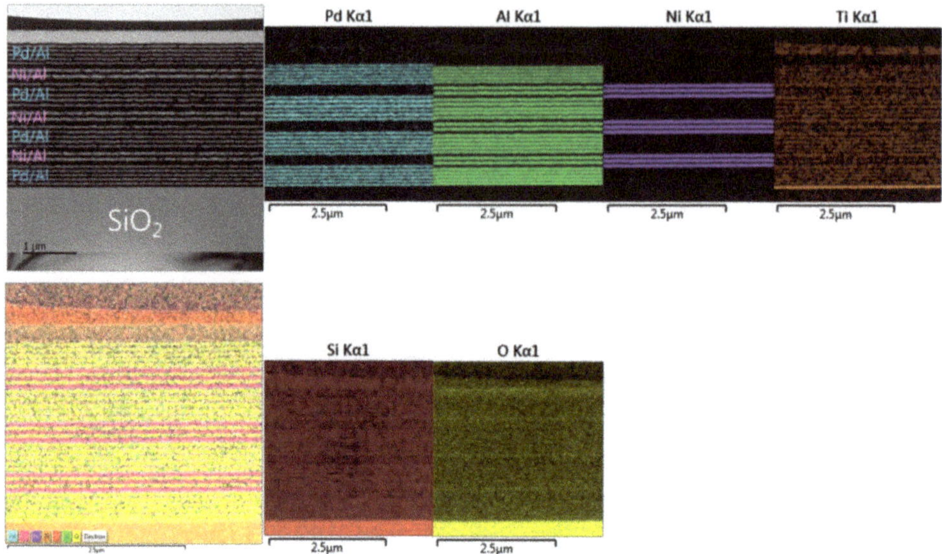

Figure 15. TEM cross section image (**upper left**) and the corresponding EDS elemental analysis (**upper right** and **lower left and right**) for the microstructural characterization of the as-deposited Pd/Al-Ni/Al MS-iRMS on the SiO$_2$-coated Si-wafer.

Regarding the electrical ignition of this reactive multi-section structure, it was found that it could easily be initiated at room temperature by applying an electric DC pulse of 10 V/max. 1 A. After initiation, the reaction front spread evenly through the structured Pd/Al-Ni/Al MS-iRMS in a self-sustained manner without any further energy supply (no preheating). The images in Figure 16 show the expansion progress of a bright glow representing the steady self-propagation of the reaction front along Pd/Al-Ni/Al MS-iRMS patterned in the linked square frame structure. The patterned Pd/Al-Ni/Al MS-iRMS frame width presented here was 500 μm, and it was also demonstrated that the minimum bond frame width showing a self-propagating exothermic reaction could be reduced to 25 μm. For the self-propagating reaction front to pass through complex frame geometries, the iRMS patterning frames must be connected to each other to assure the reaction propagation on the whole patterned wafer. Figure 17 shows an example of interconnected patterning design allowing the propagation reaction to run through the entire iRMS on the wafer. In this example, the ignition could be initiated at the edge of the wafer with the resulting reaction guided to the wafer's center from which it propagates to all wafer regions. Furthermore, it should be noted that the ignited reaction in Pd/Al-Ni/Al MS-iRMS propagated in an explosive manner, causing the ejection of small debris particles from the reacted product. However, the main reacted multi-section structure showed good adhesion to the substrate.

3.3.2. Thermal Characterization of the as-Deposited Pd/Al-Ni/Al MS-iRMS

Figure 18 shows the DSC curve of the Pd/Al-Ni/Al MS-RMS freestanding nano-foil grown on the SiO$_2$/Si-wafer coated with a Cu layer of 250 nm and released by immersion in an acidic solution bath. Three peaks that spread out towards high temperatures are observed. Such behavior is similar to the case of the usual DSC trace measured on a single Pd/Al RMS, in which the activation energy of the diffusion of Pd in Al is on the order of 190.6 kJ/mol [41], but in contrast with the case for a Ni/Al reactive system. For this, the exothermic peaks typically appear at lower temperatures in a range from 240 °C to 400 °C [1,12,16,18,42], since the activation energy of the diffusion of Ni in Al is slightly lower on the order of 145.8 kJ/mol [41]. Thence, the measured exothermic reaction of the Pd/Al-Ni/Al MS-RMS would be a superposition of the exothermic peaks of Pd/Al sections with those of the Ni/Al sections. The shift of the Ni/Al exothermic peaks to higher temperatures to superpose with those of the Pd/Al system could be related to the expansion difference of the two alternating Pd/Al and Ni/Al reactive systems ($Ni_{(CTE)} > Pd_{(CTE)}$). Thus, internal compressive stress is induced into the Ni/Al sections. Such a stress state reduces the Ni diffusion rate in the Al and consequently increases the Ni diffusion activation energy [37,38], which constrains the Ni mobility such that it takes place at a higher temperature for the reaction. The extracted total exothermic heat released from the reactive Pd/Al-Ni/Al system was determined from the DSC measurement to be equal to 1191.7 J/g, an amount that is slightly higher than the theoretically predicted enthalpy of formation for the AlNi reaction product ($\Delta H_{AlNi} = 1120$ J/g) [36] and slightly lower than that for the AlPd reaction product ($\Delta H_{AlPd} = 1260$ J/g) [36]. Hence, the energy produced in the Pd/Al-Ni/Al system confirms that the stacking of combined hRMS/mRMS reactive systems, with different reaction enthalpy, is an efficient tool for heat tuning (either heat storing or heat releasing; both must be addressed in reactive bonding applications).

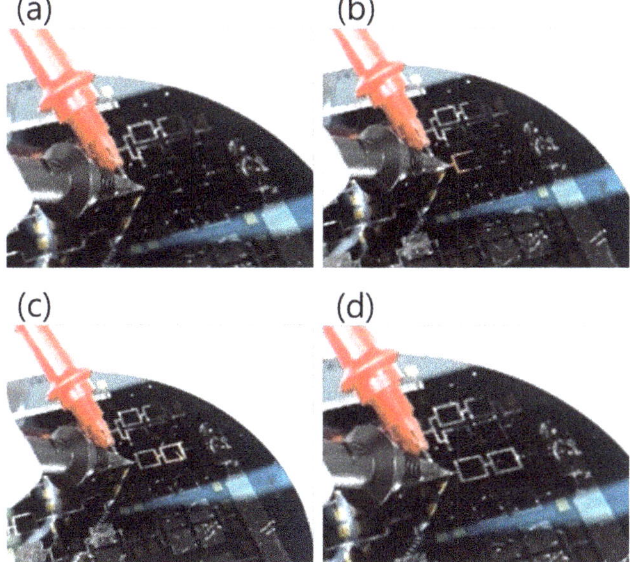

Figure 16. Images showing reaction front propagation steps across a patterned Pd/Al-Ni/Al MS-iRMS of squared frames grown on a SiO$_2$/Si-wafer. (**a**) Initiation step; (**b,c**) Self-maintained propagating reaction steps; (**d**) Reaction product morphology after the reaction step.

3.3.3. Microstructural Characterization of the Reacted Pd/Al-Ni/Al MS-iRMS

Figure 19 shows an optical microscopic image of the top-view surface morphology of the reacted Pd/Al-Ni/Al MS-iRMS. The reaction was ignited with an electrical DC pulse (10 V/max. 1 A) at room temperature, which resulted in a self-maintained propagating reaction that spread over the whole patterned RMS area within a few milliseconds. The reaction product surface morphology exhibited a spherulitic-like dendrite crystallite structure [43]. The dendrite orientations were seen to have emerged from an origin and extended toward the reaction front propagation direction to form maple leaf-like patterns. The manifestation of the reaction product in dendritic form is a common solid crystallization grown from the melt under a fast-cooling rate [44,45]. The steeper temperature gradients, systemically established behind the RMS reaction front propagation, would spontaneously introduce significant thermal stress gradients associated with the solidification process [46]. Further, thermal stress states could also increase more if the thermal expansion coefficients of the dual section materials are significantly different [38]. These factors all generated stresses that enhanced the reacted product's susceptibility to cracking or disintegration through the associated strains. Note that the self-sustained propagating reaction in our electrical ignited sample occurred in an explosive manner, ejecting reacted product in debris particles detached from the coarsened dendrite trunks and, hence, inter-dendritic regions were formed, as seen in the magnified view in Figure 19 (right). From this exhibited behavior, it is evident that, in the multi-section stack of the iRMS, adopting a section type composed of reactive reactants that have a high reaction product ductility property at room temperature would attenuate the propagating reaction effect, causing disintegrated product.

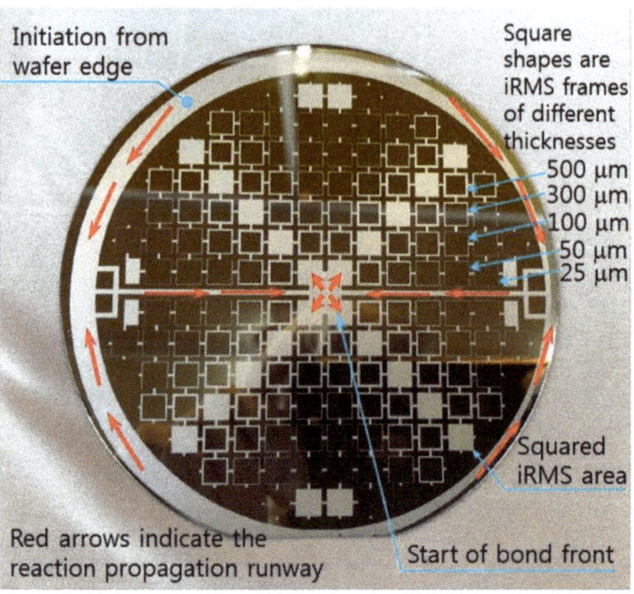

Figure 17. Example of an interconnected patterning design allowing the propagating reaction to run through the entire iRMS on the wafer.

Sample preparation by FIB for a cross-sectional analysis of the Pd/Al-Ni/Al MS-iRMS reaction product was not possible. The intense induced thermal residual stresses in the grown dendrite trunks made them very friable and fragile and, thus, the prepared lamella structure easily collapsed by the ion beam milling. Notwithstanding this obstacle, some cross-sectional analyses were carried out in order to gain more insights into the effects controlling the reaction propagation and the reaction product formation. The inset optical microscopic image in Figure 20 indicates different area zones where FIB-prepared cross section specimens were cut from a sample of the Pd/Al-Ni/Al MS-iRMS reacted product. Figure 20a depicts a TEM image of an FIB cross section of a bare area located between the inter-dendritic regions (zone A). The image confirmed that the ejected reaction product was accompanied by the local fragment detachment of the SiO_2 thermal isolating layer. Figure 20b (left) depicts a TEM image of a FIB cross section undertaken in a region where a partial reacted product was detached/ejected during the reaction propagation (zone B). The remaining product was found to belong to the first deposited Pd/Al section on the Ti/SiO_2/Si-substrate. This has been confirmed by an EDS analysis, which revealed only the formation of Al_xPd_y intermetallic compound (traces of Pd and Al elements) and the absence of nickel, which should structurally be in the subsequent Ni/Al section if it were present (Figure 20b, right). Figure 20c (left) depicts a TEM image of a FIB cross section of a region from where a partially reacted product was detached/ejected (zone C); the image shows that it concerns a remaining underneath part that corresponds to unreacted Pd/Al multilayer stack residue of the first section in direct contact with the Ti/SiO$_2$/Si-substrate. The elemental chemical composition of the intact Pd/Al layers in this unreacted residue was confirmed by the EDS analysis in Figure 20c (right).

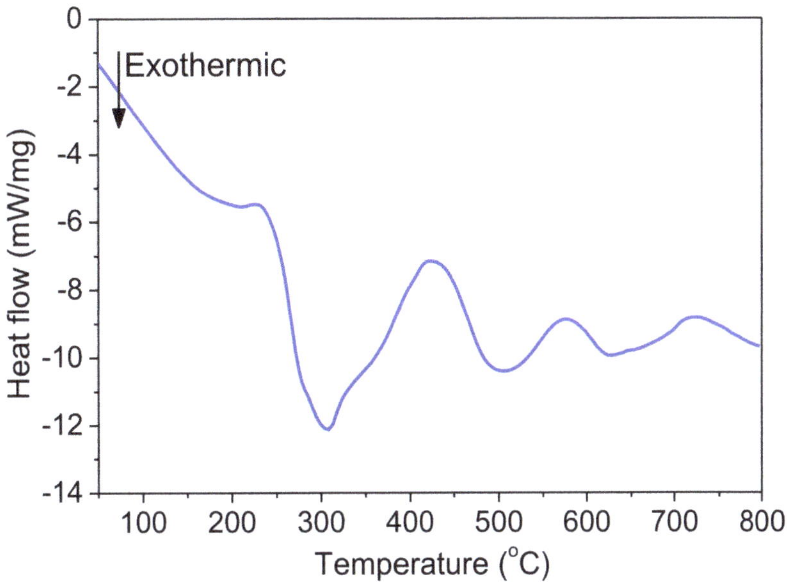

Figure 18. DSC characterization of the as-deposited Pd/Al-Ni/Al MS-iRMS.

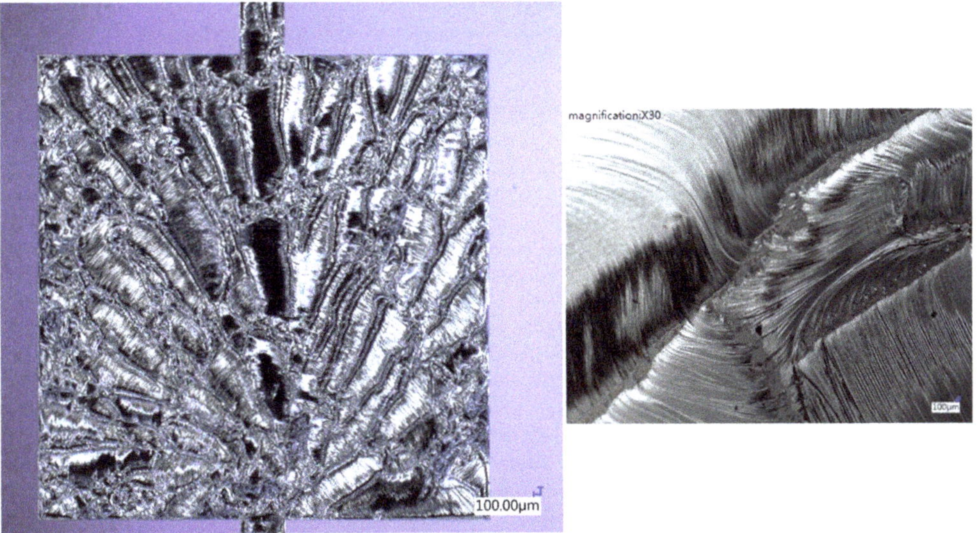

Figure 19. Optical microscopic image showing the microstructural morphology of the reacted Pd/Al-Ni/Al MS-iRMS. The right image is a magnified view depicting inter-dendritic regions formed during reaction propagation.

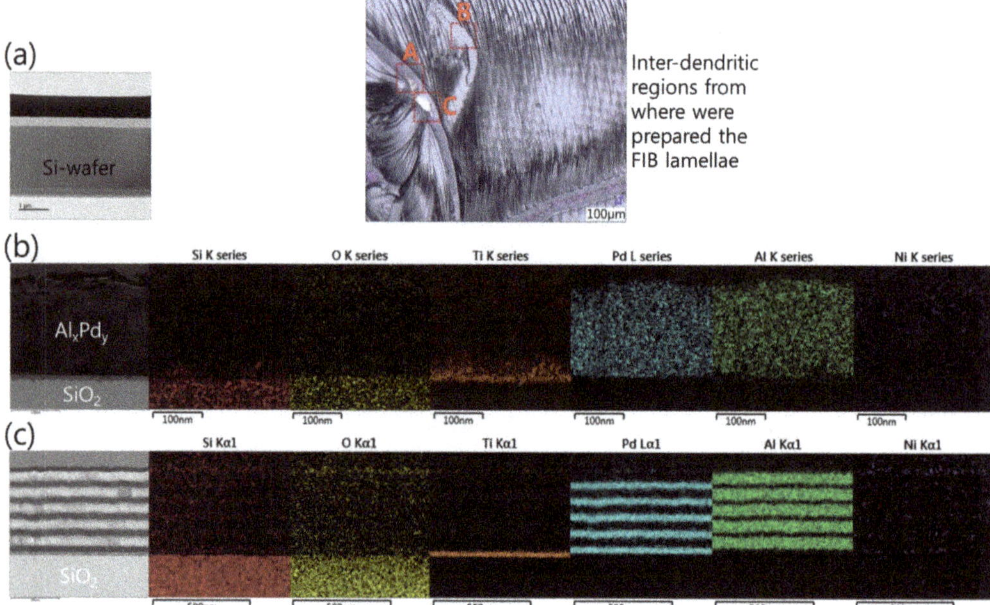

Figure 20. TEM cross section images and EDS analyses performed on different FIB cross sections selected from different regions of the upper surface of the reacted Pd/Al-Ni/Al MS-iRMS (see the inset optical microscopic image). (**a**) FIB cross section from a bare inter-dendritic area (zone A); (**b**) FIB cross section from a region where a partial reacted product was detached (zone B); (**c**) FIB cross section from a region where a partial unreacted multilayer stack residue was still attached to the substrate (zone C).

One more interesting point to raise is that it was found that the dendritic structure of the reaction product took place only when the iRMS was ignited in an open-air atmosphere or vacuum chamber. However, when the iRMS was ignited under a high punch pressure, the reaction product was a dense nodular-like structure. Figure 21 shows an optical microscopic image of the reaction product microstructure resulting from a multi-section Pd/Al-Ni/Al iRMS for which the substrate plate was set, at the ignition process step, in a press machine that was as follows: One part of the substrate plate was in a free cantilever position, and the second remaining part was covered by a SiO_2/Si-plate. Together, these were interposed between the press flat punches where they were uniaxially compressed under a pressure of 3 MPa. After the ignition, the propagation of the reaction extended, both, over the entire un-pressed MS-RMS free part and through the entire pressed MS-RMS part sandwiched between the substrate plates. The results demonstrate that pressure, which is commonly used in reactive bonding, can prevent the detachment/ejection of the reaction product in bonding processes.

Since a full intact cross-sectional lamella of the reaction product of the multi-section Pd/Al-Ni/Al nano-reactive multilayer system could not be obtained to perform structural and compositional analyses on, a complementary analysis by XRD was carried out on both the as-deposited and reacted Pd/Al-Ni/Al MS-iRMS. Thus, the crystalline structure of the multilayer materials and the phase composition of reacted multi-section materials were examined. The XRD patterns of as-deposited and reacted Pd/Al-Ni/Al MS-iRMS on the SiO_2/Si-substrate are shown in Figure 22.

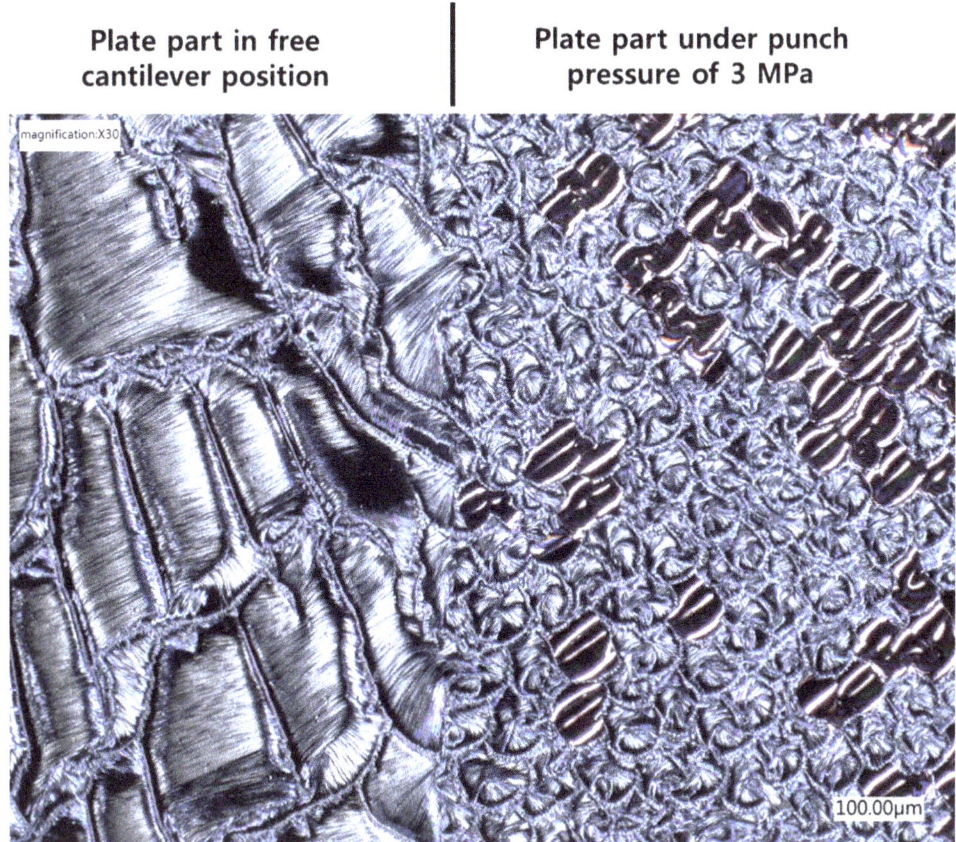

Figure 21. Optical microscope image showing the effect of the ignition under pressure on the reaction product microstructure configuration for Pd/Al-Ni/Al MS-iRMS.

For the as-deposited Pd/Al-Ni/Al MS-iRMS (Figure 22a), the diffraction peaks indexed to Al(111), Al(200), Al(311), Al(222), Pd(111), Pd(200), Pd(220), Pd(311), Pd(222), Ni(111), Ni (200), and Ni(220) indicate that the Al, Pd, and Ni elements forming the Pd/Al and Ni/Al multilayers in iRMS sections are all present in the crystalline phase (the used ICDD cards for peaks indexing are: #04-0787 for Al, #00-046-1043 for Pd, and #00-004-0850 for Ni). For the Pd/Al-Ni/Al MS-iRMS reaction product (Figure 22b), the new groups of major diffraction peaks indexed to AlPd-r(410), AlPd-r(461), AlPd-r(523), AlPd-r(182) and AlNi-c(110), AlNi-c(200), and AlNi-c(211) (r and c, respectively, denote rhombohedral and cubic) indicate that AlPd and AlNi are the major resultant intermetallic compounds of the final products of the exothermic reaction originating, respectively, from the Pd/Al and Ni/Al multilayer films. The indexed peaks were found to match the rhombohedral crystalline structure for the AlPd reaction product and cubic crystalline structure for the AlNi reaction product (corresponding ICDD cards are #00-031-0027 for r-AlPd and #01-073-2594 for c-AlNi). It should also be mentioned that some additional diffraction peaks were observed in both XRD patterns, before and after the exothermic reaction. These peaks essentially originated from the SO_2/Si-substrate and the Ti adhesive layer. The XRD data of the reaction product also revealed unreacted Pd/Al multilayer stack residues that were still attached to the substrate after the reaction propagation. They were manifested by a diffraction continuity of Pd (111) and Al (111) peaks, but with a lower intensity magnitude, indicating the existence of only a small, dispersed amount of such unreacted residue resulting from the incomplete reaction of the first deposited Pd/Al section. Furthermore, the unknown observed peaks in the XRD data of the as-deposited multi-section iRMS could be related to a solid solution reaction occurring from the pre-intermixing at the multilayer interfaces during the deposition process. Consequently, when a multi-section iRMS, with alternating sections of Pd/Al and Ni/Al multilayers undergoes an exothermic reaction process, the intermetallic AlPd and AlNi compounds are found to be the dominantly formed products.

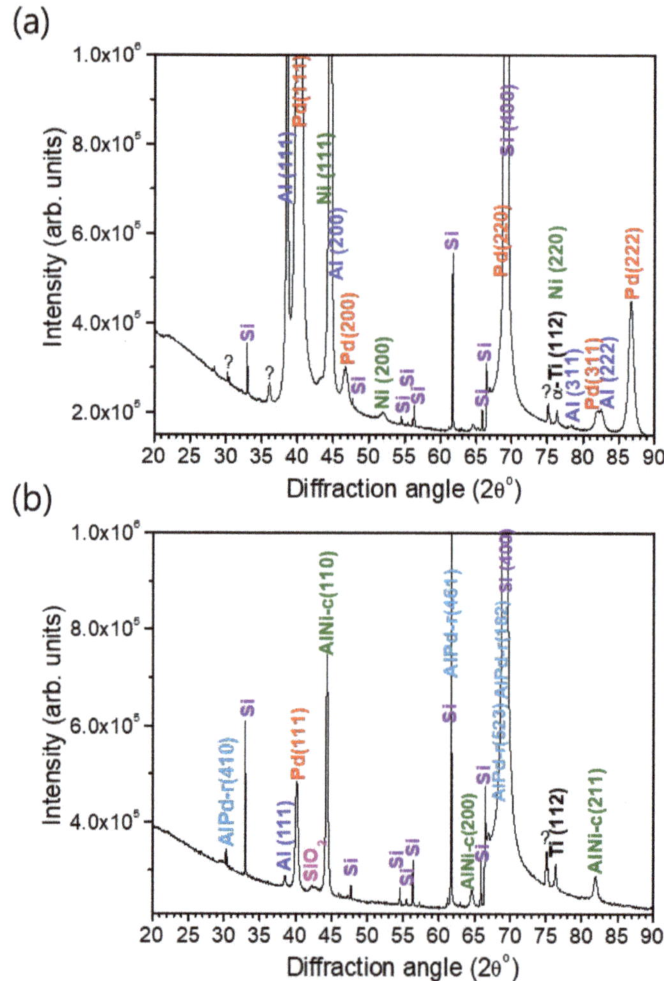

Figure 22. XRD patterns of Pd/Al-Ni/Al MS-iRMS on the SiO$_2$/Si-substrate. (**a**) As-deposited sample; (**b**) Reaction product after electrical ignition.

To briefly summarize the main findings of this investigation, the combined hRMS/mRMS reactive systems demonstrated two important beneficial impacts on the integrated multi-section reactive system: first, avoiding the spontaneous self-ignition reaction during the deposition process; and second, assuring iRMS ignition with a sustained reaction propagation simply by electrical triggering at room temperature.

4. Conclusions

We proved experimentally that, for the integration of a high energetic RMS on Si-wafer substrates for reactive bonding, two main conflicting issues should be overcome: (i) the heat accumulation and consequent spontaneous self-ignition reaction and (ii) the heat loss through the sink substrate and consequent reaction propagation quenching.

In this work, we investigated the integration of a highly energetic Pd/Al nano-reactive multilayer system on silicon wafers. To address the heat control difficulties in and around the integrated reactive system, Pd/Al-iRMS was deposited on different wafers of different thermal conductivities: a bare Si-wafer, RuO_x or PdO_x layers buffering Si-wafers, and a SiO_2-coated Si-wafer. The growth of Pd/Al-iRMS on a bare silicon wafer of higher thermal conductivity could reach a total thickness over 1 μm without the upsurging of a spontaneous self-ignition reaction during the deposition process. Nonetheless, for this structure, due to the efficient heat sink by the intimate thermal contact of RMS with the Si-wafer substrate, in order to be electrically ignitable at RT with a stable reaction front propagation, the effective Pd/Al-iRMS thickness must start from 5 μm and over, which is technically not desirable (due to high internal stress generation in the iRMS and lift-off pattering limitations). However, for the other coated Si-wafers, those with an instant thermal barrier made of metal oxide RuO_x or PdO_x buffers causing moderate thermal conductivity and the other with a SiO_2 layer of lower thermal conductivity, the Pd/Al-iRMS growth on all of them, specifically with a bilayer thickness of 100 nm or less, underwent systematic spontaneous self-ignition surging during the deposition process upon reaching a thickness of around 1 μm.

To overcome these encountered issues, we investigated a solution based on the tuning of the reaction heat release by combining RMS of a different exothermic reaction enthalpy. This type of heat modulation has direct control over the desired generation of the local temperature and reaction propagation characteristics. The materialization of this solution was accomplished by alternately stacking sections of metallic reactive multilayer Pd/Al and Ni/Al systems that, respectively, have a high and medium exothermic enthalpy of reaction. The grown heterostructure was successfully deposited directly on a SiO_2-coated Si-wafer with a bilayer of 100 nm and a total thickness of 3 μm without any spontaneous upsurge of self-ignition. Moreover, the ignition of the alternating multi-section Pd/Al-Ni/Al iRMS structure was successfully initiated with a DC pulse of 10 V/max. 1 A at RT ambient without requiring any external preheating supplement, and this was possible for the whole patterned Pd/Al-Ni/Al MS-iRMS units on the wafer. Furthermore, the reaction ignition in this heterogeneous multilayer, multi-section structure showed a stable self-sustained propagating reaction front along the whole patterned Pd/Al-Ni/Al MS-iRMS path without any local propagating reaction quenching.

The DSC analyses confirmed the predicted heat output modulation in the combined Pd/Al-Ni/Al MS-iRMS structure compared to the single structured Pd/Al iRMS. The specific heat output by the combined reactive systems effectively resulted in a heat amount (1191.7 J/g) that was relatively lower than the theoretical exothermic heat (1260 J/g) related to a pure single reactive Pd/Al system. DSC results also revealed that the phase transformation mechanisms to promote the formation of intermetallic compounds took place simultaneously, in both Pd/Al- and Ni/Al-stacked RMS, at the same temperatures and along the same temperature range. This thermal uniformity is important during the reaction of different stacked reactive systems; it avoids the reaction decoupling effect on the heat wave and the reaction propagations in each section of the stacked systems.

The analysis by X-ray diffraction of the Pd/Al-Ni/Al MS-iRMS reacted structure confirmed the formation of intermetallic compounds: mainly an AlPd reaction product with a rhombohedral crystalline structure, formed through the Pd/Al stack system, and an AlNi reaction product with a cubic crystalline structure, formed through the PNi/Al stack system. Both structures were mostly dominant in the formed reaction products.

This approach of alternating reactive sections with interfaces of different reactive affinity and different output energy demonstrated an efficient potential for the integration of the reactive system. It allowed the limitation of heat accumulation during the film growth and the tuning of heat source without changing the iRMS structural architecture. However, further investigation regarding the design parameters of the Pd/Al-Ni/Al MS-iRMS is still needed to optimize the reactive system integration to supply enough output energy capable of metal solder melting as well as compensating the heat conduction into the bonding partners. Also, given that the intermetallic compounds are highly brittle, to preserve the integrity and mechanical reliability of the reaction product's material after the reaction propagation through the iRMS, research on improving the ductility of the reaction product at room temperature is necessary. The ductility dampens the residual internal stress in the reacted iRMS product and, thus, avoids the explosive reaction with particle ejection during the formation of the reaction product. Finally, we can conclude that the proposed solution based on stacking alternating reactive systems with different exothermic heat of reactions provides an efficient potential method that could overcome the different issues of integrated RMS for bonding applications in microelectronics and microsystems technology.

Author Contributions: E.-M.B., I.-S.K. and H.Y.K. selected the investigated materials. E.-M.B. and I.-S.K. conceived the sample structures, drew and patterned the integrated structures, designed experiments and set up the experimental tools. E.-M.B. performed the experiments, analyzed the data and prepared the manuscript. All authors have read and agreed to the published version of the manuscript.

Funding: This research was supported by the National Research Council of Science & Technology (NST) grant by the Korean government (MSIT) (No. CRC-19-02-ETRI).

Acknowledgments: Authors would like to thank the National NanoFab Center engineers in the Department of Nanostructure Development for thin film deposition and sample preparation.

Conflicts of Interest: The authors declare no conflict of interest.

References

1. Boettge, B.; Braeuer, J.; Wiemer, M.; Petzold, M.; Bagdahn, J.; Gessner, T. Fabrication and characterization of reactive nanoscale multilayer systems for low-temperature bonding in microsystem technology. *J. Micromech. Microeng.* **2010**, *20*, 064018. [CrossRef]
2. Braeuer, J.; Besser, J.; Wiemer, M.; Gessner, T. Room-temperature reactive bonding by using nano scale multilayer systems. In Proceedings of the 2011 16th International Solid-State Sensors, Actuators and Microsystems Conference, Beijing, China, 5–9 June 2011; pp. 1332–1335. [CrossRef]
3. Qiu, X.; Tang, R.; Liu, R.; Guo, S.; Yu, H. A micro initiator realized by reactive Ni/Al nanolaminates for MEMS applications. In Proceedings of the 2011 16th International Solid-State Sensors, Actuators and Microsystems Conference, Beijing, China, 5–9 June 2011; pp. 1665–1668. [CrossRef]
4. Braeuer, J.; Besser, J.; Wiemer, M.; Gessner, T. Integrated nano scale multilayer systems for reactive bonding in microsystems technology. In Proceedings of the 2012 4th Electronic System-Integration Technology Conference, Amsterdam, The Netherlands, 17–20 September 2012; pp. 1–4. [CrossRef]
5. Braeuer, J.; Gessner, T. A hermetic and room-temperature wafer bonding technique based on integrated reactive multilayer systems. *J. Micromech. Microeng.* **2014**, *24*, 115002. [CrossRef]
6. Schumacher, A.; Gaiß, U.; Knappmann, S.; Dietrich, G.; Braun, S.; Pflug, E.; Roscher, F.; Vogel, K.; Hertel, S.; Kähler, D.; et al. Assembly and packaging of micro systems by using reactive bonding processes. In Proceedings of the 2015 European Microelectronics Packaging Conference (EMPC), Friedrichshafen, Germany, 14–16 September 2015; pp. 1–5.
7. Braeuer, J.; Besser, J.; Tomoscheit, E.; Klimm, D.; Anbumani, S.; Wiemer, M.; Gessner, T. Investigation of different nano scale energetic material systems for reactive wafer bonding. *ECS Trans.* **2013**, *50*, 241–251. [CrossRef]
8. Braeuer, J.; Besser, J.; Wiemer, M.; Gessner, T. A novel technique for MEMS packaging: Reactive bonding with integrated material systems. *Sens. Actuators A Phys.* **2012**, *188*, 212–219. [CrossRef]
9. Fisher, D.J. *Bonding by Self-Propagating Reaction*; Materials Research Forum LLC: Millersville, PA, USA, 2019; ISBN 978-1-64490-009-3.
10. Weihs, T.P. Self-Propagating Reactions in Multilayer Materials. In *Handbook of Thin Film Process Technology*; Glocker, D.A., Shah, S.I., Eds.; Institute of Physics Publishing: Bristol, UK, 1998; pp. F7:1–F7:13.
11. Wang, J.; Besnoin, E.; Duckham, A.; Spey, S.J.; Reiss, M.E.; Knio, O.M.; Powers, M.; Whitener, M.; Weihs, T.P. Room-temperature soldering with nanostructured foils. *Appl. Phys. Lett.* **2003**, *83*, 3987–3989. [CrossRef]
12. Wang, J.; Besnoin, E.; Duckham, A.; Spey, S.J.; Reiss, M.E.; Knio, O.M.; Weihs, T.P. Joining of stainless-steel specimens with nanostructured Al/Ni foils. *J. Appl. Phys.* **2004**, *95*, 248–256. [CrossRef]

13. Danzi, S.; Menétrey, M.; Wohlwend, J.; Spolenak, R. Thermal management in Ni/Al reactive multilayers: Understanding and preventing reaction quenching on thin film heat sinks. *CS Appl. Mater. Interfaces* **2019**, *11*, 42479–42485. [CrossRef]
14. Manesh, N.A.; Basu, S.; Kumar, R. Experimental flame speed in multi-layered nano-energetic materials. *Combust. Flame* **2010**, *157*, 476–480. [CrossRef]
15. Wang, J.; Besnoin, E.; Knio, O.M.; Weihs, T.P. Effects of physical properties of components on reactive nanolayer joining. *J. Appl. Phys.* **2005**, *97*, 114307. [CrossRef]
16. Namazu, T.; Takemoto, H.; Fujita, H.; Nagai, Y.; Inoue, S. Self-propagating explosive reactions in nanostructured Al/Ni multilayer films as a localized heat process technique for mems. In Proceedings of the 19th International Conference on Micro Electro Mechanical Systems, Istanbul, Turkey, 22–26 January 2006; pp. 286–289. [CrossRef]
17. Mann, A.B.; Gavens, A.J.; Reiss, M.E.; Van Heerden, D.; Bao, G.; Weihs, T.P. Modeling and characterizing the propagation velocity of exothermic reactions in multilayer foils. *J. Appl. Phys.* **1997**, *82*, 1178–1188. [CrossRef]
18. Gavens, A.J.; Van Heerden, D.; Mann, A.B.; Reiss, M.E.; Weihs, T.P. Effect of intermixing on self-propagating exothermic reactions in Al/Ni nanolaminate foils. *J. Appl. Phys.* **2000**, *87*, 1255–1262. [CrossRef]
19. Ma, E.; Thompson, C.V.; Clevenger, L.A.; Tu, K.N. Self-propagating explosive reactions in Al/Ni multilayer thin films. *Appl. Phys. Lett.* **1990**, *57*, 1262–1264. [CrossRef]
20. Wickersham, C.E., Jr.; Poole, J.E. Explosive crystallization in zirconium/silicon multilayers. *J. Vac. Sci. Technol. A* **1988**, *6*, 1699–1702. [CrossRef]
21. Overdeep, K.R.; Livi, K.J.T.; Allen, D.J.; Glumac, N.G.; Weihs, T.P. Using magnesium to maximize heat generated by reactive Al/Zr Nanolaminates. *Combust. Flame* **2015**, *162*, 2855–2864. [CrossRef]
22. Zapata, J.; Nicollet, A.; Julien, B.; Lahiner, G.; Esteve, A.; Rossi, C. Self-propagating combustion of sputter-deposited Al/CuO nanolaminates. *Combust. Flame* **2019**, *205*, 389–396. [CrossRef]
23. Movchan, B.A.; Demchishin, A.V. Structure and properties of thick condensates of nickel, titanium, tungsten, and aluminum oxides, and zirconium dioxide in vacuum. *Phys. Met. Metallogr.* **1969**, *28*, 653–660.
24. Thornton, J.A. The microstructure of sputter-deposited coatings. *J. Vac. Sci. Technol. A* **1986**, *4*, 3059–3065. [CrossRef]
25. Fischer, S.H.; Grubelich, M.C. Theoretical energy release of thermites, intermetallics, and combustible metals. In Proceedings of the 24th International Pyrotechnics Seminar, Monterey, CA, USA, 1 July 1998. [CrossRef]
26. Vogel, K.; Braun, S.; Hofmann, C.; Weiser, M.; Wiemer, M.; Otto, T.; Kuhn, H. Reactive Bonding. Chapter 14; In *3D and Circuit Integration of MEMS*; Esashi, M., Ed.; Wiley-VCH: Weinheim, Gemany, 2021; pp. 309–329. [CrossRef]
27. Beeby, S.; Ensell, G.; Kraft, M.; White, N. *MEMS Mechanical Sensor*; Artech House, Inc.: Norwood, MA, USA, 2004; pp. 1–5. ISBN 978-1580535366.
28. Pierson, H.O. *Handbook of Chemical Vapor Deposition (CVD): Principles, Technology, and Applications*, 2nd ed.; Noyes Publications/William Andrew Publishing, LLC Norwich: New York, NY, USA, 1999; pp. 279–283. ISBN 978-0815514329.
29. Ferizović, D.; Hussey, L.K.; Huang, Y.-S.; Muñoz, M. Determination of the room temperature thermal conductivity of RuO_2 by the photothermal deflection technique. *Appl. Phys. Lett.* **2009**, *94*, 131913.
30. Brandes, E.A.; Brook, G.B. *Smithells Metals Reference Book*; Butterworth-Heinemann, Seventh Edition: Oxford, UK, 1998; pp. 14-3–14-4. ISBN 978-0750636247.
31. Hung, M.-T.; Ju, Y.S. Process dependence of the thermal conductivity of image reversal photoresist layers. *J. Vac. Sci. Technol. B* **2007**, *25*, 224–228. [CrossRef]
32. Bell, W.E.; Tagami, M. High-temperature chemistry of the ruthenium-oxygen system. *J. Phys. Chem.* **1963**, *67*, 2432–2436. [CrossRef]
33. Miradji, F.; Souvi, S.; Cantrel, L.; Louis, F.; Vallet, V. Thermodynamic Properties of Gaseous Ruthenium Species. *J. Phys. Chem. A* **2015**, *119*, 4961–4971. [CrossRef]
34. Chaston, J.C. The Oxidation of the Platinum Metals; A Descriptive survey of the reactions involved. *Platin. Met. Rev.* **1975**, *19*, 135–140.
35. Bayer, G.; Wiedemann, H.G. Formation, dissociation and expansion behavior of platinum group metal oxides (PdO, RuO_2, IrO_2). *Thermochim. Acta* **1975**, *11*, 79–88. [CrossRef]
36. Meschel, S.V.; Kleppa, O.J. Thermochemistry of alloys of transition metals and lanthanide metals with some IIIB and IVB elements in the periodic table. *J. Alloys Compd.* **2001**, *321*, 183–200. [CrossRef]
37. Ardell, A.J.; Prikhodko, S.V. Coarsening of γ' in Ni–Al alloys aged under uniaxial compression: II. Diffusion under stress and retardation of coarsening kinetics. *Acta Mater.* **2003**, *51*, 5013–5019. [CrossRef]
38. Grapes, M.D.; Weihs, T.P. Exploring the reaction mechanism in self-propagating Al/Ni multilayers by adding inert material. *Combust. Flame* **2016**, *172*, 105–115. [CrossRef]
39. Jeong, G.-U.; Park, C.S.; Do, H.-S.; Park, S.-M.; Lee, B.-J. Second nearest-neighbor modified embedded-atom method interatomic potentials for the Pd-M (M=Al, Co, Cu, Fe, Mo, Ni, Ti) binary systems. *Calphad* **2018**, *62*, 172–186. [CrossRef]
40. Meschel, S.V.; Nash, P.; Chen, X.-Q. The standard enthalpies of formation of binary intermetallic compounds of some late 4d and 5d transition metals by high temperature direct synthesis calorimetry. *J. Alloys Compd.* **2010**, *492*, 105–115. [CrossRef]
41. Neumann, G.; Tuijn, C. *Self-Diffusion and Impurity Diffusion in Pure Metals: Handbook of Experimental Data*, Pergamon Materials Series 14, 1st ed.; Elsevier Ltd.: San Diego, CA, USA, 2009; pp. 121–148. ISBN 9780080560045.
42. Qiu, X.; Wanga, J. Bonding silicon wafers with reactive multilayer foils. *Sens. Actuators A Phys.* **2008**, *141*, 476–481. [CrossRef]

43. Gránásy, L.; Pusztai, T.; Tegze, G.; Warren, J.A.; Douglas, J.F. Growth and form of spherulites. *Phys. Rev. E* **2005**, *72*, 011605. [CrossRef]
44. Sinha, A.K. *Physical Metallurgy Handbook*; McGraw-Hill: New York, NY, USA, 2003; p. 3. ISBN 0-07-057986-5.
45. Glicksman, M.E. *Principles of Solidification: An Introduction to Modern Casting and Crystal Growth Concepts*; Springer Science Business Media: New York, NY, USA, 2011; pp. 213–235. [CrossRef]
46. Shtukenberg, A.G.; Punin, Y.O.; Gunn, E.; Kahr, B. Spherulites. *Chem. Rev.* **2012**, *112*, 1805–1838. [CrossRef] [PubMed]

Review

Polymer-Based Biocompatible Packaging for Implantable Devices: Packaging Method, Materials, and Reliability Simulation

Seonho Seok

Center for Nanoscience and Nanotechnology (C2N), University-Paris-Saclay, 91120 Palaiseau, France; seonho.seok@u-psud.fr

Abstract: Polymer materials attract more and more interests for a biocompatible package of novel implantable medical devices. Medical implants need to be packaged in a biocompatible way to minimize FBR (Foreign Body Reaction) of the implant. One of the most advanced implantable devices is neural prosthesis device, which consists of polymeric neural electrode and silicon neural signal processing integrated circuit (IC). The overall neural interface system should be packaged in a biocompatible way to be implanted in a patient. The biocompatible packaging is being mainly achieved in two approaches; (1) polymer encapsulation of conventional package based on die attach, wire bond, solder bump, etc. (2) chip-level integrated interconnect, which integrates Si chip with metal thin film deposition through sacrificial release technique. The polymer encapsulation must cover different materials, creating a multitude of interface, which is of much importance in long-term reliability of the implanted biocompatible package. Another failure mode is bio-fluid penetration through the polymer encapsulation layer. To prevent bio-fluid leakage, a diffusion barrier is frequently added to the polymer packaging layer. Such a diffusion barrier is also used in polymer-based neural electrodes. This review paper presents the summary of biocompatible packaging techniques, packaging materials focusing on encapsulation polymer materials and diffusion barrier, and a FEM-based modeling and simulation to study the biocompatible package reliability.

Keywords: biocompatible packaging; implantable; reliability; Finite element method (FEM); simulation

Citation: Seok, S. Polymer-Based Biocompatible Packaging for Implantable Devices: Packaging Method, Materials, and Reliability Simulation. *Micromachines* **2021**, *12*, 1020. https://doi.org/10.3390/mi12091020

Academic Editor: Woon-Hong Yeo

Received: 5 August 2021
Accepted: 26 August 2021
Published: 27 August 2021

Publisher's Note: MDPI stays neutral with regard to jurisdictional claims in published maps and institutional affiliations.

1. Introduction

Evolution in IC (Integrated Circuit) packaging technology has been driven by the need for higher speed and higher density devices enabling smaller form factor and lower power consumption. For example, HBM (High Bandwidth Memory) has been developed by stacking memory die based on TSV (Through Silicon Vias) and stacking with micro-bump bonding in order to achieve higher bandwidth and thus lower power consumption [1–4]. Packaging of an implantable device is critical as it determines hermeticity and compatibility of the implant system in biological environment. Thus, reliability and life-time of the implant system highly depend on both packaging materials and technology. In general, implantable device packaging houses the electronic or mechanical system through polymer encapsulation [5–12], welding or bonding of metal [13,14], and ceramics [15]. Materials of the polymer encapsulation package include epoxies, silicones, polyurethanes, polyimides, silicon-polyimides, parylenes, polycyclic-olefins, silicon-carbons, benzocyclobutenes (BCB), and liquid crystal polymers. Conventionally, titanium (Ti) box has been used for packaging of an implantable electric device such as a pacemaker in order to ensure hermetic and biocompatible packaging of the microelectronic device. While it is a well-proven hermetic implant package, the Ti-box is rather large and rigid, which evokes a pronounced foreign body reaction (FBR) upon implantation, resulting in a thick fibrous tissue encapsulation, which might decrease the implant's sensor sensitivity. Furthermore, mechanical mismatch of the Ti-box and local tissue might cause chronic discomfort for the patient [16]. Therefore, different packaging approaches have been reported to replace the existing Ti box package for implantable devices. Cardiac monitoring system has been implemented with commercial 3-axis accelerometer mounted on PCB as shown in Figure 1a. To be suitable for implant, parylene has been first coated on the surface of the accelerometer and PCB and then it is fixed on Teflon support providing stitching of the sensor on the cardiac wall with epoxy resin. Finally, a soft encapsulation in medical grade PDMS (NUSIL MED-6015) was fabricated around the device [17,18]. Pressure sensor device mounted on a stent graft has been flip-chip bonded to flexible, biocompatible polymer which has predefined metal feedthrough. After filling silicone gel around the bonded pressure sensor, a polymer has been laminated to seal the pressure sensor. A biocompatible silicone gel links the same with the thinned substrate layer in order to transfer the pressure within the aneurysm to the sensitive area of the sensor as shown in Figure 1b [19]. The flip-chip technology has been used to ensure miniaturization and flexibility of the device, compared to the commonly used wire bonding. Bare die assembly techniques such as flip-chip technology are the preferred choice, as they provide thin, small, and lightweight features and can be assembled on ceramic, laminate, Molded Interconnect Devices (MID) molded, and flexible substrates. Flip-chip technology can substitute and complement conventional surface-mounted devices (SMD) or wire bonding processes for an even higher degree of miniaturization [20]. An implantable retina stimulator is implemented by MFI (MicroFlex Interconnection) technology. The MFI technology is based on the common thermosonic ball-wedge bonding process. A gold ball is bonded by force, temperature, and ultrasound through the hole in the substrate on the IC pad as shown in Figure 1c. The metal pair is then welded together, resulting in mechanically and electrically stable interconnects [21,22]. Emergence of new implantable devices such as retina prothesis requires an innovative packaging as conventional wire-bonding techniques would not be applicable to implement massive electrical interconnect, for example, 1000 electrodes (see Figure 1d) [23]. In this case, technological barrier exists in substantial scale difference between Si chips and polymeric stimulation device for mechanical interconnection. As a solution, standard silicon wafer having through-holes of $2.51 \times 2.63 \text{ mm}^2$ is used as a temporary packaging platform. The 260-μm-thick chips are inserted from the backside and planarized using a tape on the front of the wafer. Photoresist sacrificial layer on the perimeter of the Si chip with help of anchoring parylene layer on the backside. From the frontside of the Si chip, parylene and metal interconnect has been fabricated to finalize the desired retina prothesis. The packaging and integration have been finished by separate the parylene with Si chips from the temporary Si platform through photoresist sacrificial etch [23]. Emergence of UTC (Ultra Thin Chip) opens new pathway of miniaturization of biocompatible package causing minimal neural tissue damage upon implantation for neural electrodes as shown in Figure 1e [24,25]. The initiation of UTC is made by need for embedding Si chip into packaging carrier substrate, which could reduce packaging cost by suppressing certain step of conventional packaging, for example, EMC (Epoxy Mold Compound) for flip-chip technology.

From the examples shown in Figure 1, it is found that the biocompatible packaging starts to use conventional packaging technology such as wire-bonding, flip-chip bonding, PCB chip carrier and advances to use microfabrication technology due to the need for maximum miniaturization and lots of electrical wiring. Implantable microsensors have been packaged in a conventional method such as EMC (Epoxy Mold Compound), wire bonding, PCB, and they are bonded and encapsulated with soft or biocompatible material to be suitable in biological environment. Furthermore, advanced medical device such as neural prosthesis is implemented with flexible material, its electrical interconnection with silicon IC is the essence of packaging technology. Thinned Si IC can be embedded into polymer material and it can be integrated with the polymer-based neural prothesis. Such a thin silicon chip is ideally best approach to achieve maximum miniaturization of an active implant. This paper presents polymer-based biocompatible packaging techniques for implantable devices. Biocompatible packaging approaches, focusing on materials and the packaging process, have been summarized in Section 2. Section 3 addresses the reliability issues of the biocompatible packaging as well as FEM modeling and simulation based on interfacial fracture mechanics.

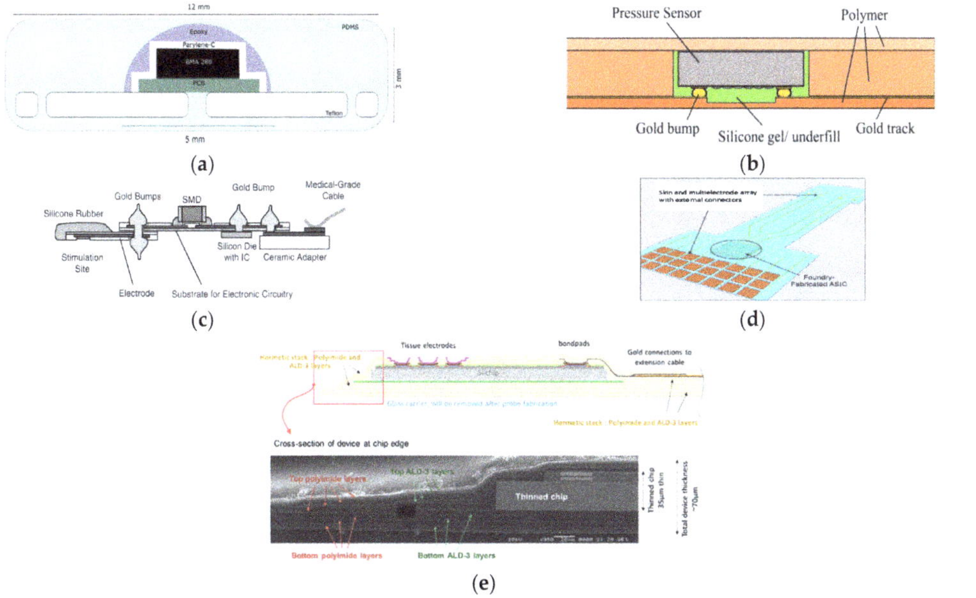

(a)

(b)

(c)

(d)

(e)

Figure 1. Examples of implantable device packaging. (**a**) PDMS encapsulation of wire-bonded chip; (**b**) Polyimide encapsulation of flip-chipped Si chip; (**c**) MicroFlex Interconnect (MFI); (**d**) MicroFlex Interconnect (MFI); (**e**) Ultra-Thin-Chip (UTC) packaging.

2. Biocompatible Packaging Methods

The objective of the biocompatible packaging is to provide a protection for the implanted electronic device to be tolerating the harsh biological environment in order to increase life-time of implanted device. Sealing of the implanted device is one of the critical aspects of long-term reliable biocompatible package. Materials and process will determine the type of sealing of the biocompatible packaging; hermetic, watertight, or permeable. Particular difficulty of the biocompatible packaging is the need for feedthrough as the implanted device should interact with the biological medium in different ways; electrically, chemically, mechanically, or optically. It is required that the feedthrough should withstand mechanical stress due to biological environment such as muscle activity. In addition, it should not add significantly to the mechanical load of the implant, as in the case of tethered neural implants whose cable weight and flexibility may negatively affect tissue response [26,27]. As shown in Figure 1, it can be said that the biocompatible packaging is mainly being achieved in two different ways: (1) polymer encapsulation of conventional circuit board (2) chip-level packaging.

2.1. Polymer Encapsulation

Figure 2 shows conceptual drawing of polymer encapsulation of Si chip wire-bonded on PCB board. The role of polymer encapsulation is to protect the packaged circuit preventing biofluid from penetrating during its operation. Such a polymer encapsulation is implemented through a molding process which uses predefined mold to encapsulate the implanted electronics. To be well-bonded and guaranteed their mechanical properties, the polymer materials should be cured at designated temperature for suitable time duration. It could make additional thermal stress on the implantable device in conventional package and sometimes newly-formed interfaces between encapsulated device and encapsulation polymer suffers from delamination. In addition, this packaging approach has drawbacks in view of miniaturization of implantable devices as it houses conventional package with biocompatible polymer encapsulation. Most of the biocompatible packaging uses this approach even for advanced implantable neural devices. Material properties and process conditions of frequently-used polymer materials are summarized in Table 1 [28]. Polymer material has lower mechanical modulus, which reduces mechanical stress to surrounding tissues. The polymer materials are utilized through coating for molding or bonding for lamination to encapsulate the package inside.

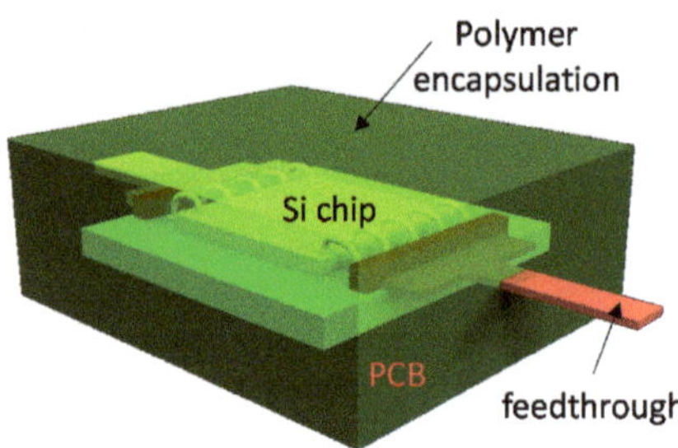

Figure 2. Polymer encapsulation of conventional package for biocompatible package.

Table 1. Summarizes material properties and process conditions of polymer materials.

Properties	Polyimide +	Epoxy	Parylene-C	PDMS ++	SU8 +++
Possible thickness (µm)	3–20		1–100	10–100 with spincoating	1–300
Moisture absorption (%)	-		0.06	<1	0.55–0.65
Glass transition temperature (°C)	-	≥40 *	-	-	200–210
Thermal coefficient of expansion (ppm/K)	35	52 (below Tg) * 191 (above Tg) *	35	-	52
Tensile strength (MPa)	200	34 **	69	6.2	60
Elastic modulus (MPa)	3400	4800 **	3200	0.1–0.5	2000

+ HD microsystem HD4100 Series, * EPOTEK-302 data sheet, ** Araldite 2014, ++ Nusil MED-1000, +++ Microchem SU8-2000 and SU8-3000 series.

Most of the polymer materials need to be cured at certain temperature in order to ensure stable mechanical properties. For example, epoxy will have different mechanical properties depending on curing conditions. Epoxy, cured at room temperature for 24 h (equivalently around 50% curing rate), shows yield strength of 24 MPa and elastic modulus of 2 GPa, while it has 35 MPa yield strength and 5 GPa Young's modulus after 4 h curing at 64 °C [29]. In addition, hydrophobic surface treatment of the polymer is one of good solutions to improve the reliability of the biocompatible package [30–32].

2.2. Chip-Level Packaging

The important perspectives in implantable electronic package are biocompatibility, hermeticity, and miniaturization. Biocompatibility typically refers to the way the body tolerates the presence of an implant material and thus polymer materials given in Table 1 and Ti (titanium) box are frequently used for the biocompatible package as explained in previous section Hermeticity of a package is referred as the integrity of sealed packages to resist gas and liquids penetrating the seal or an opening (crack) in the package, especially critical to the reliability and longevity of a packaged electronics. A diffusion barrier based on thin-film passivation can be deposited onto a Si chip or a polymer encapsulation in order to prevent biofluid penetration or diffusion of IC materials into the body. Table 2 summarizes material properties of frequently-used diffusion barrier of the biocompatible package. Diffusion barrier is a dielectric film deposited on a Si surface or an encapsulating polymer to avoid liquid passage through biocompatible package [33–40]. The dielectric layer, Al_2O_3 deposited by ALD (Atomic Layer Deposition) is frequently used as moisture barrier between polymer layers to increase adhesion strength between the polymer layers or enhance the life-time of a neural electrode bilayered with parylene material [41,42]. Ultra-thin thermally-grown silicon dioxide transferred to flexible substrate has shown high robustness as biofluid barrier compared with conventional approaches such as LCP and Al_2O_3/Parylene-C [43].

Table 2. Material properties of diffusion barrier layer of the biocompatible package *.

Material Properties	SiO$_2$	Si$_3$N$_4$	SiC	Al$_2$O$_3$
Density (g/cm^3)	2.65	3.44	3.21	3.95
Thermal coefficient of expansion (10^{-5} K^{-1})	0.05	0.28	0.44	0.70
Elastic modulus (GPa)	66.3	310	90	330
Poisson ratio	0.15	0.27	0.35	0.22
Tensile strength (MPa)	45	400	240	240

* Table 2 is partially taken from reference [21].

Figure 3 shows a concept of UTC-based chip-scale biocompatible package. It has merits of miniaturization compared with polymer encapsulation of conventional package. Miniaturization of the biocompatible package is important because mechanical load of the implanted device may negatively affect tissue response. Therefore, chip-scale biocompatible packaging may be more suitable for advanced implantable devices compared to the previous polymer encapsulation. It is realized, with microfabrication technologies, combining Si chips and polymer-based devices such as neural probe, retina prostheses, etc. It would make it possible to integrate Si chip into soft materials for neural probes without utilizing a chip carrier such as PCB. Technological difficulties of the integration process with two different materials rise in size mismatch between Si chips and polymer devices and thus innovative integration techniques, including metallization, are highly demanded. The Si chips are typically in a range of 200 μm after fab-out, while the polymer devices have total thickness up to tens of micrometers even if multiple polymer layers have been used. In general, typical Si chip having a few hundred micrometers in thickness is assembled to flexible circuit through flip-chip bonding with metallic bumps such as copper, gold, etc. Anisotropic Conductive Adhesive (ACA) is a good approach to integrate Si chips onto flexible substrate [44–46]. The advantages of ACA are reduced processing steps, lower processing temperature, and fine pitch capability. Stability of ACA's electrical or mechanical performance depends on associated adhesive types; thermoplastics, such as polymer, or thermosetting, such as epoxies and silicones. ACA flip-chip technology is used to assemble bare chips where the pitch is extremely fine, normally less than 120 μm. ACA flip-chip bonding exhibits better reliability on flexible chip carriers because the ability of flex provides compliance to relieve stresses. For example, the internal stress generated during resin curing can be absorbed by the deformation of the chip carrier. ACA joint stress analysis indicated that the residual stress is larger on rigid substrates than on flexible substrates after bonding [47]. The thickness of ACA has ranged from 20 μm to 75 μm depending on the associated materials, and the process temperature is usually less than 200 °C. The drawback of ACA technique exists in that it needs a bonding process that is a reason of low throughput, and the finest pitch it can provide is limited down to hundred micrometers, as mentioned earlier. In case of retina protheses, it requires a great of number of interconnects for 1000 electrodes [48]. The packaging methods relying on bonding would not be desirable due to its low throughput and thus an innovative way of integration is necessary to achieve high density (fine pitch) chip-scale integrated interconnect packaging. Standard microfabrication-based chip-level packaging enables the density of interconnects to scale to the limits of photolithography used to define the etch holes over the on-chip pads [48]. However, integration packaging process is not simple because it is based on sacrificial layer release of the temporary guiding substrate. The guide substrate is required to hold relatively thicker Si chips compared with the parylene-based neural interface device during the microfabrication interconnect process. Such non-conventional processes sometimes create process errors such as misalignment between pad on Si chip and neural electrode, which eventually deteriorates the process yield as well as process cost. Ultra-thin-chip (UTC), defined as less than 20 μm in thickness, packaging could be a solution to tackle the process barrier related with thickness mismatch between Si chip and polymer materials. Fabrication of UTC is a big challenge and it can be implemented in different ways; grinding, epitaxial growth, SOI, silicon wafer with buried cavities, etc. Furthermore, UTC having 10–50 μm thickness shows good flexibility and good mechanical stability and it provides excellent flexibility and unconditional stability when its thickness is less than 10 μm. Embedded UTC in a polymer encapsulation may have great advantage because it provides low mechanical stress as well as biocompatibility. Such a thin Si chips may also be beneficial in view of process compatibility between polymer materials and silicon.

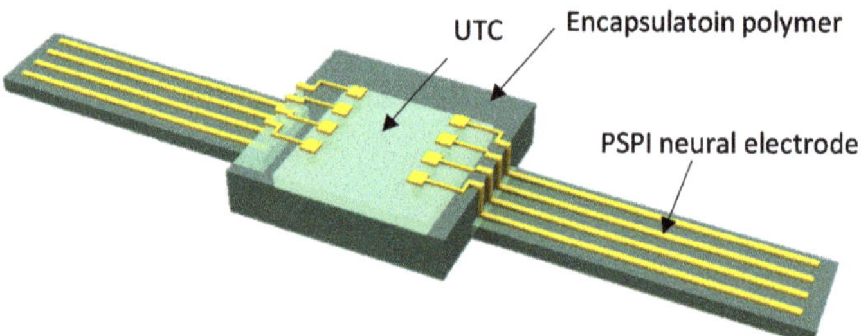

Figure 3. Conceptual drawing of chip-scale biocompatible package.

3. Reliability of Biocompatible Package

The biocompatible packages should be tested to estimate the reliability and life time as conventional packages. It is generally carried out through an accelerated aging test, which uses aggravated environmental conditions, such as temperature and humidity, to predict the expected life time of test devices or packages. In electronic package, the flip-chipped Si IC must use organic underfill to protect solder bumps by substantially reducing the mechanical stress. However, the underfill may create a new failure mode of the package due to CTE (Coefficient of Thermal Expansion) mismatch between the materials in joint. Shear or peeling stress could result in delamination of imperfect underfill with voids or microcracks under temperature cycling conditions. Such delamination is considered as mixed mode interfacial fracture and thus it is studied using FEM modeling and simulation [49–53]. Concerning biocompatible package, it is also similar case to the underfill of flip-chipped Si chip because polymer encapsulation creates multiple interfaces with the encapsulated Si chips, as explained earlier. Therefore, the theory of interfacial fracture mechanics is briefly explained, and then, an example of FEM modeling and simulation based on fracture mechanics is presented.

3.1. Estimation of Package Life-Time through Acceleration Aging Test

The acceleration aging test simulates real-time aging using elevated temperatures to artificially speed up the aging process. This test enables to get the expected life-time of the device under test. The estimation of the life-time can be calculated by using Arrhenius equation as follows [54].

$$k = Ae^{\frac{-E_a}{k_B T}}$$

(1)

where k is rate constant, A is constant, E_a is the activation energy, k_B is the Boltzmann constant and T is absolute temperature.

If an acceleration aging test is running at T_2 instead of T_1, destruction will occur at a rate k_2 where

$$\log \frac{k_2}{k_1} = \frac{E_a}{k_B}\left(\frac{1}{T_1} - \frac{1}{T_2}\right) \tag{2}$$

Suppose rate doubles between 32 °C and 42 °C,

$$\log 2 = \frac{E_a}{k_B}\left(\frac{1}{305\,°K} - \frac{1}{315\,°K}\right) \tag{3}$$

Therefore, $\frac{E_a}{k_B} = 6663\,°K$ (*Hence* $E = 0.58$ eV). Suppose the test temperature is 67 °C, whereas the temperature in life is 37 °C.

$$\log \frac{k_2}{k_1} = 6663\left(\frac{1}{310\,°K} - \frac{1}{340\,°K}\right) \tag{4}$$

Therefore, $\frac{k_2}{k_1} = 6.66$, the speed up factor is achieved.

Therefore, the test period for a 5-year life should be $\frac{5\,years}{6.66} = 274\,days$.

3.2. Interfacial Fracture Mechanics

Conventional crack has been dealt in assumption that the material is homogeneous, but packaging for electronics or implantable device should be considered as non-homogeneous materials, creating different interfaces. In fracture mechanics, there are three types of cracks, referred to as mode I, II, and III, as shown in Figure 4. Mode I is a normal opening mode, while mode II and III are shear sliding mode and shear tearing mode, respectively. In case of homogeneous material, any fracture mode may be described by one of the three basic modes or their combination. The stresses near crack tip in the crack plane (xz-plane) for these three modes can be expressed as ($y = 0$, x->0+) [55],

$$\sigma_{yy} = \frac{K_I}{\sqrt{2\pi x}} + O(\sqrt{x}),\ \sigma_{xy} = \sigma_{yz} = 0 \tag{5}$$

$$\sigma_{xy} = \frac{K_{II}}{\sqrt{2\pi x}} + O(\sqrt{x}),\ \sigma_{yy} = \sigma_{yz} = 0 \tag{6}$$

$$\sigma_{yz} = \frac{K_{III}}{\sqrt{2\pi x}} + O(\sqrt{x}),\ \sigma_{yy} = \sigma_{xy} = 0 \tag{7}$$

respectively, where the three parameters K_I, K_{II}, and K_{III} are named stress intensity factors corresponding to the opening, sliding, and tearing (anti-plane shearing) modes of fracture, respectively. These equations shows that stress tend to be infinity as x approaches to crack tip.

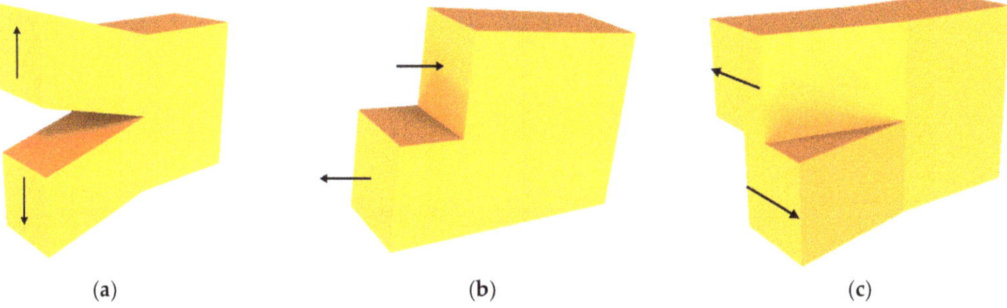

(a)   (b)   (c)

Figure 4. Basic fracture modes. (**a**) mode I (opening); (**b**) mode II (shearing); (**c**) mode III (tearing).

Bimaterial interfacial cracks tend to exhibit mode mixity with coupling between mode I and mode II [54]. Stress along the interface ahead of crack tip is given

$$\sigma_{yy} + i\sigma_{xy} = \frac{Kx^{i\varepsilon}}{\sqrt{2\pi x}} \tag{8}$$

where $K = K_I + iK_{II}$, x is the distance from crack tip, σ_{yy} and σ_{xy} are stress component normal and parallel to crack surface, respectively. The oscillatory index, ε, a function of the material properties

$$\varepsilon = \frac{1}{2\pi} ln \frac{\frac{\kappa_1}{\mu_1} + \frac{1}{\mu_2}}{\frac{\kappa_2}{\mu_2} + \frac{1}{\mu_1}} \tag{9}$$

where μ is shear modulus, $\kappa = 3\text{--}4\,v$ for plane strain or $= (3 - v)/(1 + v)$ for plane stress. The subscripts 1 and 2 represent each material associated to build the interface.

Phase angle, a measure of mode mixity, is given as follows.

$$\psi = tan^{-1}\left(\frac{K_{II}}{K_I}\right) \tag{10}$$

Another important parameter for interface fracture, energy release rate G, can be found

$$G = \frac{1}{cosh^2(\pi\varepsilon)} \frac{|K|^2}{E^*} \tag{11}$$

where $\frac{2}{E^*} = \frac{1}{E_1} + \frac{1}{E_2}$

The delamination will occur when the stress energy release rate G falls in following condition

$$G \geq G_c(\psi) \tag{12}$$

where Gc is critical stress release rate of an interface, which is determined through experimental characterization.

3.3. FEM Simulation of a Biocompatible Package

Finite element method (FEM) simulation, with appropriate mechanics theories, becomes useful to find a solution of the reliability issues of different package structures. For example, underfill has significant impact on flip chip package reliability due to delamination driven by coefficient of thermal expansion (CTE) mismatch between organic substrate and silicon die. Such a delamination problem can be solved through FEM study of the effects of various design variables, including underfill material properties, fillet dimensions, and die overhang on underfill delamination fracture parameters. Since the delamination can be considered as a bimaterial interfacial crack, the FEM study is based on interfacial fracture mechanics, fundamentally a mixed mode, including both energy release rate and phase angle [38]. Likewise, the biocompatible package can be simulated through FEM modeling to study its reliability issues concerning delamination of encapsulating polymer. Figure 5a shows a conceptual drawing of biocompatible package for FEM simulation. The Si chip is attached to PCB carrier with glue and then wire-bonded for electrical connection. Flexible cables are connected with typical flexible cable connector at I/O ports of the Si chip. The conventional package has been encapsulated with a biocompatible polymer. Failure of such biocompatible package could be caused by imperfect encapsulation material with voids or microcracks as indicated in Figure 5b. Therefore, the failure mode, due to the initial crack, is of interest for the FEM simulation. Figure 5c presents 2D FEM model for fracture analysis due to the initial crack. The initial crack has been defined at the interface between Si chip and encapsulation polymer, epoxy. As it is a half model of the package, boundary condition has been correspondingly defined at symmetric line; ux = uy = 0 at x = y = 0, ux = 0 at x = 0. Table 3 summaries material properties and dimension of the model. The 2D element behavior is defined as axisymmetric or plane strain and applied temperature load is $-100\ ^\circ\text{C}$ in assumption that the package is under the acceleration aging test for its life-time estimation. Initial crack with crack tip is defined starting from Si edge at the interface with encapsulation epoxy. As the objective of this simulation is to find stress parameters related with the failure mode, finer mesh has been defined at the crack tip as shown in Figure 5d. The applied thermal load from the acceleration aging test makes the biocompatible package deformed due to CTE difference between materials. Thus, the deformation of the package has been first found in the simulation as shown in Figure 6a. The package has deformation of out-of-plane bending due to the thermal loading and thus Si chip is under tensile stress as expected.

Next, the crack modes at the crack tip have been checked. Figure 6b,c clearly demonstrates the direction of opening and shearing modes of an initially closed crack interface. It is important to understand the change of fracture parameters, including both energy release rate and phase angle, during crack propagation, i.e., increase in crack length.

Figure 7 shows stress intensity factor, phase angle, and strain energy release rate as function of crack length, respectively. Opening mode (K_I) and shear mode (K_{II}) increase as the crack length increases. After 0.5 mm crack length, K_{II} is slightly reduced and K_I is still increasing, which is confirmed in phase angle change as function of crack length. Phase angle starts from near 90° (shear mode) and shows 72° at 1 mm crack length. Phase angle of 0° represents opening mode. Therefore, shear mode strength of epoxy is important when crack length is small, while tensile mode strength becomes important when crack length is substantial. SERR in mode 2 (G_2) has maximum at 0.5 mm crack length, while SERR in mode 1 (G_1) increases in the crack length of interest.

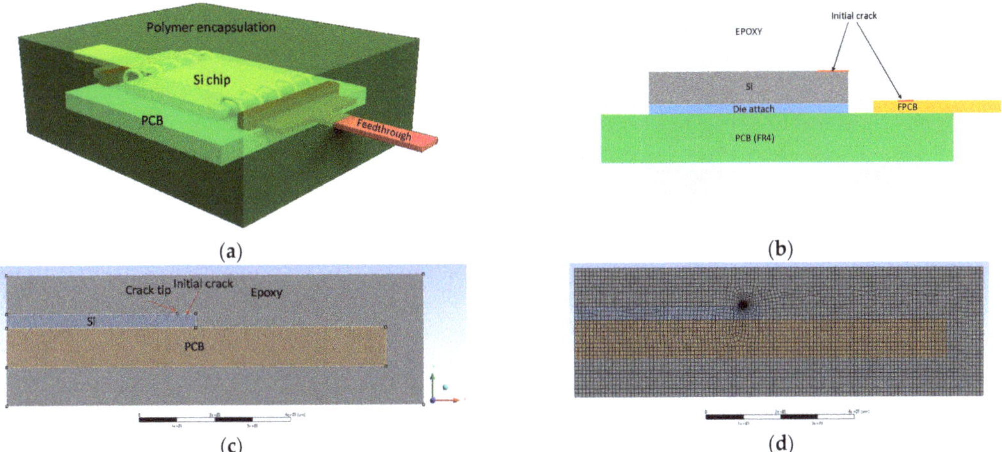

Figure 5. Finite element method (FEM) model of the biocompatible package. (**a**) Concept; (**b**) Cross-sectional view; (**c**) 2D model; (**d**) Meshed model.

Table 3. Material properties and model dimension.

Name	Elastic Modulus (GPa)	Poisson Ratio	Coefficient of Thermal Expansion (/°C)	Dimension Width (mm)	Height (µm)
Epoxy *	1.58	0.4	60.7×10^{-6}	11	3350
Silicon **	190	0.28	3.1×10^{-6}	5	350
PCB (FR4) ***	24	0.15	14.5×10^{-6}	10	1000

* Material properties has been extracted from Ref. [29]. ** Material properties have been taken from Ref [56]. *** Material properties have been extracted from Ref [57].

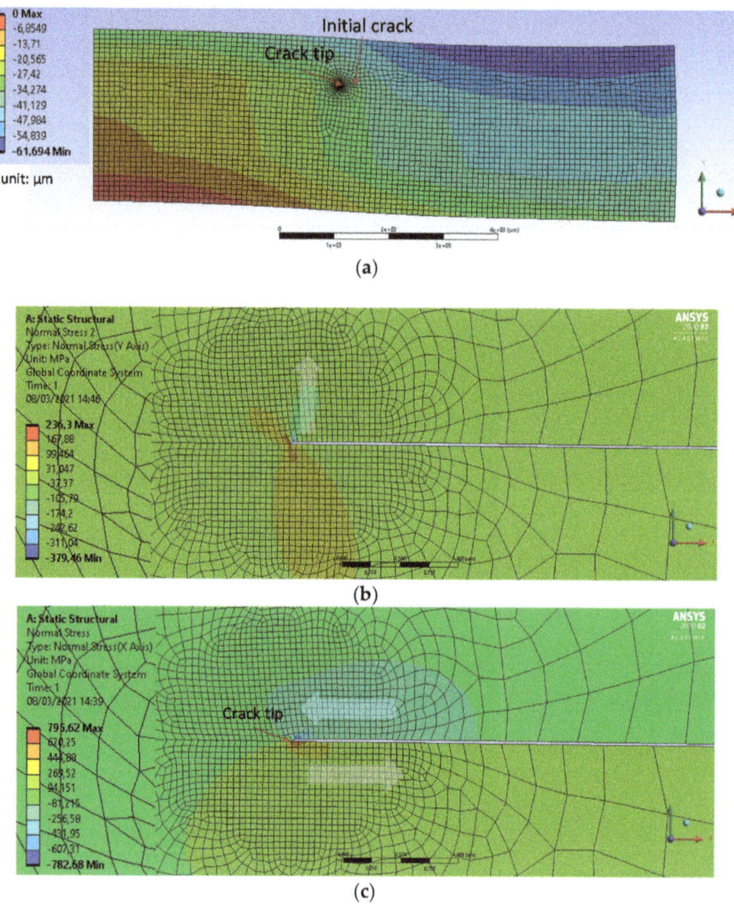

(a)

(b)

(c)

Figure 6. Package deformation and corresponding fracture mode at crack tip. (**a**) Deformation of the package; (**b**) Opening mode at crack tip; (**c**) Shear mode at crack tip.

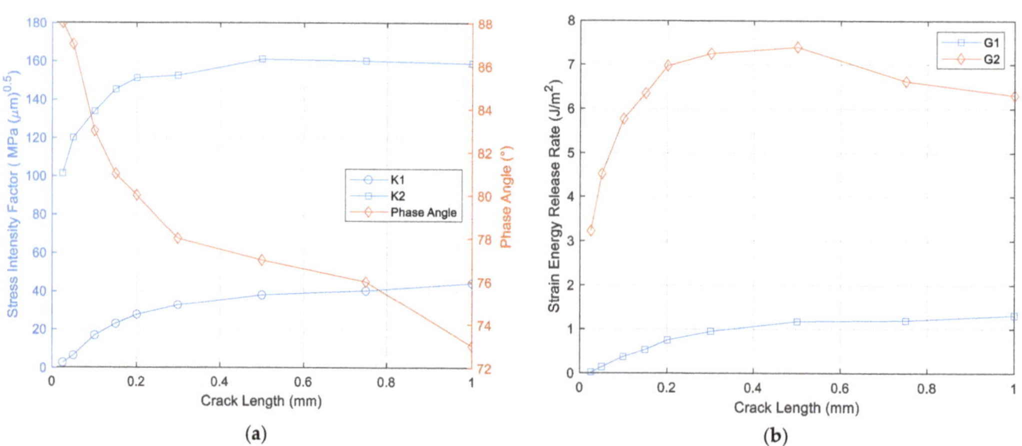

(a)

(b)

Figure 7. (**a**) Stress intensity factor, phase angle as function of crack length, and (**b**) strain energy release rate as function of crack length.

In case of small crack length, mode 2 is a dominant factor, as validated with phase angle and thus, critical toughness (G_c) in mode 2 becomes important. When the crack length becomes substantial, critical toughness (G_c) in mode 1 plays an important role. In general, G_c in mode 2 has bigger value that that of in mode 1 [53].

The effect of material property of the packaging material has been studied as it is one of major critical parameters. The encapsulation polymer is the packaging material for biocompatibility, so its elasticity has been changed to check its effect to the packaging reliability. As seen in Figure 8, strain energy release rate (SERR) increases with elastic modulus of encapsulation material, which indicates that stiffer encapsulation will generate a larger crack driving force and accelerate crack propagation once the crack is initiated.

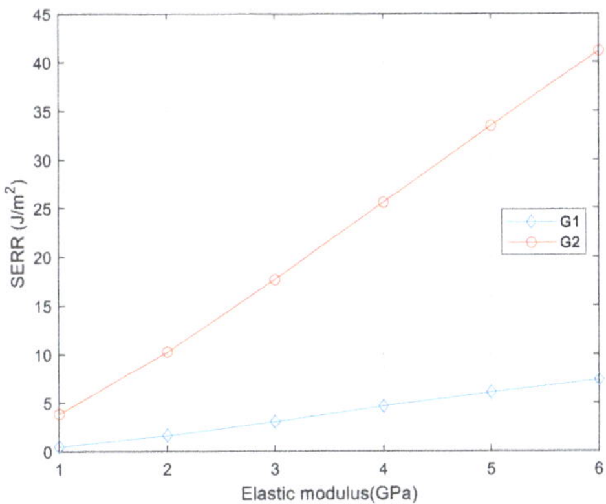

Figure 8. Strain energy release rate (SERR) as function of elasticity of packaging polymer.

Given with fracture analysis results, geometric parameters of the packaging have been studied to find optimal package dimension. The dimension parameters of interest in the package design are top epoxy height, bottom epoxy height as indicated in Figure 9. Initial thickness of top and bottom epoxy is 100 µm and the crack length is fixed to 0.5 mm. Figure 10 shows the thickness effect on fracture parameters, strain energy release rate (SERR) and phase angle. SERR increases with top epoxy thickness, while it reduces with bottom epoxy thickness. Closer examination on the bending mode of the package reveals that bending modes are dependent on the ratio between top and bottom epoxy height, as shown in Figure 11. When top epoxy height becomes thicker than that of bottom epoxy, opening the fracture mode becomes dominant due to the packaging bending mode. However, increasing bottom epoxy thickness maintains the initial bending mode making shear fracture mode far more dominant. Thus, it is recommended that bottom epoxy height should be always thicker than that of top epoxy as shear mode adhesion has greater than that of opening mode. It makes the Si chip under tensile stress, which reduces Si chip delamination from carrier substrate.

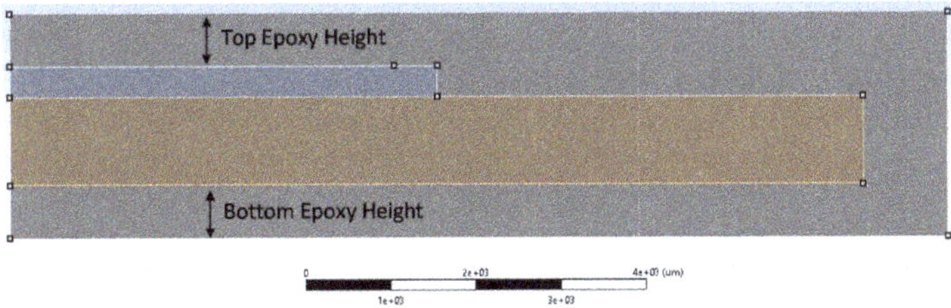

Figure 9. Dimension parameters for parametric analysis.

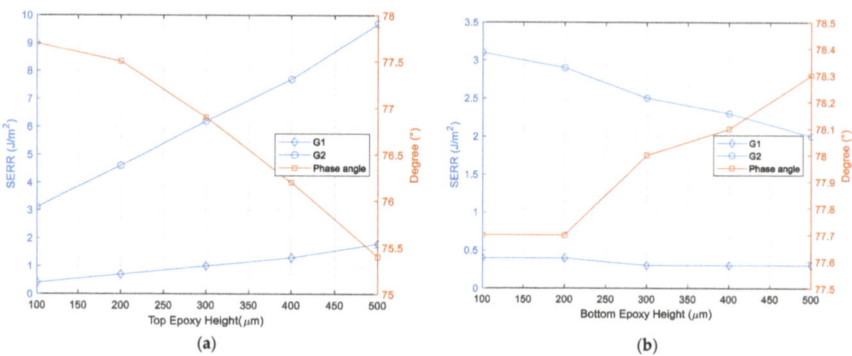

Figure 10. (**a**) Fracture parameters as function of top epoxy height. (**b**) Fracture parameters as function of bottom epoxy height.

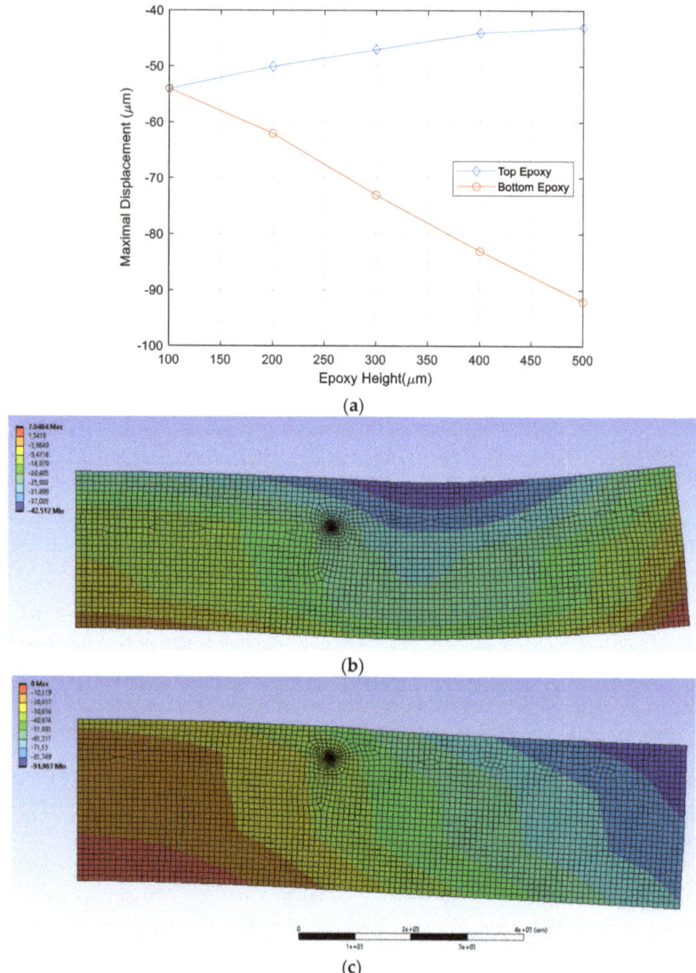

Figure 11. Bending mode as a function of epoxy height. (**a**) Maximal displacement vs. Epoxy height (**b**) Deformation of package when top epoxy height is 400 μm; (**c**) Deformation of package when bottom epoxy height is 500 μm.

4. Conclusions

Biocompatible packaging plays a more and more important role in implantable medical devices due to emergence of new technology such as neural prosthesis. Evolution of the biocompatible packaging has been recently reported; it has been recognized that flip-chip bonding with a bare chip is one of best way of packaging in terms of miniaturization as is conventional electronics packaging. However, microfabrication-like packaging has been frequently reported as advanced UTC (Ultra-Thin Chip) technology is introduced. The major advantage of UTC technology is the capability of maximum miniaturization and thus, biocompatible packaging can take advantage of this new technology as package mechanical stress can cause undesirable FBR (Foreign Body Reaction) during implant. The biocompatible packaging approach can be categorized in two different ways, polymer encapsulation of conventional package and chip-level packaging. Material properties for encapsulation polymers and diffusion barrier have been presented in view of long-term biocompatible packaging. Diffusion barrier is a dielectric or ceramic layer to prevent liquid penetration or leaching of toxic chemical out of packaged Si chip, while encapsulating polymer is a protection polymer with lower elasticity causing smaller mechanical stress to surrounding tissue. As encapsulating polymer has multiple interfaces with the packaged objects such as sensors, Si chip, bumps, and bonding wire, there would be interfacial delamination caused by imperfect manufacturing of polymer coating, resulting in voids or microcracks. FEM modeling and simulation, based on fracture mechanics, is an efficient way to understand of the failure mode due to the interfacial delamination. At FEM modeling and simulation, phase angle based on stress intensity factor of mode I and mode II is used to find principal failure mode of encapsulation polymer for biocompatible package. Package design can be also made through FEM-based parametric study of package geometric parameters. In conclusion, this paper addresses that it is of high importance that packaging technology, material selection, modeling, and simulation are essential for long-term reliable biocompatible package.

Funding: This research received no external funding.

Conflicts of Interest: The authors declare no conflict of interest.

References

1. Lee, D.U.; Kim, K.W.; Kim, K.W.; Kim, H.; Kim, J.Y.; Park, Y.J.; Kim, J.H.; Kim, D.S.; Park, H.B.; Shin, J.W.; et al. 25.2 A 1.2 V 8 Gb 8-channel 128 GB/s high-bandwidth memory (HBM) stacked DRAM with effective microbump I/O test methods using 29 nm process and TSV. In Proceedings of the 2014 IEEE International Solid-State Circuits Conference Digest of Technical Papers (ISSCC), San Francisco, CA, USA, 9–13 February 2014; pp. 432–433.
2. Jun, H.; Cho, J.; Lee, K.; Son, H.-Y.; Kim, K.; Jin, H.; Kim, K. HBM (High Bandwidth Memory) DRAM Technology and Architecture. In Proceedings of the 2017 IEEE International Memory Workshop (IMW), Monterey, CA, USA, 14–17 May 2017; pp. 1–4.
3. Jun, H.; Nam, S.; Jin, H.; Lee, J.-C.; Park, Y.J.; Lee, J.J. High-Bandwidth Memory (HBM) Test Challenges and Solutions. *IEEE Des. Test* **2016**, *34*, 16–25. [CrossRef]
4. Kim, S.; Kim, S.; Cho, K.; Shin, T.; Park, H.; Lho, D.; Park, S.; Son, K.; Park, G.; Kim, J. Processing-in-memory in High Bandwidth Memory (PIM-HBM) Architecture with Energy-efficient and Low Latency Channels for High Bandwidth System. In Proceedings of the 2019 IEEE 28th Conference on Electrical Performance of Electronic Packaging and Systems (EPEPS), Montreal, QC, Canada, 6–9 October 2019; pp. 1–3.
5. Kim, S.J.; Lee, D.S.; Kim, I.G.; Sohn, D.W.; Park, J.Y.; Choi, B.K.; Kim, S.W. Evaluation of the biocompatibility of a coating mate-rial for an implantable bladder volume sensor. *Kaohsiung J. Med. Sci.* **2012**, *28*, 123–129. [CrossRef] [PubMed]
6. Teo, A.; Mishra, A.; Park, I.; Kim, Y.J.; Park, W.-T.; Yoon, Y.-J. Polymeric Biomaterials for Medical Implants and Devices. *ACS Biomater. Sci. Eng.* **2016**, *2*, 454–472. [CrossRef] [PubMed]
7. Pérez-Merino, P.; Dorronsoro, C.; Llorente, L.; Durán, S.; Jiménez-Alfaro, I.; Marcos, S. In vivo chromatic aber-ration in eyes implanted with intraocular lenses. *Invest. Ophthalmol. Vis. Sci.* **2013**, *54*, 2654–2661. [CrossRef] [PubMed]
8. Kirsten, S.; Uhlemann, J.; Braunschweig, M.; Wolter, K.J. Packaging of electronic devices for long-term implan-tation. In Proceedings of the 35th International Spring Seminar on Electronics Technology (ISSE), Bad Aussee, Austria, 9–13 May 2012; IEEE: Piscataway, NJ, USA, 2012; pp. 123–127.
9. Hogg, A.; Aellen, T.; Uhl, S.; Graf, B.; Keppner, H.; Tardy, Y.; Burger, J. Ultra-thin layer packaging for implanta-ble electronic devices. *J. Micromech. Microeng.* **2013**, *23*, 075001. [CrossRef]

10. Hassler, C.; von Metzen, R.P.; Ruther, P.; Stieglitz, T. Characterization of parylene C as an encapsulation mate-rial for implanted neural prostheses. *J. Biomed. Mater. Res. Part B* **2010**, *93*, 266–274.

11. Lee, S.W.; Min, K.S.; Jeong, J.; Kim, J.; Kim, S.J. Monolithic Encapsulation of Implantable Neuroprosthetic Devices Using Liquid Crystal Polymers. *IEEE Trans. Biomed. Eng.* **2011**, *58*, 2255–2263. [CrossRef]

12. Min, K.S.; Lee, C.J.; Jun, S.B.; Kim, J.; Lee, S.E.; Shin, J.; Chang, J.W.; Kim, S.J. A Liquid Crystal Polymer-Based Neuromodulation System: An Application on Animal Model of Neuropathic Pain. *Neuromodul. Technol. Neural Interface* **2013**, *17*, 160–169. [CrossRef] [PubMed]

13. Kramar, T.; Michalec, I.; Kovacocy, P. The laser beam welding of titanium grade 2 alloy. *GRANT J.* **2012**, *1*, 77–79.

14. Schuettler, M.; Ordonez, J.S.; Santisteban, T.S.; Schatz, A.; Wilde, J.; Stieglitz, T. Fabrication and test of a hermetic miniature implant package with 360 electrical feedthroughs. In Proceedings of the 2010 Annual International Conference of the IEEE Engineering in Medicine and Biology, Buenos Aires, Argentina, 31 August–4 September 2010; pp. 1585–1588. [CrossRef]

15. Chlebowski, A.L.; Chow, E.Y.; Ellison, C.; Irazoqui, P.P. Integrated LTCC packaging for use in biomedical devices. *Bio-Med Mater. Eng.* **2012**, *22*, 361–372. [CrossRef]

16. Op de Beeck, M.; O'Callaghan, J.; Qian, K.; Malachowski, K.; Vanfleteren, J.; Van Hoof, C. Biocompatible packaging solutions for implantable electronic systems for medical applications. In Proceedings of the IEEE Circuits and Systems Society Forum on Emerging and Selected Topics, Seoul, Korea, 20 May 2012.

17. Brancato, L.; Weydts, T.; Oosterlinck, W.; Herijgers, P.; Puers, R. Biocompatible packaging of an epicardial accel-erometer for real-time assessment of cardiac motion. *Procedia Eng.* **2016**, *168*, 80–83. [CrossRef]

18. Brancato, L.; Weydts, T.; De Clercq, H.; Dimiaux, T.; Herijgers, P.; Puers, R. Biocompatible Packaging and Testing of an Endocardial Accelerometer for Heart Wall Motion Analysis. *Procedia Eng.* **2015**, *120*, 840–844. [CrossRef]

19. Kirsten, S.; Schubert, M.; Braunschweig, M.; Woldt, G.; Voitsekhivska, T.; Wolter, K.-J. Biocompatible packaging for implantable miniaturized pressure sensor device used for stent grafts: Concept and choice of materials. In Proceedings of the 2014 IEEE 16th Electronics Packaging Technology Conference (EPTC), Singapore, 3–5 December 2014; pp. 719–724.

20. Velten, T.; Ruf, H.; Barrow, D.; Aspragathos, N.; Lazarou, P.; Jung, E.; Malek, C.; Richter, M.; Kruckow, J.; Wackerle, M. Packaging of bio-MEMS: Strategies, technologies, and applications. *IEEE Trans. Adv. Packag.* **2005**, *28*, 533–546. [CrossRef]

21. Meyer, J.-U.; Stieglitz, T.; Scholz, O.; Haberer, W.; Beutel, H. High density interconnects and flexible hybrid assemblies for active biomedical implants. *IEEE Trans. Adv. Packag.* **2001**, *24*, 366–374. [CrossRef]

22. Stieglitz, T.; Schuetter, M.; Koch, K.P. Implantable biomedical microsystems for neural prostheses. *IEEE Eng. Med. Boil. Mag.* **2005**, *24*, 58–65. [CrossRef] [PubMed]

23. Rodger, D.C.; Tai, Y.-C. Microelectronic packaging for retinal prostheses. *IEEE Eng. Med. Boil. Mag.* **2005**, *24*, 52–57. [CrossRef] [PubMed]

24. De Beeck, M.O.; Verplancke, R.; Schaubroeck, D.; Cuypers, D.; Cauwe, M.; Vandecasteele, B.; O'Callaghan, J.; Braeken, D.; Andrei, A.; Firrincieli, A.; et al. ltra-thin biocompatible implantable chip for bidirectional communication with peripheral nerves. In Proceedings of the 2017 IEEE Biomedical Circuits and Systems Conference, Turin, Italy, 19–21 October 2017; pp. 1–4.

25. Barz, F.; Lausecker, R.; Wallrabe, U.; Ruther, P.; Paul, O. Wafer-level shellac-based interconnection process for ultrathin silicon chips of arbitrary shape. In Proceedings of the 2016 IEEE 29th International Conference on Micro Electro Mechanical Systems (MEMS), Shanghai, China, 24–28 January 2016; pp. 520–523. [CrossRef]

26. Biran, R.; Martin, D.C.; Tresco, P.A. The brain tissue response to implanted silicon microelectrode arrays is increased when the device is tethered to the skull. *J. Biomed. Mater. Res.* **2007**, *82*, 169. [CrossRef]

27. Du, Z.J.; Kolarcik, C.L.; Kozai, T.D.; Luebben, S.D.; Sapp, S.A.; Zheng, X.S.; Nabity, J.A.; Cui, X.T. Ultrasoft microwire neural electrodes improve chronic tissue integration. *Acta Biomater.* **2017**, *53*, 46–58. [CrossRef] [PubMed]

28. Inmann, A.; Hodgins, D. *Implantable Sensor Systems for Medical Applications*; Woodhead Publishing Series in Biomaterials; Elsevier: Amsterdam, The Netherlands, 2013.

29. Lapique, F.; Redford, K. Curing effects on viscosity and mechanical properties of a commercial epoxy resin adhesive. *Int. J. Adhes. Adhes.* **2002**, *22*, 337–346. [CrossRef]

30. Dy, E.; Ho, C.-M. Development of a cytomic force transducer for experimentalmechanobiology. In Proceedings of the 2009 IEEE 22nd International Conference on Micro Electro Mechanical Systems, Sorrento, Italy, 25–29 January 2009; pp. 391–394.

31. Erismis, M.A.; Neves, H.P.; De Moor, P.; Puers, R.; Van Hoof, C. A water-tight packaging of MEMS electrostatic actuators for biomedical applications. *Microsyst. Technol.* **2010**, *16*, 2109–2113. [CrossRef]

32. Chang, K.-C.; Hsu, M.-H.; Lu, H.-I.; Lai, M.-C.; Liu, P.-J.; Hsu, C.-H.; Ji, W.-F.; Chuang, T.-L.; Wei, Y.; Yeh, J.-M.; et al. Room-temperature cured hydrophobic epoxy/graphene composites as corrosion inhibitor for cold-rolled steel. *Carbon* **2014**, *66*, 144–153. [CrossRef]

33. Phan, H.-P. Implanted Flexible Electronics: Set Device Lifetime with Smart Nanomaterials. *Micromachines* **2021**, *12*, 157. [CrossRef]

34. Song, E.; Li, R.; Jin, X.; Du, H.; Huang, Y.; Zhang, J.; Xia, Y.; Fang, H.; Lee, Y.K.; Yu, K.J.; et al. Ultrathin Trilayer Assemblies as Long-Lived Barriers against Water and Ion Penetration in Flexible Bioelectronic Systems. *ACS Nano* **2018**, *12*, 10317–10326. [CrossRef]

35. Pham, T.A.; Nguyen, T.K.; Vadivelu, R.K.; Dinh, T.; Qamar, A.; Yadav, S.; Phan, H.P. A versatile sacrificial layer for transfer printing of wide bandgap materials for implantable and stretchable bioelectronics. *Adv. Funct. Mater.* **2020**, *30*, 2004655. [CrossRef]

36. Li, J.; Li, R.; Du, H.; Zhong, Y.; Chen, Y.; Nan, K.; Won, S.M.; Zhang, J.; Huang, Y.; Rogers, J.A. Ultrathin, Transferred Layers of Metal Silicide as Faradaic Electrical Interfaces and Biofluid Barriers for Flexible Bioelectronic Implants. *ACS Nano* **2019**, *13*, 660–670. [CrossRef] [PubMed]
37. Phan, H.P.; Zhong, Y.; Nguyen, T.K.; Park, Y.; Dinh, T.; Song, E.; Nguyen, N.T. Long-lived, transferred crystal-line silicon carbide nanomembranes for implantable flexible electronics. *ACS Nano* **2019**, *13*, 11572–11581. [CrossRef] [PubMed]
38. Fang, H.; Yu, K.J.; Gloschat, C.; Yang, Z.; Song, E.; Chiang, C.-H.; Zhao, J.; Won, S.M.; Xu, S.; Trumpis, M.; et al. Capacitively coupled arrays of multiplexed flexible silicon transistors for long-term cardiac electrophysiology. *Nat. Biomed. Eng.* **2017**, *1*, 1–12. [CrossRef]
39. Reiher, A.; Günther, S.; Krtschil, A.; Witte, H.; Krost, A.; Opitz, T.; Voigt, T. In vitro stimulation of neurons by a planar Ti–Auelectrode interface. *Appl. Phys. Lett.* **2005**, *86*, 103901. [CrossRef]
40. Jiang, Y.; Li, X.; Liu, B.; Yi, J.; Fang, Y.; Shi, F.; Tian, B. Rational design of silicon structures for optically con-trolled multiscale biointerfaces. *Nat. Biomed. Eng.* **2018**, *2*, 508–521. [CrossRef]
41. Lee, C.D.; Meng, E. Mechanical properties of thin-film Parylene-Metal-Parylene devices. *Front. Mech. Eng.* **2015**, *1*, 10. [CrossRef]
42. Xie, X.; Rieth, L.; Williams, L.; Negi, S.; Bhandari, R.; Caldwell, R.; Sharma, R.; Tathireddy, P.; Solzbacher, F. Long-term reliability of Al2O3 and Parylene C bilayer encapsulated Utah electrode array based neural interfaces for chronic im-plantation. *J. Neural. Eng.* **2014**, *11*, 026016. [CrossRef]
43. Fang, H.; Zhao, J.; Yu, K.J.; Song, E.; Farimani, A.B.; Chiang, C.H.; Jin, X.; Xue, Y.; Xu, D.; Du, W.; et al. Ultrathin, transferred layers of thermally grown silicon dioxide as biofluid barriers for biointegrat-ed flexible electronic systems. *Proc. Natl. Acad. Sci. USA* **2016**, *113*, 11682–11687. [CrossRef]
44. Park, D.; Oh, T.S. Comparison of Flip-Chip Bonding Characteristics on Rigid, Flexible, and Stretchable Sub-strates: Part I. Flip-Chip Bonding on Rigid Substrates. *Mater. Trans.* **2017**, *58*, 1212–1216. [CrossRef]
45. Balde, J.W. *Foldable Flex and Thinned Silicon Multichip Packaging Technology*; Springer Science Business Media: New York, NY, USA, 2003.
46. Haberland, J.; Becker, M.; Lutke-Notarp, D.; Kallmayer, C.; Aschenbrenner, R.; Reichl, H. Ultrathin 3D ACA FlipChip-in-Flex Technology. In Proceedings of the IMAPS Device Packaging Conference, Scottsdale, AZ, USA, 9 March 2010.
47. Wu, C.; Liu, J.; Yeung, N. Reliability of ACF in flip-chip with various bump heights. In Proceedings of the 4th International Conference on Adhesive Joining and Coating Technology in Electronics Manufacturing. Presented at Adhesives in Electronics 2000 (Cat. No.00EX431), Espoo, Finland, 18–21 June 2000; pp. 101–106.
48. Rodger, D.C.; Weiland, J.D.; Humayun, M.S.; Tai, Y.C. Scalable Flexible Chip-level Parylene Package for High Lead Count Retinal Prosthesis. In Proceedings of the 13th International Conference on Solid-State Sensors, Actuators and Microsystems, 2005. Digest of Technical Papers, Seoul, Korea, 5–9 June 2005; pp. 1973–1976.
49. Ayhan, A.O.; Nied, H.F. Finite element analysis of interface cracking in semiconductor packages. *IEEE Trans. Compon. Packag. Technol.* **1999**, *22*, 503–511. [CrossRef]
50. Ayhan, A.O. Finite Element Analysis of Semiconductor Package Debonding Due to Thermal Cycling. Master's Thesis, Lehigh University, Bethlehem, PA, USA, 1997.
51. Zhong, Z.; Yip, P.K. Finite element analysis of a three-dimensional package. *Solder. Surf. Mt. Technol.* **2003**, *15*, 21–25. [CrossRef]
52. Fan, X.; Wang, H.; Lim, T. Investigation of the underfill delamination and cracking in flip-chip modules under temperature cyclic loading. *IEEE Trans. Compon. Packag. Technol.* **2003**, *24*, 84–91. [CrossRef]
53. Zhai, C.J.; Sidharth; Blish, R.C., II; Master, R.N. Investigation and Minimization of Underfill Delamination in Flip Chip Packages. *IEEE Trans. Device Mater. Reliab.* **2004**, *4*, 86–91. [CrossRef]
54. Donaldson, P.E.K.; Sayer, E. A technology for implantable hermetic packages. Part 2: An implementation. *Med. Biol. Eng. Comput.* **1981**, *19*, 403–405. [CrossRef] [PubMed]
55. Sun, C.T.; Jin, Z.-H. Chapter 3 the elastic stress field around a crack tip. In *Fracture Mechanics*; Elsevier Inc.: Amsterdam, The Netherlands, 2012.
56. Seok, S. Experiment and Analysis of the Effect of BCB Sealing Ring Flatness on BCB Cap Transfer Pack-aging. *Microsyst. Technol.* **2021**, *27*, 263–268. [CrossRef]
57. Zhang, T.; Choi, K.K.; Rahman, S.; Cho, K.; Baker, P.; Shakil, M.; Heitkamp, D. A hybrid surrogate and pattern search optimization method and application to microelectronics. *Struct. Multidiscip. Optim.* **2006**, *32*, 327–345. [CrossRef]

 *micromachines*

Article

Geometrical Effects on Ultrasonic Al Bump Direct Bonding for Microsystem Integration: Simulation and Experiments

Jun-Hao Lee [1], Pin-Kuan Li [1], Hai-Wen Hung [1], Wallace Chuang [2], Eckart Schellkes [2], Kiyokazu Yasuda [3] and Jenn-Ming Song [1,3,4,5,*]

[1]   Department of Materials Science and Engineering, National Chung Hsing University, Taichung 402, Taiwan; r840722@dragon.nchu.edu.tw (J.-H.L.); klausli@spil.com.tw (P.-K.L.); hung523@yahoo.com.tw (H.-W.H.)

[2]   Automotive Electronics Department, Robert Bosch Taiwan Co., Ltd., Taipei 104, Taiwan; Wallace.Chuang@tw.bosch.com (W.C.); Eckart.Schellkes@tw.bosch.com (E.S.)

[3]   Division of Materials and Manufacturing Science, Graduate School of Engineering, Osaka University, Osaka 565-0871, Japan; yasuda@mapse.eng.osaka-u.ac.jp

[4]   Research Center for Sustainable Energy and Nanotechnology, National Chung Hsing University, Taichung 402, Taiwan

[5]   Innovation and Development Center of Sustainable Agriculture, National Chung Hsing University, Taichung 402, Taiwan

*   Correspondence: samsong@nchu.edu.tw

Abstract: This study employed finite element analysis to simulate ultrasonic metal bump direct bonding. The stress distribution on bonding interfaces in metal bump arrays made of Al, Cu, and Ni/Pd/Au was simulated by adjusting geometrical parameters of the bumps, including the shape, size, and height; the bonding was performed with ultrasonic vibration with a frequency of 35 kHz under a force of 200 N, temperature of 200 °C, and duration of 5 s. The simulation results revealed that the maximum stress of square bumps was greater than that of round bumps. The maximum stress of little square bumps was at least 15% greater than those of little round bumps and big round bumps. An experimental demonstration was performed in which bumps were created on Si chips through Al sputtering and lithography processes. Subtractive lithography etching was the only effective process for the bonding of bumps, and Ar plasma treatment magnified the joint strength. The actual joint shear strength was positively proportional to the simulated maximum stress. Specifically, the shear strength reached 44.6 MPa in the case of ultrasonic bonding for the little Al square bumps.

Keywords: finite element analysis; ultrasonic bonding; metal direct bonding; microsystem integration

Citation: Lee, J.-H.; Li, P.-K.; Hung, H.-W.; Chuang, W.; Schellkes, E.; Yasuda, K.; Song, J.-M. Geometrical Effects on Ultrasonic Al Bump Direct Bonding for Microsystem Integration: Simulation and Experiments. *Micromachines* **2021**, *12*, 750. https://doi.org/10.3390/mi12070750

Academic Editor: Seonho Seok

Received: 23 May 2021
Accepted: 23 June 2021
Published: 26 June 2021

Publisher's Note: MDPI stays neutral with regard to jurisdictional claims in published maps and institutional affiliations.

1. Introduction

Recent research has promoted the application of ultrasonic bonding, which is conventionally used in the bonding of metal/plastic or plastic/plastic, to metal direct bonding. Ultrasonic bonding was recently developed for hermetic sealing in microelectromechanical system (MEMS) packaging. Since 2002, ultrasonic bonding has been used in homogenous and heterogenous metal bonding, including Al/Al, In/Au, and Au/Al [1,2]. Since 2007, studies on the application of ultrasonic bonding to tinned Cu columns have emerged [3–5], and the first report of successful direct Cu bump bonding appeared in 2013. At present, metal direct bonding, including Cu rim sealing in sensors or flip chip bonding of Au or Cu bumps, is under development in the packaging processes of microelectromechanical systems [6–8]. All these progresses are due to the advantages of ultrasonic bonding, i.e., the extremely short bonding time (in seconds or less than 1 s) and the operability at room temperature in ambience.

In ultrasonic bonding [9], fast horizontal vibration is generated using ultrasound to bond two metal surfaces through rapid friction at the bonding interface. The friction damages the surface oxidation layers, removes impurities, and generates a high thermal energy, thereby drastically reducing the bonding time. This method does not require the use of a flux. The short bonding time generates little byproducts (e.g., intermetallic compounds) and therefore allows the resulting component to maintain its favorable electrical properties. Despite advantages such as short bonding time, high bonding strength, high electrical conductivity, and low processing temperature, ultrasonic bonding is less effective when the bonding surface is large.

ANSYS (v.19.2, 2018, Ansys, Inc., Canonsburg, PA, USA) is a program capable of simulating the stress of a material subject to force. Regarding the use of ANSYS in the simulation of ultrasonic bonding, Wang et al. [10] simulated the stress and strain generated by the bonding of composite polymer materials and Al, and compared the simulated joint shear strength with the value obtained in an experimental demonstration. Arai et al. [11] simulated horizontal and vertical deformation in bumps subject to ultrasonic bonding and suggested that high deformation is associated with high bonding strength. Sasaki et al. [12] simulated the effect of different welding heads on ultrasonic bonding and reported a positive correlation between bonding strength and the depth of stress distribution. In ultrasonic bonding, friction exerts a substantial effect on the stress and strain distributions. Increasing the frequency of friction increases the equivalent strain of the corner surface. This indicates that when the equivalent strain is highly concentrated at the corner of a workpiece, the resulting deformation becomes less even [13]. Myung et al. [14] considered thermal cycle parameters in finite element analysis (FEA) of Cu/Cu bonds, the formation of cracks could be predicted.

This study aimed to improve the feasibility of ultrasonic bonding in microsystem integration and employ metal bump direct bonding as an alternative to conventional polymer and solder joining techniques, such as die attachment, flip chip, and ball grid array packaging. As for flip chip, the arrays of bumps with the functions of mechanical support and electrical and thermal conductance usually consist of solders. Studs of gold, copper, silver, and their alloys have also been applied [15–19]. In such cases, intermetallic compounds usually form at the interface with Al pads and likely lead to reliability problems, especially when subject to a harsh environment. Al bump arrays made from bonding wires were recently developed to join with Al pads for high temperature applications such as SiC power devices [20]. This joint structure is free from the formation of interfacial intermetallic compounds and the subsequent problems. To optimize the ultrasonic Al bump bonding process, we attempted to conduct experiments on and simulate the use of ultrasonic bonding to integrate Al bump arrays with Al thin film. In addition to using ANSYS to simulate the stress distribution of bumps subject to force, this study includes an experimental demonstration to determine the effects of bump shape and size on simulated maximum stress and actual joint strength.

2. Experimental Procedures

2.1. Finite Element Analysis

To explore the effect of bump shape and size on the effectiveness of ultrasonic bonding, FEA was performed to simulate the stress exerted on the bump surface during the bonding process. Figure 1 presents the setup for simulation. First, 3D software was used to create a model and generate three types of bumps, namely little square bumps (LSB), little round bumps (LRB), and big round bumps (BRB). The bumps were stacked on a chip coated with metal thin films, and the workpiece was placed on a platform with a heating rod. Model meshing was performed based on the bump size. Next, the static structural model was selected in ANSYS. The steady-state thermal model was then selected and linked with the static structural model to configure the properties of the materials, namely Al, Cu, and Ni/Pd/Au. Owing to limitations in computation time and capability, the total bump height was set as 5 μm and 10 μm, and the thickness of Pd and Au was set as 1 μm. With respect to the meshing of samples, the adaptive sizing mode was used. Minimum edge length was the bump height, i.e., 5 μm for 5 μm-thick bumps, and 10 μm for 10 μm-thick bumps. As in the example given in Figure 2a, the 5 μm and 10 μm-thick bumps were all divided into 44 cells. As for the Pd or Au with the thickness of 1 μm (Figure 2b), the minimum edge length was 1 μm and meshing cell number was 36.

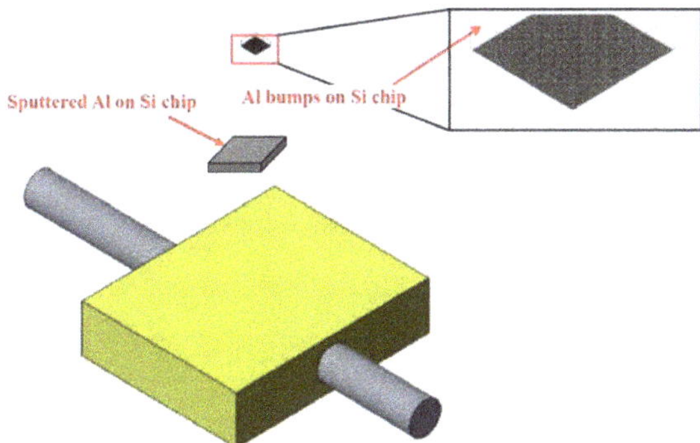

Figure 1. Simulation setup for ultrasonic bonding in this study.

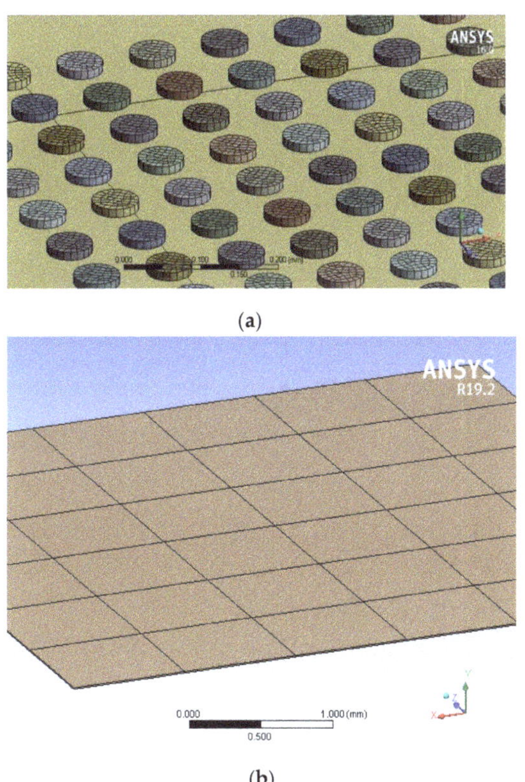

(**a**)

(**b**)

Figure 2. Meshing for simulation: (**a**) 10 μm-thick big round bumps (BRB), (**b**) 1 μm-thick Pd for Ni/Pd/Au.

All material parameters were set according to theoretical values (Table 1); for example, the density, Young's modulus, and Poisson's ratio of Al were 2.7 g/cm^3, 70 GPa, and 0.35, respectively. The configured model was input into ANSYS to determine the model materials. Figure 2 illustrates the model meshing. The boundary condition, set using the fixed support model, of force was determined as the upper surface of the die surface, with an input force of 200 N. Vibration displacement time and distance were input in the bump movement. For the steady-state thermal model setting, the heating rod temperature was set as 200 °C. Notably, frictional displacement was considered in the simulation to accurately reflect the actual process of ultrasonic bonding. This was achieved by configuring the material friction coefficient (Table 1), frictional displacement (6 μm), and frictional frequency (35 kHz). Finally, the equivalent stress was input to solve the final stress value.

Table 1. Material property parameters in finite element analysis (FEA).

Bump	Al	Cu	Ni	Pd	Au
Density (g/cm^{-3})	2.7	8.96	8.908	12.023	19.3
Poisson's Ratio	0.35	0.34	0.31	0.39	0.44
Young's Modulus (GPa)	70	110	200	121	79
Coefficient of Friction	1.4	0.2	0.53	0.36	0.26

2.2. Preparation of Al Bumps

Figure 3 presents the photomasks designed for the BRB, LRB, and LSB. The chip size was 2 × 2 mm^2, and the total contact area of the bumps was kept constant at 1.12–1.13 mm^2. Lithography and sputtering fabrication techniques were used according to the sample design. Ti-coated Si chips were used following subtractive lithography etching and additive lithography sputtering to create the three types of bump array. Figure 4a shows the subtractive method sequence to form Al bumps, in which the chip was sputtered with an Al film, followed by photo resist coating and lithography. Ion etching was applied to form Al bumps. Figure 4b displays the additive sequence, in which lithography was performed to etch out the defined patterns on the coated photoresist. Subsequently, sputtering was performed to deposit Al bumps, and after that the photoresist was removed.

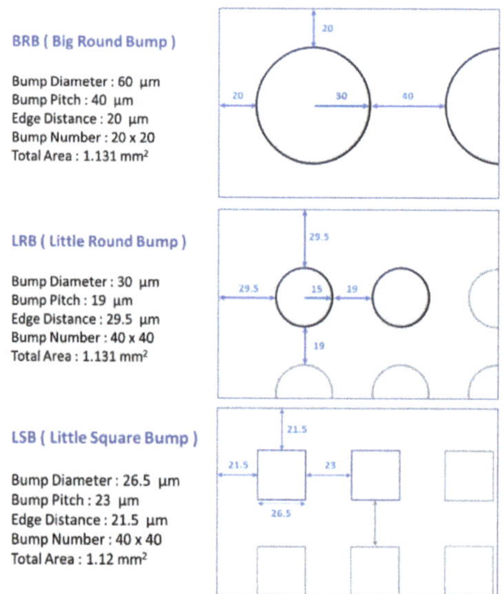

Figure 3. Specifications of the three bump types.

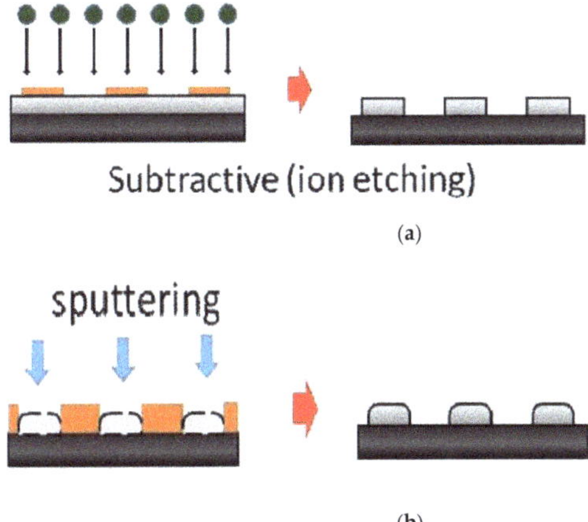

Figure 4. Illustrations of Al bump fabrication: (**a**) subtractive method and (**b**) additive method.

2.3. Surface Pretreatment and Surface Energy Measurement

Surface modification of the Al bump surface was performed using Ar plasma prior to bonding. The conditions of Ar plasma were as follows: power of 200 W, gas flow of 50 sccm, and treating time of 20 min.

Surface energy was estimated according to contact angle measurement. The surface energy γ_s can be divided into two parts, dispersive part γ_s^D and polar part γ_s^P. Deionized water (γ_l^P = 46.8 mN/m and γ_l^D = 26 mN/m) and CH_2I_2 (γ_l^P = 6.7 mN/m, γ_l^D = 44.1 mN/m) were adopted to calculate γ_s using Equations (1) and (2) [21].

$$\gamma_{sl} = \gamma_s - \gamma_l \cos\theta \tag{1}$$

$$\gamma_{sl} = \gamma_s + \gamma_l - 2\left[\left(\gamma_s^D + \gamma_l^D\right)^{\frac{1}{2}} + \left(\gamma_s^P + \gamma_l^P\right)^{\frac{1}{2}}\right] \tag{2}$$

2.4. Ultrasonic Bonding and Joint Strength Measurement

The substrate specimens were square Al-coated Si chips with a length of 3 mm. The substrate was bonded with a Si chip with Al bumps on it through ultrasonic bonding. Figure 5 is a schematic of the customized ultrasonic bonding machine used in this study. The vibration frequency and amplitude of ultrasonic bonding were 35 kHz and 6 µm, respectively. The bonding process was performed in air or nitrogen atmosphere. The bonding temperature was 200 °C, the bonding load was 200 N, and the ultrasonic vibration time was 5 s.

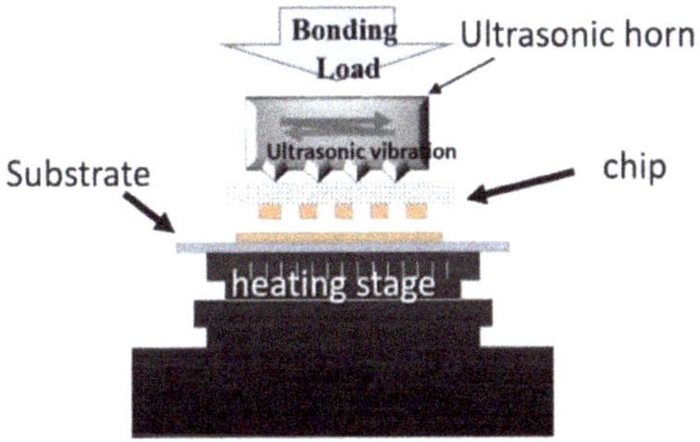

Figure 5. Schematic of the ultrasonic bonding machine.

Figure 6 illustrates the shear testing for obtaining the joint strength. The shear rate was 0.2 mm/min. A thermosetting glue was used to fix the joint specimen, followed by pushing in the knife shear to adjust the shear plane for the shear test.

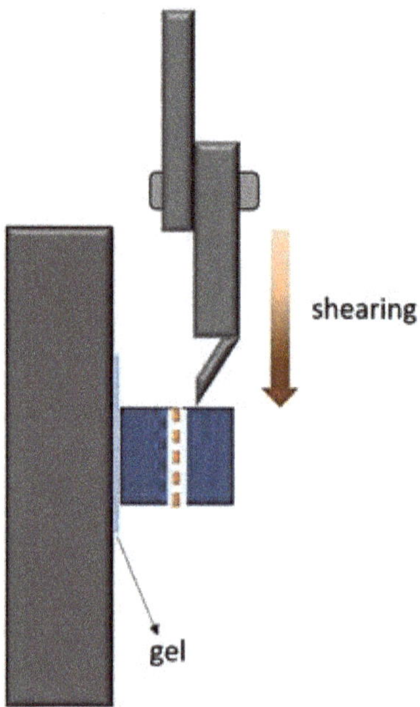

Figure 6. Illustration of the shear testing.

3. Results and Discussion

3.1. Simulation of Stress Distribution

FEA was performed to simulate the stress distribution during ultrasonic bonding. Figures 7 and 8, respectively, present the simulated stress distributions when the bump height was 10 and 5 μm. The maximum stresses were also labeled. The three bump types exhibited similar stress distributions, with the compressive stress concentrated on the edges of the specimens. Particularly, the BRB specimen exhibited relatively lower stress at the center, whereas the stress distributions of the LRB and LSB specimens were more even. For each material, the LSB exhibited a greater maximum compressive stress than the other two bump types, while LRB and BRB did not differ greatly. In addition, the 10 μm-thick bumps showed higher compressive stresses than the 5 μm-thick bumps for all the bump types. The statistics for the stress and corresponding area fraction illustrated in Figure 9 support the above observation about the stress distribution. Overall, stress was concentrated at the peripheral areas of the chips, and the bump shape influenced the stress distribution more than the bump size.

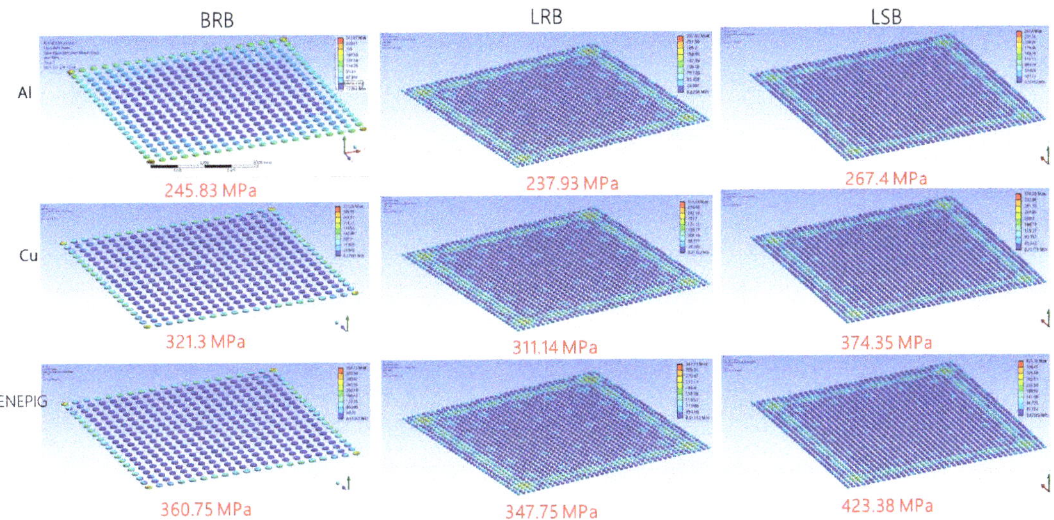

Figure 7. Stress distributions at the bonding interface when bump thickness was 10 µm.

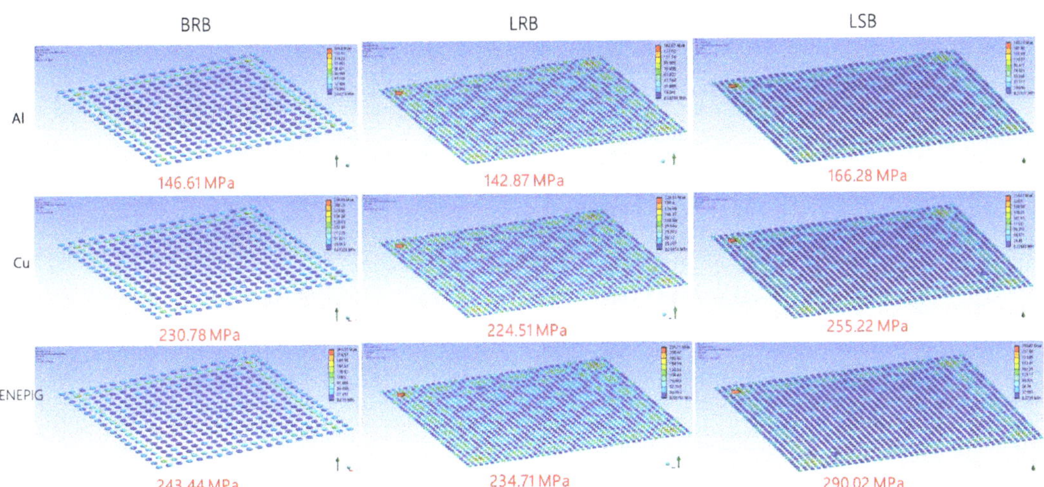

Figure 8. Stress distributions at the bonding interface when bump thickness was 5 µm.

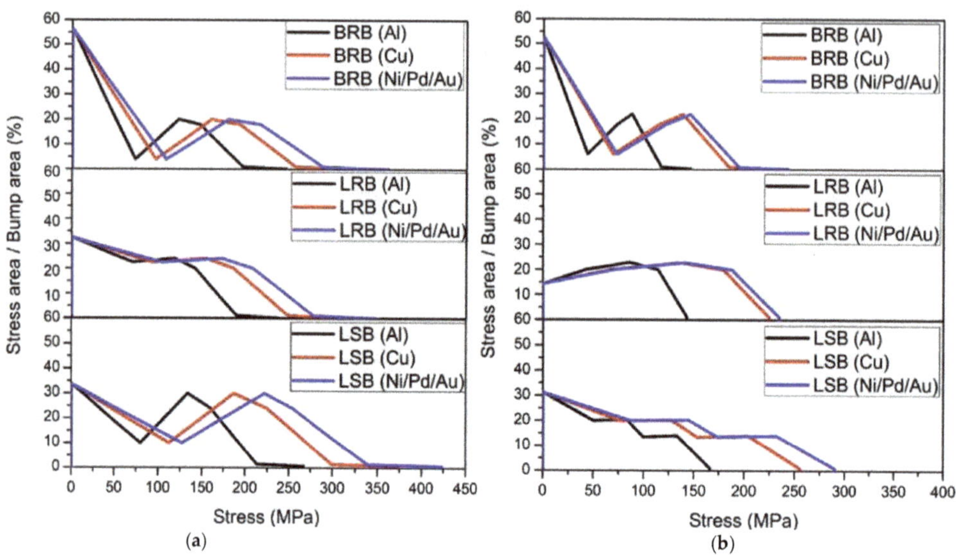

Figure 9. Quantitative simulated stress distribution on bonding surface and corresponding area fraction: (**a**) 10 μm bumps and (**b**) 5 μm bumps.

3.2. Simulated MAXIMUM STRESS with Respect to Materials and Bump Geometry

Figures 10 and 11 compile the maximum stress determined in the simulation. For the 10 μm-thick bumps (Figure 10), the maximum stress comparisons all reveal that Ni/Pd/Au exhibited the largest stress, followed by Cu and Al. The maximum stress of the BRB was only slightly higher than that of the LRB for all materials, by approximately 5%; presenting little difference caused by size dissimilarity (Figure 10a). In contrast, the maximum stress of the LSB was approximately 15–20% greater than that of the LRB (Figure 10b). Figure 11 presents the simulation results when the bump height was 5 μm. The maximum thickness of the BRB was approximately 5% greater than that of the LRB, and that of the LSB was 10–20% greater than that of the LRB. The maximum stress of the 10-μm thick specimens was greater than that of the 5-μm thick specimens, and the increase in maximum stress depended on material type. Specifically, the Al specimens exhibited the largest increase (60%), followed by Ni/Pd/Au (45%) and Cu (40%).

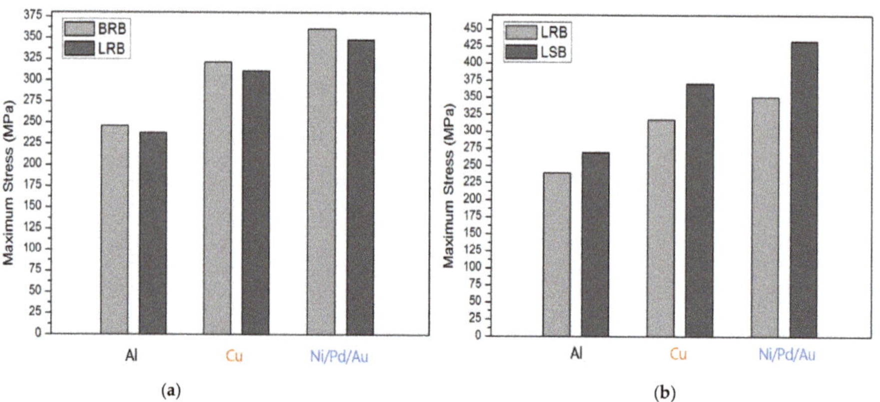

Figure 10. Comparison of the maximum stress of 10-μm thick specimens: (**a**) BRB and LRB; (**b**) LRB and LSB.

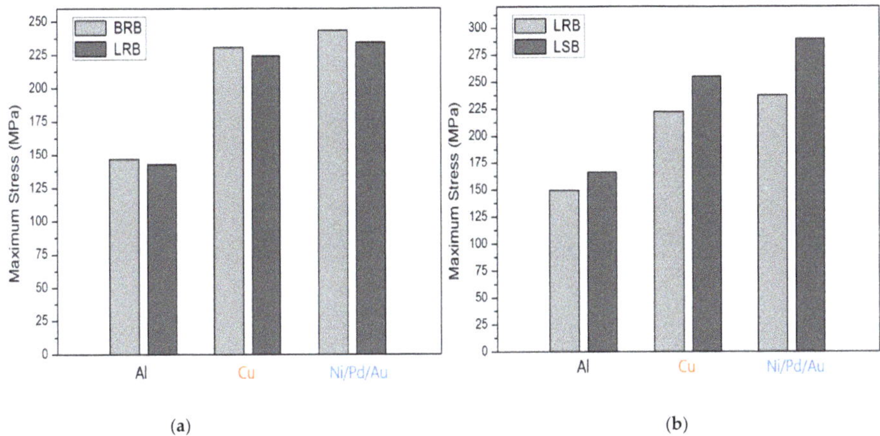

(a)   (b)

Figure 11. Comparison of the maximum stress of 5-μm thick specimens: (**a**) BRB and LRB; (**b**) LRB and LSB.

3.3. Experimental Demonstration of Al/Al Joint Strength

Figure 12a,b present the top views of LSB Al bumps prepared using the subtractive and additive methods, respectively. The contours of the Al bumps shown in Figure 12c,d suggest than the bump height of these two samples was about 1 μm. The additive bumps exhibited an arched surface. On the contrary, the subtractive bumps showed a flat surface, with an effective contact surface area at least 40% larger than that of the additive bumps. The arched surface of additive bumps might have resulted from the shadowing effect of sputtering due to the photoresists. The data given in Table 2 indicate that when ultrasonic bonding was performed in air without surface pre-treatment, the subtractive LSB exhibited an average shear strength of 22.5 MPa, whereas the bonding of the additive LSB was unsuccessful. When ultrasonic bonding was performed in a nitrogen atmosphere, the average shear strength of the subtractive LSB increased slightly to 23.7 MPa, whereas the bonding of the additive LSB still failed. When surface activation right before bonding was performed using Ar plasma, the bonding of the additive LSB remained unsuccessful, but the shear strength of the subtractive LSB increased considerably to 44.6 MPa. The enhancement by Ar plasma bombardment could be ascribed to the significant increase in surface energy, indicating sufficient surface activation [22,23]. The significant increase in surface energy could be verified by the data shown in Figure 13, which were derived from the contact angles of untreated surface using Equations (1) and (2), which were 74.9° for H_2O and 45.9° for CH_2I_2, while those of the Ar plasma-bombarded surface were 4.6° for H_2O and 31.5° for CH_2I_2.

Table 2. Comparison of the ultrasonic bonding results of the LSB bumps created using subtractive lithography etching and additive lithography sputtering (the conditions of Ar plasma: 200 W−50 sccm−20 min).

Photolithography Type	Pre-Treatment	Bonding Environment	Joint Strength
Additive	Without pre-treatment	Air	Failed
		N_2 atmosphere	Failed
	Ar plasma	N_2 atmosphere	Failed
Subtractive	Without pre-treatment	Air	22.5 MPa
		N_2 atmosphere	23.7 MPa
	Ar plasma	N_2 atmosphere	44.6 MPa

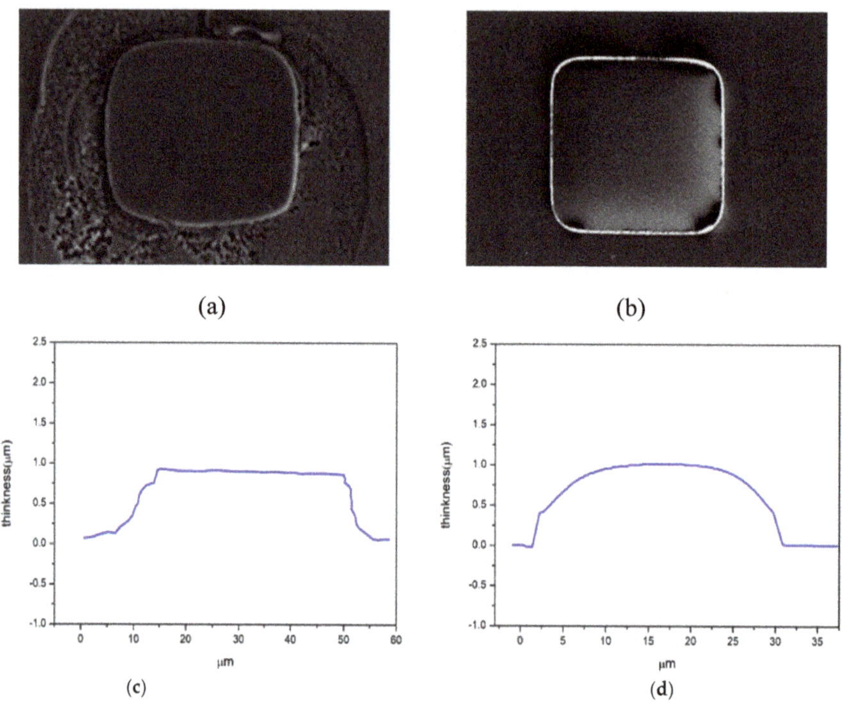

Figure 12. LSB: (a) top-down view of the subtractive Al bump; (b) top-down view of the additive Al bump; and cross-sectional atomic force microscopy scans of the (c) subtractive Al bump and (d) additive Al bump.

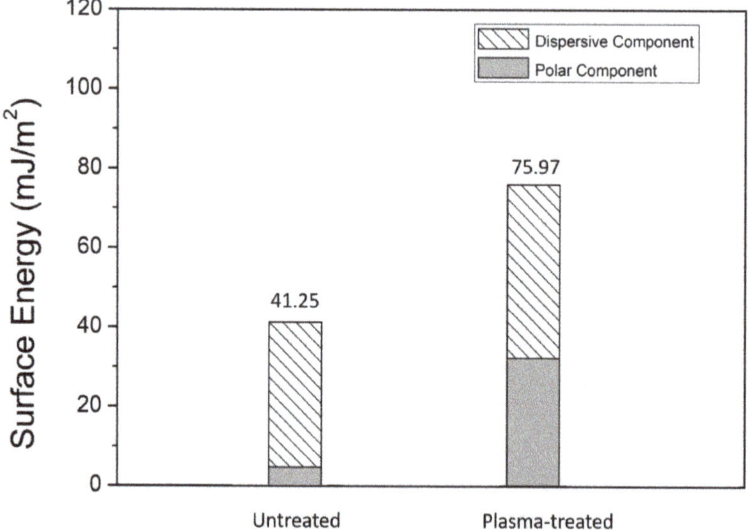

Figure 13. Surface energy of untreated and Ar-plasma treated Al surfaces.

3.4. Geometrical Effects on Joint Strength

Figure 14 compiles actual joint strength and the maximum stress simulated using the aforesaid FEA on Al/Al bonding. When the bump thickness was 10 μm, the maximum stress of the LSB, LRB, and BRB was 267.4, 237.93, and 245.8 MPa, respectively. When the bump thickness was 5 μm, the maximum stress of the LSB, LRB, and BRB was 166.28, 142.87, and 146.61 MPa, respectively. As tabulated in Table 3, and also Figure 14, the shear test results reveal that the actual shear strength of the LSB, LRB, and BRB with surface pre-treatment was 44.6, 28.5, and 30.1 MPa, respectively, whose sequence approximated the maximum stresses estimated through simulation. This again indicates that under the same contact area, bump shape exerted a more notable effect on ultrasonic bonding than bump size. The test results for both the actual joint strength and the simulated maximum stress supported this inference.

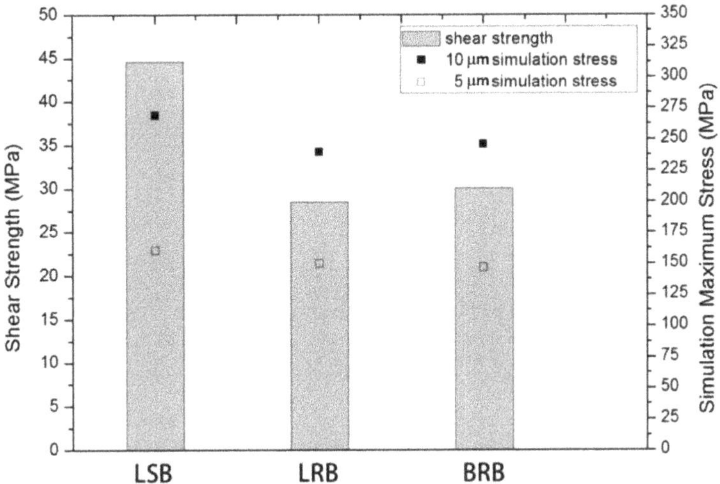

Figure 14. Simulated maximum stress and actual joint shear strength of the LSB, LRB, and BRB.

Table 3. Actual joint strength of the BRB, LRB, and LSB subjected to Ar plasma pretreatment (200 W, 50 sccm, and 20 min) and ultrasonic bonding.

	Pre-Treatment	Bonding Environment	Joint Strength
BRB			30.1 MPa
LRB	Ar plasma	N_2 atmosphere	28.5 MPa
LSB			44.6 MPa

Diffusion under pressure and stress has been studied thermodynamically [24]. The estimation of diffusivity under stress conditions is based on the following equation.

$$\ln\left(\frac{D}{D_0}\right) = (\sigma - \sigma_0)\frac{V^*}{kT} \tag{3}$$

where D is the diffusivity under the set conditions, D_0 denotes the diffusivity at as-received state, V^* is atomic volume, and 0 respectively represents the stresses with or free from stresses. Accordingly, it can be inferred that an increased compressive stress brought about an increase in D/D_0. Therefore, the enhanced self-diffusivity and thus accelerated atom diffusion resulting from the compressive stress accounts for the greater bonding strength of the square bumps.

4. Conclusions

In this study, ultrasonic bonding was utilized to perform direct metal bonding for bump arrays. FEA was adopted to explore the effects of bump shape, size, height, and material on the stress distribution on the bonding surface. An experimental demonstration of Al/Al joint strength was also carried out. The simulation results reveal that the stress distribution of the large bumps was more uneven than that of the small bumps, while the bump shape and height influenced the maximum stress more significantly. Material type did not exert a notable effect on stress distribution, but the maximum stress was positively proportional to the Young's modulus. When ultrasonic bonding was applied to bond Al bumps and thin film, only the subtractive bump underwent successful bonding, and Ar plasma pretreatment magnified the actual joint strength by multiple folds: the actual joint strength of the subtractive LSB specimens with Ar plasma pretreatment reached 44.6 MPa. A comparison of the experimental and FEA results verifies that the tendency of the joint strength observed in the experimental demonstration was consistent with that of the maximum stress estimated using FEA.

Author Contributions: J.-M.S. conceived the research idea, led this cooperative work, and prepared the draft. P.-K.L. carried out the main experiments. J.-H.L. and H.-W.H. carried out the simulations and related analysis. W.C. contributed to methodology and result discussion. E.S. supervised the project and research. K.Y. supervised the ultrasonic bonding technique. All authors have read and agreed to the published version of the manuscript.

Funding: This research was supported by the Ministry of Science and Technology (Taiwan) under contracts MOST 109-2221-E-005-040-MY3.

Acknowledgments: This work was supported by the Ministry of Science and Technology (Taiwan) under contracts MOST 109-2221-E-005-040-MY3, for which the authors are grateful. This work was also supported by the "Innovation and Development Center of Sustainable Agriculture" from the Featured Research Center Program within the framework of the Higher Education Sprout Project of the Ministry of Education (Taiwan).

Conflicts of Interest: The authors declare no conflict of interest.

References

1. Kim, J.; Chiao, M.; Lin, L. Ultrasonic bonding of In/Au and Al/Al for hermetic sealing of MEMS packaging. In Proceedings of the Fifteenth IEEE International Conference on Micro Electro Mechanical Systems (Cat. No.02CH37266), Las Vegas, NV, USA, 20–24 January 2002; pp. 415–418.
2. Kim, J.; Jeong, B.; Chiao, M.; Lin, L. Ultrasonic Bonding for MEMS Sealing and Packaging. *IEEE Trans. Adv. Packag.* **2009**, *32*, 461–467. [CrossRef]
3. Li, J.; Han, L.; Duan, J.; Zhong, J. Interface mechanism of ultrasonic flip chip bonding. *Appl. Phys. Lett.* **2007**, *90*, 242902. [CrossRef]
4. Li, M.; Li, Z.; Xiao, Y.; Wang, C. Rapid formation of Cu/Cu3Sn/Cu joints using ultrasonic bonding process at ambient temperature. *Appl. Phys. Lett.* **2013**, *102*, 094104. [CrossRef]
5. Masumoto, M.; Arai, Y.; Tomokage, H. Effect of Direction of Ultrasonic Vibration on Flip-Chip Bonding. *Trans. Jpn. Inst. Electron. Packag.* **2013**, *6*, 38–42. [CrossRef]
6. Takigawa, R.; Kawano, H.; Shuto, T.; Ikeda, A.; Takao, T.; Asano, T. Room-temperature vacuum packaging using ultrasonic bonding with Cu compliant rim. In Proceedings of the 2014 4th IEEE International Workshop on Low Temperature Bonding for 3D Integration (LTB-3D), Tokyo, Japan, 15–16 July 2014; p. 6886183.
7. Qiu, L.; Ikeda, A.; Noda, K.; Nakai, S.; Asano, T. Room-Temperature Cu Microjoining with Ultrasonic Bonding of Cone-Shaped Bump. *Jpn. J. Appl. Phys.* **2013**, *52*, 4. [CrossRef]
8. Arai, Y.; Nimura, M.; Tomokage, H. Cu-Cu Direct Bonding Technology Using Ultrasonic Vibration for Flip-chip Intercon-nection. In Proceedings of the 2015 International Conference on Electronics Packaging and iMAPS All Asia Conference (ICEP-IAAC), Kyoto, Japan, 14–17 April 2015; pp. 468–472.
9. Yasuda, K. Ultrasonic Bonding Technology for Micro System Integration. *J. Jpn. Inst. Electron. Packag.* **2019**, *22*, 395–399. [CrossRef]
10. Wang, H.; Hao, X.; Yan, K.; Zhou, H.; Lin, H. Ultrasonic vibration-strengthened adhesive bonding of CFRP-to-aluminum joints. *J. Mater Process Technol.* **2018**, *257*, 213–226. [CrossRef]
11. Arai, Y.; Miyamoto, Y.; Nimura, M.; Tomokage, H. Evaluation of ultrasonic vibration energy for copper-to-copper bonding by flip-chip technology. In Proceedings of the 2014 International Conference on Electronics Packaging (ICEP), Toyama, Japan, 23–25 April 2014; pp. 658–661.

12. Sasaki, T.; Komiyama, K.; Pramudita, A.J. Influence of tool edge angle on the bondability of aluminum in ultrasonic bonding. *J. Mater Process Technol.* **2018**, *252*, 167–175. [CrossRef]
13. Kang, C.-G.; Kwon, H. Finite element analysis considering fracture strain of sheath material and die lubricant in extrusion process of Al/Cu clad composites and its experimental investigation. *Int. J. Mech. Sci.* **2002**, *44*, 247–267. [CrossRef]
14. Myung, W.; Kim, K.Y.; Kim, Y.; Jung, S.B. The reliability of ultrasonic bonded Cu to Cu electrode for 3D TSV stacking. *J. Mater. Sci. Mater. Electron.* **2017**, *28*, 16467–16475. [CrossRef]
15. Lim, M.R.; Sauli, Z.; Aris, H.; Retnasamy, V.; Lo, C.; Muniandy, K.; Khan, N.; Foong, C.S. First level interconnection based on optimization of Cu stud bump for chip to chip package. In *AIP Conference Proceedings*; AIP Publishing: Melville, NY, USA, 2018; Volume 2045, p. 020095.
16. Rong, L.M.; Sauli, Z.; Aris, H.; Retnasamy, V. Reliability comparison between solder and solderless flip chip interconnection in terms of high temperature storage. In *AIP Conference Proceedings*; AIP Publishing LLC: Melville, NY, USA, 2018; Volume 2203, p. 020094.
17. Chuang, T.-H.; Hsu, S.-W.; Chen, C.-H. Intermetallic Compounds at the Interfaces of Ag–Pd Alloy Stud Bumps with Al Pads. *IEEE Trans. Compon. Packag. Manuf. Technol.* **2020**, *10*, 1657–1665. [CrossRef]
18. Fujiwara, S.; Harada, M.; Fujita, Y. Interconnection reliability and interfacial structure between Au alloy bump and Al pad using ultrasonic bonding. *Weld. Int.* **2016**, *30*, 1–9. [CrossRef]
19. Kim, H.-J.; Cho, J.-S.; Park, Y.-J.; Lee, J.; Paik, K.-W. Effects of Pd addition on Au stud bumps/Al pads interfacial reactions and bond reliability. *J. Electron. Mater.* **2004**, *33*, 1210–1218. [CrossRef]
20. Tanisawa, H.; Kato, F.; Koui, K.; Sato, S.; Watanabe, K.; Takahashi, H.; Murakami, Y.; Sato, H. Transient thermal characteristics of high-temperature SiC power module enhanced with Al-bump technology. *Jpn. J. Appl. Phys.* **2018**, *57*, 04FR10. [CrossRef]
21. Owens, D.K.; Wendt, R.C. Estimation of surface free energy of polymers. *J. Appl. Polym. Sci.* **1969**, *13*, 1741–1747. [CrossRef]
22. Kim, S.Y.; Hong, K.; Kim, K.; Yu, H.K.; Kim, W.K.; Lee, J.L. Effect of N2, Ar, and O2 plasma treatments on surface properties of metals. *J. Appl. Phys.* **2008**, *103*, 076101. [CrossRef]
23. Kim, J.-W.; Kim, K.-S.; Lee, H.-J.; Kim, H.-Y.; Park, Y.-B.; Hyun, S. The effect of plasma pre-cleaning on the Cu-Cu direct bonding for 3D chip stacking. In Proceedings of the 18th IEEE International Symposium on the Physical and Failure Analysis of Integrated Circuits (IPFA), Incheon, Korea, 4–7 July 2011; pp. 1–4.
24. Aziz, M.J. Thermodynamics of diffusion under pressure and stress: Relation to point defect mechanisms. *Appl. Phys. Lett.* **1997**, *70*, 2810–2812. [CrossRef]

micromachines

Article

Study of the Absorption of Electromagnetic Radiation by 3D, Vacuum-Packaged, Nano-Machined CMOS Transistors for Uncooled IR Sensing

Gil Cherniak [1], Moshe Avraham [1,2], Sharon Bar-Lev [1], Gady Golan [2] and Yael Nemirovsky [1,*]

[1] Electrical Engineering Department, Technion—Israel Institute of Technology, Haifa 32000, Israel; gilcher@campus.technion.ac.il (G.C.); smoa@technion.ac.il (M.A.); sharonb@technion.ac.il (S.B.-L.)

[2] Department of Electrical Engineering and Electronics, Ariel University, Ariel 40700, Israel; gadygolan@gmail.com

* Correspondence: nemirov@technion.ac.il

Abstract: There is an ongoing effort to fabricate miniature, low-cost, and sensitive thermal sensors for domestic and industrial uses. This paper presents a miniature thermal sensor (dubbed TMOS) that is fabricated in advanced CMOS FABs, where the micromachined CMOS-SOI transistor, implemented with a 130-nm technology node, acts as a sensing element. This study puts emphasis on the study of electromagnetic absorption via the vacuum-packaged TMOS and how to optimize it. The regular CMOS transistor is transformed to a high-performance sensor by the micro- or nano-machining process that releases it from the silicon substrate by wafer-level processing and vacuum packaging. Since the TMOS is processed in a CMOS-SOI FAB and is comprised of multiple thin layers that follow strict FAB design rules, the absorbed electromagnetic radiation cannot be modeled accurately and a simulation tool is required. This paper presents modeling and simulations based on the LUMERICAL software package of the vacuum-packaged TMOS. A very high absorption coefficient may be achieved by understanding the physics, as well as the role of each layer.

Keywords: thermal sensors; TMOS sensor; finite difference time domain; optical and electromagnetics simulations

Citation: Cherniak, G.; Avraham, M.; Bar-Lev, S.; Golan, G.; Nemirovsky, Y. Study of the Absorption of Electromagnetic Radiation by 3D, Vacuum-Packaged, Nano-Machined CMOS Transistors for Uncooled IR Sensing. *Micromachines* **2021**, *12*, 563. https://doi.org/10.3390/mi12050563

Academic Editor: Seonho Seok

Received: 14 April 2021
Accepted: 12 May 2021
Published: 16 May 2021

Publisher's Note: MDPI stays neutral with regard to jurisdictional claims in published maps and institutional affiliations.

1. Introduction

There has been a great deal of interest in low-cost uncooled IR sensors in recent years, which may bolster a wide range of new applications, such as consumer electronics, smart homes, Internet of Things (IoT) devices, and mobile applications. Over the years, thermal sensors have been extensively applied to uncooled passive infrared (PIR) sensing. Thermal detection sensors are based on mechanisms that change some measurable property of a material due to the temperature rise of that material as caused by the absorption of electromagnetic radiation. Of these, the most important state-of-the-art thermal detectors are microbolometers, thermopiles, and pyroelectric IR (PIR) sensors [1–6].

Commercially available PIR sensors are usually based on decades-old pyroelectric detector technology. The main drawback of current pyroelectric PIR detectors is that the sensors can only detect moving objects and not the presence of hot objects. Furthermore, as the response times of these sensors are relatively high, fast moving targets are often not detected. They can also fail to detect intruders that move slowly or crawl. They also suffer from false events, which are particularly common at elevated temperatures.

Micromachining has been the enabling technology for sensitive thermal sensors, which require very low thermal mass and very low thermal conductivity [7]. When an optical power irradiates a micro-machined thermal sensor packaged in vacuum, its steady state temperature increases by $\Delta Tss = \eta Pir/Gth$, where Gth [W/K] is the thermal conductance of the holding arms and η is the absorbing efficiency of the radiation.

The advent of microelectromechanical systems (MEMS) and nanoelectromechanical systems (NEMS) technologies in CMOS technology has enabled the production of high performance bolometers and thermopiles. CMOS and its derivative CMOS-SOI are the prevalent microelectronics technologies and the key to a significant cost reduction in many monolithically integrated electro-optical sensors.

Microbolometers are still relatively expensive as they require additional fabrication steps, such as vanadium oxide deposition, on top of standard surface micromachining processes. Thermopiles are being compatible with standard CMOS processes that allow low-cost production with large volumes but require powerful amplifiers since the internal signal is low.

Recently, novel uncooled thermal sensors based on CMOS-SOI technology have been extensively pursued [8–14], mainly for IR and THz detection, and more recently for gas sensing [15–18]. The sensor, dubbed TMOS, is based on a suspended micro- or nano-machined transistor fabricated in standard CMOS-SOI process and released by dry etching. The thermally isolated transistor, operating at subthreshold, converts small temperature changes to electrical signals as the transistor I-V characteristics are strongly dependent upon temperature at a subthreshold level [12].

At present, the most advanced TMOS processing is based on a nanometric 0.13-μm CMOS SOI technology and is implemented with 8-inch wafers. The silicon technology is available for advanced yet standard CMOS and MEMS FABs [19], offering wafer-level processing and packaging (WLP) with integrated optical windows and filters above the vacuum. As a result, there is a considerable and exceptional potential for cost reduction. Since the TMOS may be operated at a subthreshold level, thus consuming very low power, it may be powered by a battery, enabling a wide range of applications related to mobile phones, smart homes, security, and IoT. This low power feature of the TMOS is a great advantage in comparison to the passive bolometers for example, which currently dominate the IR imaging market. In addition, since the TMOS is a transistor, which is an active device exhibiting large internal gain even at subthreshold, it exhibits unprecedented temperature sensitivity $TCV[K^{-1}] = (dV/dT)/V$ and very high responsivity in terms of voltage/wattage in comparison to the commercial passive thermal sensors such as pyroelectric sensors, thermopiles, and diodes.

This paper presents modeling and simulations based on the LUMERICAL software package [20] for the absorption of IR radiation by a vacuum-packaged TMOS sensor. Since the TMOS is processed in a CMOS-SOI FAB and is comprised of multiple thin layers, the absorbed electromagnetic radiation cannot be modeled accurately and a simulation tool is required. Section 2 discusses the choice of the simulation tool and the advantages and drawbacks of LUMERICAL. Section 3 presents the design of a TMOS pixel, as well as the vacuum-packaged TMOS. Incident IR radiation is first transmitted through the upper silicon wafer, which provides the optical window. The transmitted electromagnetic radiation is absorbed in the MEMS/NEMS released TMOS. Section 4 discusses the absorption, reflection, and transmission of the front optical window cap. Since the optical window has a simple structure, the simulation is compared with analytical numerical modeling with MATLAB. The correspondence between the simulation, modeling and measurements validates the results. Section 5 is the heart of this paper and presents the optimized absorption of the TMOS with an impedance matching layer made of TiN. Section 6 summarizes the paper. The goal of this study is to give physical insight regarding the layers and mechanisms that play primary roles in TMOS electromagnetic absorption, as well as to assist in the use of LUMERICAL for simulations.

2. The LUMERICAL Simulation Tool

The absorption simulation was carried out using the LUMERICAL finite-difference time-domain (FDTD) tool and the LUMERICAL knowledge base [20]. It is a state-of-the-art tool for solving Maxwell's equations in complex geometries, which allows solving and analyzing electromagnetics in complex photonics problems. FDTD splits the simulated region into many mesh cells and solves the equations relating to the time and space dependence of the electromagnetic fields at the cell boundaries. Useful quantities can be calculated by using these data, such as the Poynting vector and the transmission/reflection of radiation or absorbed power.

One of the main advantages of LUMERICAL FDTD is its ability to simulate very thin layers that are relative to the radiation wavelength and to use a precise material refractive index that is frequency-dependent in order to get an accurate analysis of the model. Furthermore, LUMERICAL software allows creation of time-domain field propagation simulations that contribute to the understanding of the nature of electromagnetic absorption in the studied system.

There are also several challenges associated with this software:

- The division of the space into mesh points require that a unit cell will be much smaller than the shortest wavelength and smaller than the smallest geometrical feature. This might require multiple mesh points and result in very long simulation durations;
- Introducing boundary conditions at the boundary of the mesh can result in the introduction of errors;
- Resonators with high Q factor require very long simulation times to converge, if at all.

A Brief Comparison between FEM and FDTD

Electromagnetic simulators solve Maxwell equations, which correspond to the initial conditions and boundary conditions. Due to the wide variety of types, shapes, and dimensions of problems, there is no single simulation solver method that is best suitable for all problems and applications. For 3D problems, there are two common solver methods: FDTD, which is used in this paper with LUMERICAL, and finite element methods (FEMs), which are used in many commercial solvers like HFSS and COMSOL Multiphysics.

The simulation processes for the FDTD and FEM solvers are similar. The first step is to define the physical model, which includes the geometry and the properties of the materials. The second step is to set up the simulation, which includes defining the simulation's general settings, boundary conditions and discretizing the physical model to cells. The last step is to run the simulation and postprocess the results.

Although the simulation processes are similar for both methods, the implementation difference between the methods in the way they each solve and discretize the domain may have a huge effect on the results and computation time for different applications.

In the FDTD approach, the region of the simulation is defined. The simulation domain is discretized by a rectangular Cartesian style mesh cell. For each cell, the FDTD method solves Maxwell equations on each cell and time step. The FDTD method solves the equation in the time domain, which makes it usually more suitable for time domain reflectometry. Another advantage for solving the equation in the time domain is that by one simulation the results for broadband frequencies can be achieved.

In the FEM method, the domain volume is discretized to a finite number of elements and nodes, usually by tetrahedron cell mesh. The field is approximated for each cell. Among the nodes, a piecewise polynomial solution is assumed and applying the boundary conditions and simulation properties yields a sparse matrix to determine the fields. Unlike the FDTD method, the FEM method solves the equations in the frequency domain, which makes it more suitable for example for resonators or other high-Q circuits. RF engineers and researchers prefer to use CST or HFSS commercial software while electro-optical scientists may appreciate the advantages of LUMERICAL.

3. TMOS Pixel Design

Figure 1 presents a single nano-machined pixel, including an overview of the layout and a 3D model. The layout of a typical pixel is shown in Figure 1a. The suspended TMOS is released by RIE and DRIE dry etching. For reproducible processing across 8-inch wafers, all gaps must be the same.

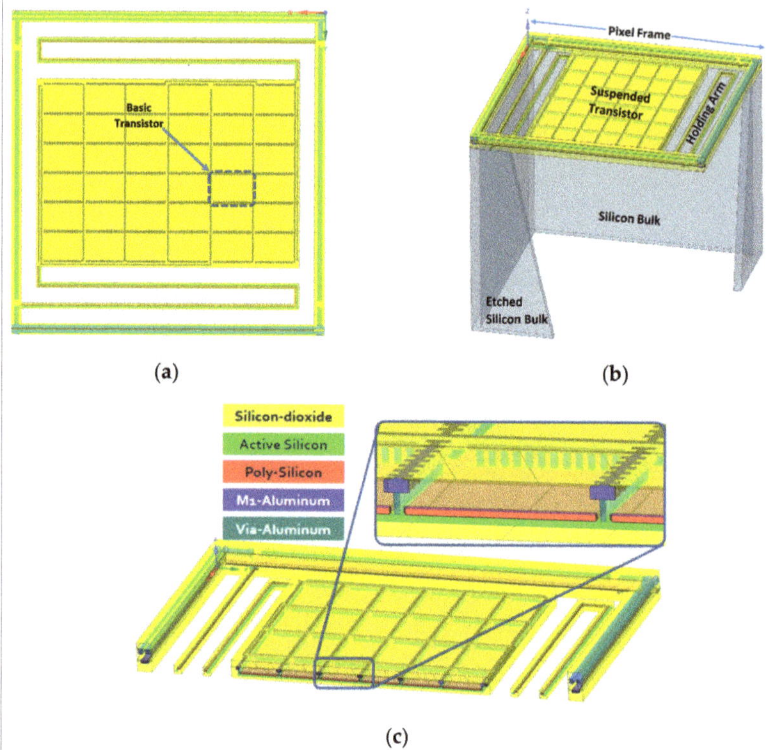

(a)

(b)

(c)

Figure 1. TMOS single nano-machined pixel. (**a**) Overview; (**b**) 3D model; (**c**) cross section of the pixel.

The TMOS released transistor of the pixel has the designed form factor W/L, where W is the width and L is the length of the transistor. It is based on a serial and parallel combination of the largest transistor that the PDK provides in order to be able to use PDK models. For example, large channel length, L is obtained by serially connecting two transistors and the required W is obtained by combining in parallel several transistors.

The vacuum-packaged device is shown in Figure 2 [21].

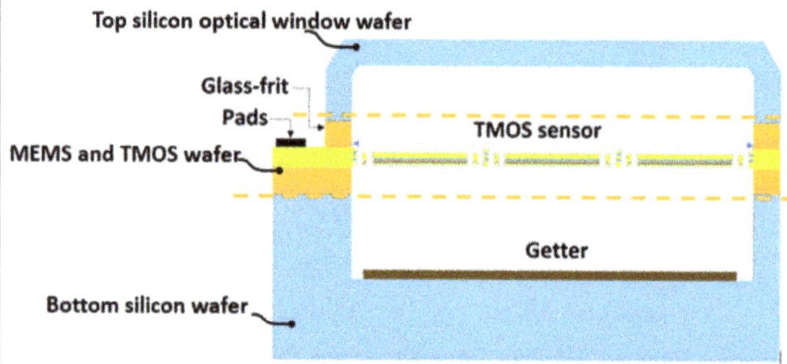

Figure 2. Schematic of the wafer-level package architecture with a getter layer for a high vacuum.

The challenges for wafer-level vacuum packages with MEMS devices are well established and reported in the literature [22]. In the case of optical MEMS sensors, such as thermal sensors, the package plays an important role in the performance, as discussed below.

4. The Absorption, Reflection, and Transmission of the Top Silicon Optical Window

4.1. Modeling

The TMOS cap is fabricated by two layers: a silicon wafer of the order of 100 μm and a much thinner silicon dioxide layer. The optical wave transmission of this Fabry–Perot cavity can be analytically calculated with transfer matrices of the electric fields [23].

Figure 3 exhibits the absorption, reflection, and transmission of the top optical window.

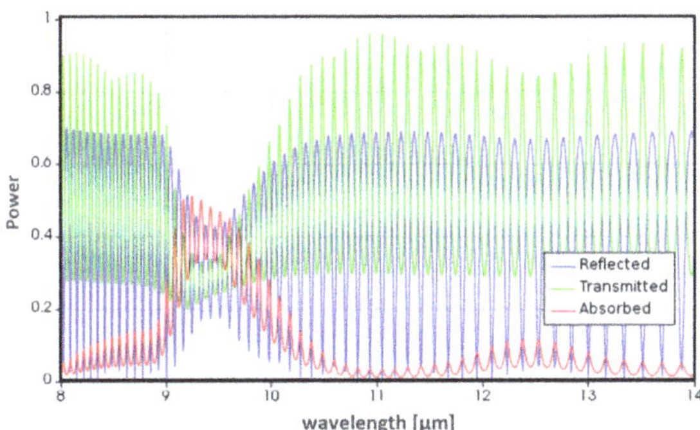

Figure 3. The modeled absorption, reflection, and transmission of the optical window as a function of the wavelength. The silicon wafer size is ~100 μm and the oxide layer is of the order of 0.1 μm.

4.2. LUMERICAL Simulation

A FDTD model has been created in LUMERICAL as described in Figure 4 in order to simulate the TMOS optical window.

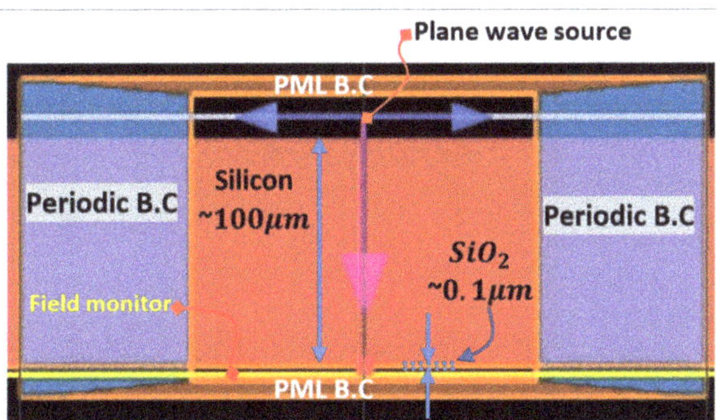

Figure 4. The simulation model and setup. The simulated area is outlined by the orange rectangle. The lateral dimensions are in the order of 0.1 μm. PML B.C.: perfectly matched layer boundary conditions.

As can be seen in Figure 4, on the sides of the FDTD region periodic boundary conditions were applied and on the top and bottom of the FDTD region a perfectly matched layer (PML) boundary conditions were applied. A radiation source with the wavelength range of 5–20 μm was set above the TMOS cap. The refractive indices of the materials were set by the software using Palik data [20,24].

The simulation results are shown in Figure 5.

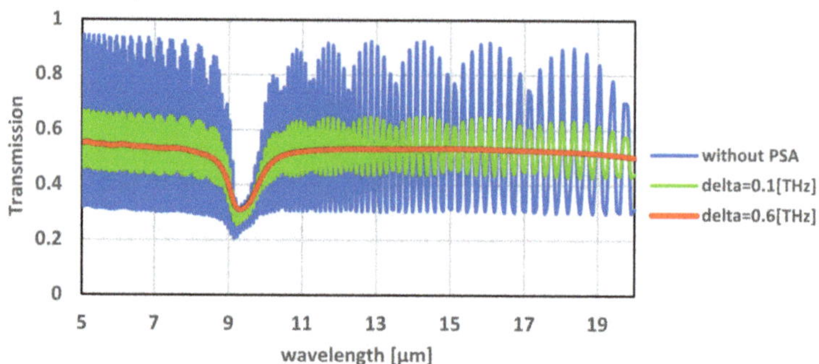

Figure 5. Simulations results for the transmission of the TMOS optical window as a function of the wavelength. PSA: partial spectral averaging (δ: delta, see Section 4.3).

The simulations and modeling show good agreement, as can be seen in Figure 6.

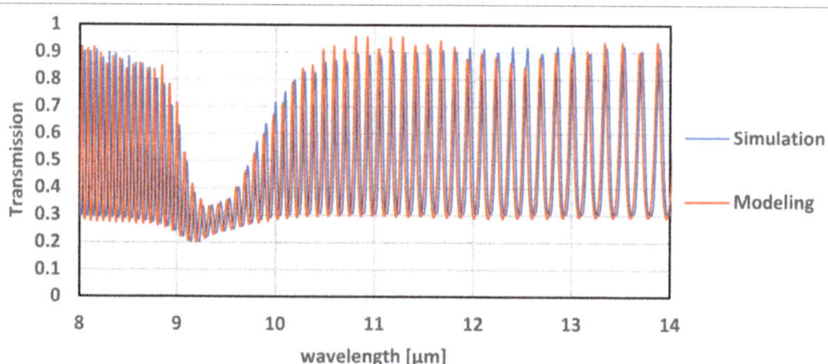

Figure 6. Simulation and modeling of transmission results as a function of the wavelength.

The measured transmission results, compared with the simulation, as shown in Figure 7, are very different and do not show the resonance lines. A good fit between measurements and simulation or modeling requires partial spectral averaging, and δ (delta) estimation, as explained below.

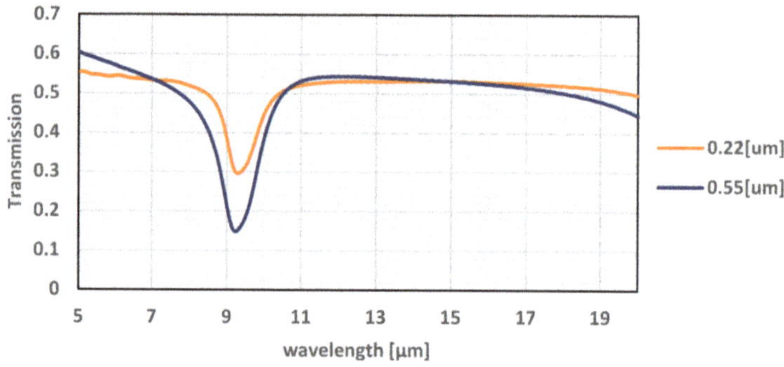

Figure 7. The simulated optical transmittance of the TMOS optical window for different silicon dioxide thickness as a function of the wavelength. The simulated results are in good agreement with measured results.

4.3. Comparison with Measurements

The optical window that was simulated was a Fabry–Perot cavity with two dielectric slab devices. One of the slabs was much wider (>100 μm) than the typical wavelength used for the measurements. Therefore, inside this slab, we expected to find a standing wave. During measurement, the transmitted power through the device was measured for a range of wavelengths. The simulation calculates the transmission for a specific single wavelength, whereas measurement is experimentally conducted around a certain bandwidth, since it is difficult to produce a monochromatic source. Therefore, the calculated simulation should be averaged by a convolution with a Lorentzian weighting function, which is a function mainly used to characterize narrow spectrum lines, such as emission by atoms, laser radiation, etc. [20].

The Lorentzian weighting function is defined by:

$$\left|h\left(\omega,\omega'\right)\right|^2 = \frac{\delta}{\left(\omega - \omega'\right) + \left(\pi\delta\right)^2} \tag{1}$$

where δ is the FWHM (full width at half maximum) and ω is the Lorentzian center frequency. The value of δ can be evaluated by differentiating the relation $\lambda = c/f$ with respect to f, resulting in $|df| = f/\lambda \; d\lambda$, where $d\lambda$ is the measurement error of the wavelength. The value of df can be used for the evaluation of the FWHM. As described in Figure 5, the value of the FWHM used in the simulation was $\delta = 0.6$ (THz), which corresponds to a value of $d\lambda = 200$ (nm). This value yields a match between the measurements and simulation when using that value.

5. Optimized Absorption of the TMOS Sensor with an Impedance Matching Layer with the Right Thickness and Location

To gain an understanding of the electromagnetic absorption of the TMOS sensor, the classical propagation of waves should be applied. The dielectric layers of the TMOS released transistor (see Figure 1c) are comprised of silicon dioxide and silicon nitride. The silicon oxide bond absorbs at ~9.3 μm, while the silicon nitride bond absorbs at ~11.5–12 μm [25]. The physical description becomes complicated since there are several built-in Fabry–Perot resonators and the LUMERICAL simulations become very useful here. Furthermore, it is well established from bolometers that an impedance matching layer with the correct thickness and placement allows optical absorption with 90% efficiency along large arrays. The concept of impedance matching layer for an ideal $\lambda/4$ optical cavity is described in Appendix A [26]. The transmission line description, discussed in Appendix B, provides an intuitive description of an ideal optical cavity.

It is apparent that in order to reduce reflection and enhance absorption, an impedance matching layer is mandatory. This layer should provide the required impedance at the frequency of the incident radiation. It should be based on a standard metallization available in the CMOS FAB. Luckily, Ti and TiN may provide this layer. At DC, the specific resistivity of these layers is too low, but at the IR frequency of interest (around 10^{13} Hz) the values increase by a factor of ~3 because of the plasma effect [25].

In the TMOS sensor case, the modeling is more complicated than that of Appendix B. We assume that a thin "impedance matching layer" is sandwiched between two optical materials (the interlevel dielectric material of the TMOS). When the skin depth into the metal is larger than its thickness, the metallic film can be considered as a regular optical layer (that may be described by the Drude model [25]). In this case, the Salisbury screen becomes a dielectrically-coated Salisbury sheet (DSS) [26–29] (see an example in Figure 8):

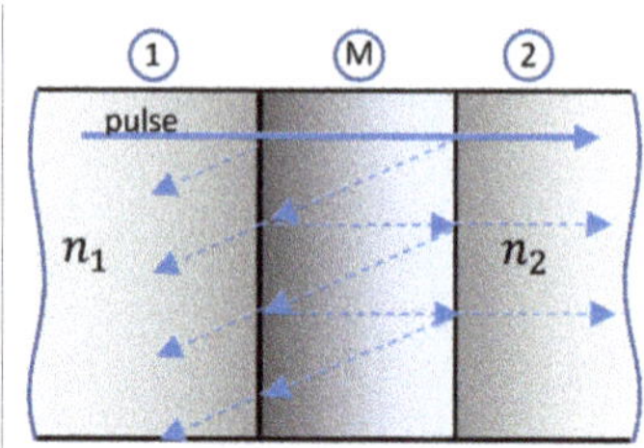

Figure 8. Schematic of an optical system consisting of an optical material (1) with n_1, thin metal layer (M), and optical material (2) with n_2.

As discussed above, the definition of "impedance matching layer" in this case is much more complicated: What should be the thickness and the location of this layer in the cross section of Figure 1c? Theoretically, Fabry–Perot or transmission line modeling may be performed, but LUMERICAL simulations are very effective and useful here.

At this point, it is important to remind the reader that the accuracy of the LUMERICAL simulation is very much dependent upon the mesh dimensions. Accuracy increases if the mesh is less than one tenth of the wavelength. Furthermore, a simple "sanity check" should be performed to validate the results. The sum of absorption, reflection, and transmission should be 1. If it is higher than 1, the mesh should be redefined and be reduced in dimensions.

We present simulation results below that demonstrate how the thickness of the layer (Section 5.1) and location (Section 5.2) affect absorption. At certain bandpass regions, high absorption of 90% may be obtain (Section 5.3).

5.1. The Effect of the Thickness of the TiN

To begin with, we wanted to examine the influence of the thickness of the TiN layer on the absorbance. We simulated the model for varying thickness from 10 nm up to 30 nm, all positioned in the same height of 2.16 µm from the bottom of the Box. The absorption, reflection, and transmission of this simulation can be seen in Figure 9.

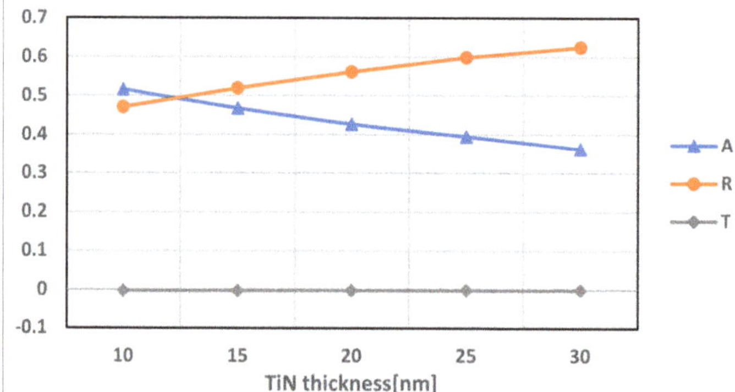

Figure 9. Absorption (A), reflection (R), and transmission (T) of the transistor with varying thicknesses of the TiN layer. Simulation is performed at 9.5 µm wavelength.

It can be seen from Figure 9 that the absorption in the transistor is more significant as the layer thickness decreases and can be improved from 36% up to 52%.

5.2. The Effect of the Location of the TiN along the Upper Silicon Oxide

After optimizing the thickness of the TiN layer, we continued with optimizing the absorbance of the sensor by changing the location of the TiN along the upper silicon oxide, which starts after the last metal layer at the height in the order of ~1.5 μm (see Figure 10).

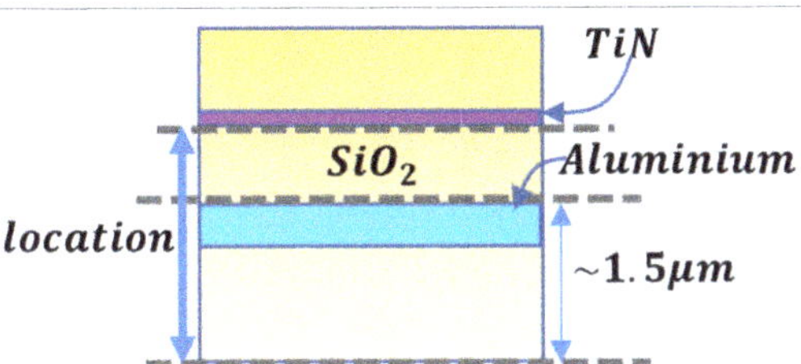

Figure 10. Schematic of the TiN location in the TMOS sensor.

The absorption, reflection, and transmission were simulated for different locations along the silicone oxide for 10-nm and 20-nm layers of TiN.

It can be seen in Figure 11 that the absorbance is higher at a lower TiN placement. As shown in Section 5.1, the higher absorption is reached by the 10–20 nm TiN layer and is slightly higher for 20 nm layer of TiN.

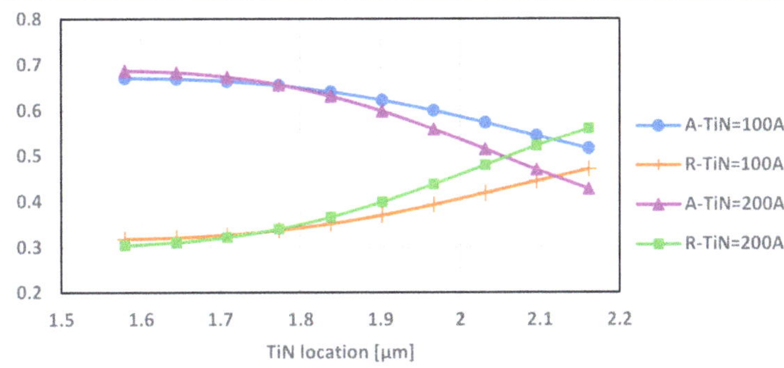

Figure 11. Absorption (A), reflection (R), and transmission (T) of the transistor with varying locations of the TiN layer along the upper silicon oxide for 10-nm and 20-nm layer thickness. This simulation is performed at a 9.5-μm wavelength.

5.3. Simulation of Absorption at an Optical Bandpass

In practice, the TMOS sensor is provided with a bandpass filter that is determined by the use case. We present the simulation below within a bandpass of 6–13 μm. The simulations were performed with default continuous wave normalization, where in this state the power monitors are normalized by the Fourier transform of the source pulse. In other words, it returns the impulse response of the system. For most applications this is the best choice, but in some applications there may be critical mismatch. For this paper, there is a good agreement between the field's results yield by the normalized power over the source spectrum to the impulse response of single frequency simulations, as can be seen in Figure 12.

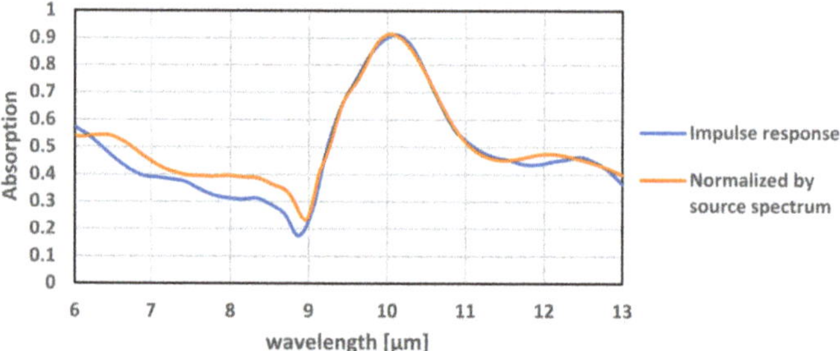

Figure 12. Absorption of the TMOS transistor with a TiN layer thickness of 20 nm located at 1.58 μm (see Figure 10) as a function of the wavelength.

For the optimized thickness and location of the TiN layer, an absorption of about 90% can be achieved. The simulation results for various TiN layer thickness and location can be seen in Figure 13.

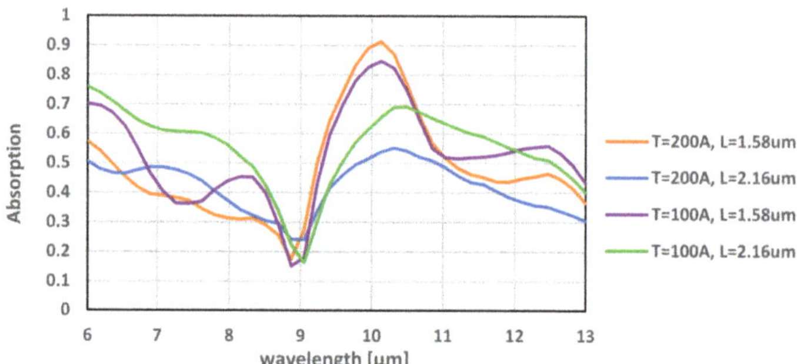

Figure 13. Absorption of the TMOS transistor for various thicknesses (T) and locations (L) as a function of the wavelength.

Figure 14 shows the simulation results for additional SiN layer with thickness of 0.1 μm for various locations. Though the maximum absorption decreases by about 10% around 10 μm for most of the bandpass spectrum, the absorption increases by about 20%.

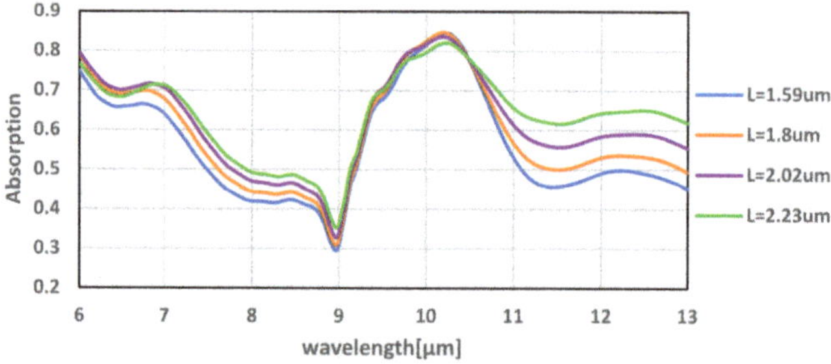

Figure 14. Absorption of the TMOS transistor with additional SiN layer with a thickness of 0.1 μm for various locations (L) as a function of the wavelength.

6. Summary

The goal of this study was to provide physical insight and an intuitive approach to the layers and mechanisms that play primary roles regarding 3D, vacuum-packaged, nano-machined TMOS electromagnetic absorption.

The main parameters that determine the overall absorption efficiency of the vacuum-packaged TMOS sensor are (i) the *transmission* of the optical window and (ii) the *absorption* of the released TMOS sensor, which should be as high as possible. The latter is significantly increased by the impedance matching layer (a dielectric-coated Salisbury sheet) [27–29].

As can be seen in Figure 15, the degrading effect of the oxide, which is part of the optical window, upon the transmission of the electromagnetic radiation in the bandpass of interest (>5 μm) is clearly observed. The average transmission is ~55% while at the resonance wavelengths (around 9.3 μm) it is significantly reduced. By integrating an antire-flection (AR) coating on the optical window, its transmission may be significantly increased.

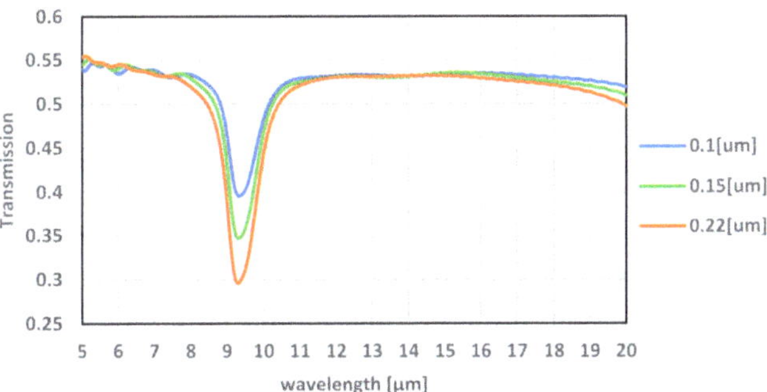

Figure 15. The effect of the thickness of the oxide on the transmission of the optical window. The simulation is achieved with PSA averaging and $\delta = 0.6$ (THz).

The absorption of the TMOS sensor may be optimized with the LUMERICAL simulations. By optimizing the thickness of the *impedance matching layer* of a thin TiN layer and its location along the upper inter-level dielectric of the TMOS, a significant absorption of 90% may be achieved. Adding more than one impedance matching layer may also enhance absorption [27] (chapter 9). Physical insight may be gained by applying transmission lines modeling and Smith chart modeling, but this is beyond the scope of this paper [27–29].

Moreover, the properties of thin layers are different from the bulk materials and depend on the deposition technology. Appendix C reports the values used in this study. Better simulations may be achieved if the actual TMOS main dielectric layers and refractive index are measured. The values of $n(\lambda)$ and $k(\lambda)$ play a role in determining the exact values of absorption. For example, the refractive index for the main TMOS material SiO_2 used in the simulation can be seen in Figure 16.

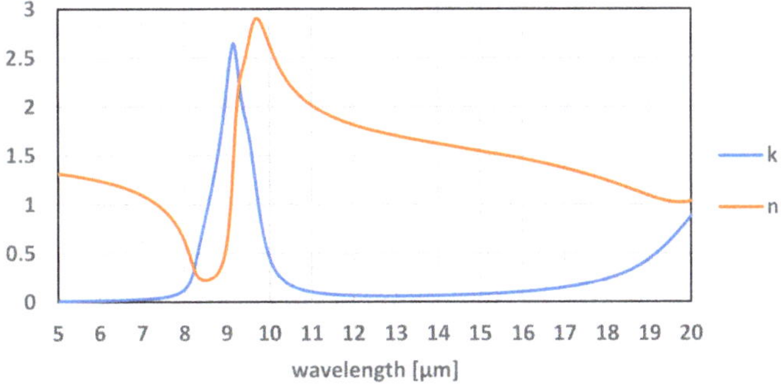

Figure 16. The refractive index of SiO_2 as a function of the TMOS bandpass wavelength where (**n**) is the real part and (**k**) is the imaginary part.

In summary, the LUMERICAL software package is an excellent tool; however, like all software, it requires understanding of the limitations of the tool and the underlying approximation of results. A designer that is also an expert in transmission line theory and Smith charts may achieve very high absorption of the order of 90% within the bandpass of interest.

Author Contributions: Conceptualization, Y.N. and G.G.; methodology, Y.N., G.C. and M.A.; software, G.C., M.A. and S.B.-L.; validation, G.C., M.A. and S.B.-L.; formal analysis, G.C. and M.A.; investigation, G.C. and M.A.; resources, Y.N.; data and M.A.; writing—review and editing, Y.N.; visualization, G.C. and M.A.; supervision, Y.N. and G.G.; project administration, Y.N.; funding acquisition, Y.N. All authors have read and agreed to the published version of the manuscript.

Funding: This research was funded by TODOS technologies, grant number 2024790.

Acknowledgments: We are grateful for the technical support provided by Greg Baethge, ANSYS LUMERICAL. The authors highly appreciate the contributions of the research team at Technion as well as at STM involved with the TMOS project. The funding by TODOS technologies is gratefully acknowledged.

Conflicts of Interest: The authors declare no conflict of interest.

Appendix A. Impedance Matching Layer: Wave Absorption with a Salisbury Sheet

It is well established that if a sheet with a thin layer of 377 $\Omega/\square$ (ohm per square), known as a Salisbury sheet [26], is placed at $\lambda/4$ in front of a perfectly conducting sheet, the incident wave in air will be completely absorbed in the sheet and dissipated as heat, without any reflection, as shown in Figure A1.

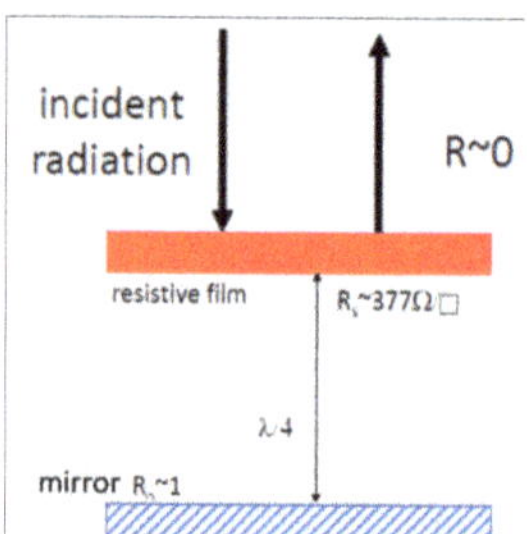

Figure A1. Ideal optical cavity based on Salisbury screen.

The 377 $\Omega/\square$ (ohm per square) is the intrinsic impedance of air or vacuum and the sheet of resistive film (the Salisbury Sheet) is known as "impedance matching layer".

In this arrangement, the impedance presented to the incident wave at the sheet is the impedance of the sheet backed by an infinite impedance, since at $\lambda/4$ the perfectly conducting layer at the bottom is transformed to an infinite impedance. As a result, the incident wave is totally absorbed; however, there is a standing wave and energy circulation between the resistive and conducting sheets and this energy is dissipated as heat in the resistive sheet.

Appendix B. Modeling *the Ideal Optical Cavity* with Transmission Lines

The easiest way to model the ideal optical cavity is by using transmission line theory. Optical waves travel through space as through an infinite transmission line. A pixel with an electrical impedance Z can be modeled as a parallel load between two halves of this line (see Figure A2). The refection coefficient is given by:

$$\Gamma = \frac{Z_0 - (Z||Z_0)}{Z_0 + (Z||Z_0)} \tag{A1}$$

For $Z = Z_0$ the reflection is $1/3$.

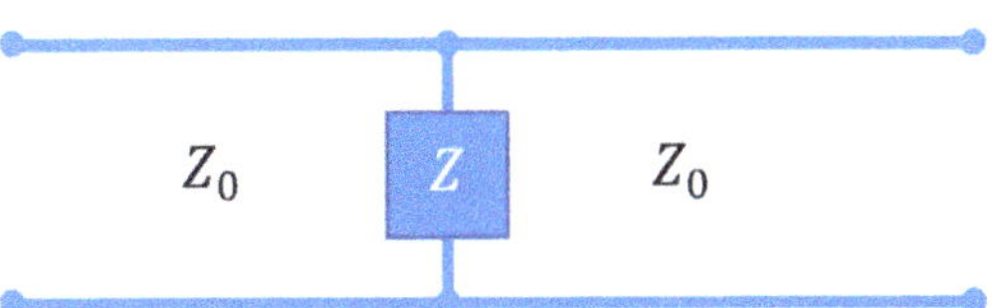

Figure A2. Schematic model of transmission line with pixel with impedance Z. (note Z_0 is the impedance of the air or vacuum).

To maximize absorption, we need to minimize reflection. This is achieved with $\lambda/4$ optical cavity. We can achieve a perfect matching by using a "$\lambda/4$ transformer" terminated by a perfect conductor ($Z_L = 0$), which transforms the impedance seen by the waves to an infinite impedance, in parallel to the impedance of the pixel (see Figure A3).

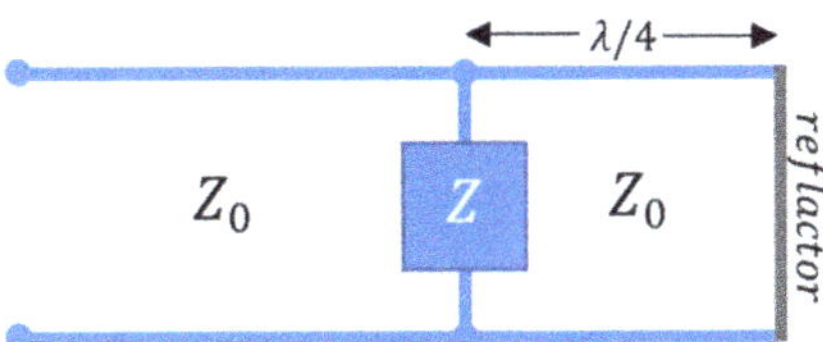

Figure A3. Schematic model of transmission line with pixel with impedance Z and $\lambda/4$ transformer for perfect matching.

Accordingly, the reflection coefficient is:

$$\Gamma = \frac{Z_0 - (Z||Z_0)}{Z_0 + (Z||Z_0)} = \frac{Z_0 - (Z||\infty)}{Z_0 + (Z||\infty)} = \frac{Z_0 - Z}{Z_0 + Z} \tag{A2}$$

Hence, if $Z = Z_0$, the reflection is canceled. Since both, the optical cavity and the reflector, do not absorb energy, all the energy is absorbed by the pixel—the only dissipative element.

Appendix C. Optical and Material Parameters of the Main Layers Used in the Simulations

Table A1. Refractive index of TiN, SiN, and SiO$_2$ at $\lambda = 9.5$ μm.

Material	N	K
TiN	8.5	16.4
SiN	1.56	0.71
SiO$_2$	2.7	1.71

Note: It is well known that the physical properties of thin layers are different from the properties of bulk materials and depend on the deposition technology.

References

1. Jha, C.M. Introduction. In *Thermal Sensors*; Springer: New York, NY, USA, 2015; pp. 1–3.
2. Iniewski, K. (Ed.) *Smart Sensors for Industrial Applications, Part 2*; CRC Press: Boca Raton, FL, USA, 2013.
3. Budzier, H.; Gerlach, G. *Thermal Infrared Sensors: Theory, Optimization and Practice*; John Wiley & Sons Ltd.: Chichester, UK, 2011; ISBN 978-0-470-87192.
4. Rogalski, A. *Infrared Detectors*, 2nd ed.; CRC Press: Boca Raton, FL, USA, 2011.
5. Kastek, M.; Madura, H.; Sosnowski, T. Passive infrared detector for security systems design, algorithm of people detection and field tests result. *Int. J. Saf. Secur. Eng.* **2013**, *3*, 10–23. [CrossRef]
6. Neumann, N.; Banta, V. P12—Comparison of Pyroelectric and Thermopile Detectors. *Proc. IRS²* **2013**, *14*, 139–143.

7. Kaajakari, V. *Practical MEMS*; Small Gear Pub: Las Vegas, NV, USA, 2009.

8. Gitelman, L.; Stolyarova, S.; Bar-Lev, S.; Gutman, Z.; Ochana, Y.; Nemirovsky, Y. CMOS-SOI-MEMS Transistor for Uncooled IR Imaging. *IEEE Trans. Electron Devices* **2009**, *56*, 1935–1942. [CrossRef]

9. Nemirovsky, Y.; Svetlitza, A.; Brouk, I.; Stolyarova, S. Nanometric CMOS-SOI-NEMS Transistor for Uncooled THz Sensing. *IEEE Trans. Electron Devices* **2013**, *60*, 1575–1583. [CrossRef]

10. Saraf, T.; Brouk, I.; Shefi, S.B.-L.; Unikovsky, A.; Blank, T.; Radhakrishnan, P.K.; Nemirovsky, Y.; Bar-Lev, S. CMOS-SOI-MEMS Uncooled Infrared Security Sensor with Integrated Readout. *IEEE J. Electron Devices Soc.* **2016**, *4*, 155–162. [CrossRef]

11. Svetlitza, A.; Blank, T.; Stolyarova, S.; Brouk, I.; Shefi, S.B.-L.; Nemirovsky, Y. CMOS-SOI-MEMS Thermal Antenna and Sensor for Uncooled THz Imaging. *IEEE Trans. Electron Devices* **2016**, *63*, 1260–1265. [CrossRef]

12. Zviagintsev, A.; Blank, T.; Brouk, I.; Bloom, I.; Nemirovsky, Y. Modeling the Performance of Nano Machined CMOS Transistors for Uncooled IR Sensing. *IEEE Trans. Electron Devices* **2017**, *64*, 1–7. [CrossRef]

13. Zviagintsev, A.; Bar-Lev, S.; Brouk, I.; Bloom, I.; Nemirovsky, Y. Modeling the Performance of Mosaic Uncooled Passive IR Sensors in CMOS–SOI Technology. *IEEE Trans. Electron Devices* **2018**, *65*, 4571–4576. [CrossRef]

14. Blank, T.; Brouk, I.; Bar-Lev, S.; Amar, G.; Meimoun, E.; Bouscher, S.; Meltsin, M.; Vaiana, M.; Maierna, A.; Castagna, M.E.; et al. Non-Imaging Digital CMOS-SOI-MEMS Uncooled Passive Infra-Red Sensing Systems. *IEEE Sensors J.* **2021**, *21*, 3660–3668. [CrossRef]

15. Nemirovsky, Y.; Stolyarova, S.; Blank, T.; Bar-Lev, S.; Svetlitza, A.; Zviagintsev, A.; Brouk, I. A New Pellistor-Like Gas Sensor Based on Micromachined CMOS Transistor. *IEEE Trans. Electron Devices* **2018**, *65*, 5494–5498. [CrossRef]

16. Shlenkevitch, D.; Stolyarova, S.; Blank, T.; Brouk, I.; Nemirovsky, Y. Novel Miniature and Selective Combustion-Type CMOS Gas Sensor for Gas-Mixture Analysis—Part 1: Emphasis on Chemical Aspects. *Micromachines* **2020**, *11*, 345. [CrossRef] [PubMed]

17. Avraham, M.; Stolyarova, S.; Blank, T.; Bar-Lev, S.; Golan, G.; Nemirovsky, Y. A Novel Miniature and Selective CMOS Gas Sensor for Gas Mixture Analysis—Part 2: Emphasis on Physical Aspects. *Micromachines* **2020**, *11*, 587. [CrossRef] [PubMed]

18. Shlenkevitch, D.; Avraham, M.; Stolyarova, S.; Blank, T.; Nemirovsky, Y. *Catalytic Gas Sensor Based on Micro Machined CMOS Transistor*; Institute of Electrical and Electronics Engineers (IEEE): New York, NY, USA, 2019; pp. 1–4.

19. STMicroelectronics. Available online: http://www.st.com (accessed on 15 May 2021).

20. FDTD Solutions 7. Available online: http://www.lumerical.com (accessed on 11 January 2012).

21. Urquia, M.A.; Allegato, G.; Paleari, S.; Tripodi, F.; Oggioni, L.; Garavaglia, M.; Nemirovsky, Y.; Blank, T. High vacuum wafer level packaging for uncooled infrared sensor. In Proceedings of the 2020 Symposium on Design, Test, Integration & Packaging of MEMS and MOEMS (DTIP), Lyon, France, 15–26 June 2020; pp. 1–5.

22. Lee, B.; Seok, S.; Chun, K. A study on wafer level vacuum packaging for MEMS devices. *J. Micromech. Microeng.* **2003**, *13*, 663–669. [CrossRef]

23. Born, M.; Wolf, E.; Bhatia, A.B. *Principles of Optics: Electromagnetic Theory of Propagation, Interference and Diffraction of Light/Max Born and Emil Wolf*; Cambridge University Press: Cambridge, UK, 1959.

24. Ghosh, G.; Palik, E.D. *Handbook of Optical Constants of Solids, Five-Volume Set: Handbook of Thermo-Optic Coefficients of Optical Materials with Applications*; Elsevier Science: Amsterdam, The Netherlands, 1997.

25. Sze, S.M.; Ng, K.K. *Physics of Semiconductor Devices*; Wiley-Interscience: New York, NY, USA, 2007.

26. Kraus, J.D.; Daniel, A.F. *Electromagnetics: With Applications*; WCB/McGraw-Hill: Boston, MA, USA, 1953.

27. Munk, B.A. *Frequency Selective Surfaces Theory and Design*; Wiley: Hoboken, NJ, USA, 2000.

28. Jung, J.Y.; Park, J.Y.; Han, S.; Weling, A.S.; Neikirk, D.P. Wavelength-selective infrared Salisbury screen absorber. *Appl. Opt.* **2014**, *53*, 2431–2436. [CrossRef] [PubMed]

29. Kröll, J.; Darmo, J.; Unterrainer, K. Metallic wave-impedance matching layers for broadband terahertz optical systems. *Opt. Express* **2007**, *15*, 6552–6560. [CrossRef] [PubMed]

 *micromachines*

MDPI

Article

Reliability Evaluation of Fan-Out Type 3D Packaging-On-Packaging

Pao-Hsiung Wang, Yu-Wei Huang and Kuo-Ning Chiang *

Department of Power Mechanical Engineering, National Tsing Hua University, Hsinchu 300, Taiwan;
s101033812@gmail.com (P.-H.W.); s106033852@m106.nthu.edu.tw (Y.-W.H.)
* Correspondence: knchiang@pme.nthu.edu.tw

Abstract: The development of fan-out packaging technology for fine-pitch and high-pin-count applications is a hot topic in semiconductor research. To reduce the package footprint and improve system performance, many applications have adopted packaging-on-packaging (PoP) architecture. Given its inherent characteristics, glass is a good material for high-speed transmission applications. Therefore, this study proposes a fan-out wafer-level packaging (FO-WLP) with glass substrate-type PoP. The reliability life of the proposed FO-WLP was evaluated under thermal cycling conditions through finite element simulations and empirical calculations. Considering the simulation processing time and consistency with the experimentally obtained mean time to failure (MTTF) of the packaging, both two- and three-dimensional finite element models were developed with appropriate mechanical theories, and were verified to have similar MTTFs. Next, the FO-WLP structure was optimized by simulating various design parameters. The coefficient of thermal expansion of the glass substrate exerted the strongest effect on the reliability life under thermal cycling loading. In addition, the upper and lower pad thicknesses and the buffer layer thickness significantly affected the reliability life of both the FO-WLP and the FO-WLP-type PoP.

Keywords: fan-out wafer-level package; finite element; glass substrate; reliability life; packaging-on-packaging

Citation: Wang, P.-H.; Huang, Y.-W.; Chiang, K.-N. Reliability Evaluation of Fan-Out Type 3D Packaging-On-Packaging. *Micromachines* **2021**, *12*, 295. https://doi.org/10.3390/mi12030295

Academic Editor: Seonho Seok

Received: 5 February 2021
Accepted: 5 March 2021
Published: 10 March 2021

Publisher's Note: MDPI stays neutral with regard to jurisdictional claims in published maps and institutional affiliations.

1. Introduction

As electronics technology progresses, 3D integrated systems are becoming increasingly important in the realization of lightweight devices with higher performance and better miniaturization. These systems not only effectively reduce the package footprint and weight, but also improve system performance by reducing the system circuit communication length. The fan-out package structure has good electrical and thermal performance and is used in many applications for system integration. Moreover, packaging-on-packaging (PoP) technology allows packages to be stacked three-dimensionally, thus achieving high-density integration and improving chip-to-chip performance (e.g., in applications with high-frequency data exchange between application processes and memory) [1–3].

Glass is an insulating material, and hence its electrical characteristics are more favorable than those of silicon. In addition, the thickness of a glass substrate can be modified without additional thinning and polishing processes, and through glass via (TGV) technology does not require an additional barrier layer, greatly reducing the production cost. Moreover, the coefficient of thermal expansion (CTE) of glass can be optimized to reduce warpage [4,5] and improve the reliability life of stacked fan-out wafer-level packaging (FO-WLP). Hence, a fan-out structure with a glass substrate is favorable for high-frequency applications [4,6]. However, before mass production, packaging must pass reliability life testing under thermal cycling loading; in the JEDEC(Joint Electron Device Engineering Council) standard (JESD22-A104D), the thermal range is −40 to 125 °C. Finite element analysis is widely used to optimize package structures [7–12]. In this study, we proposed an FO-WLP with a glass substrate architecture and used it as a PoP. We fabricated FO-WLP test samples, subjected them to onboard thermal cycling testing (OBTCT), and verified the results against those of the finite element models described below. In the fabricated FO-WLP structure, the corner solder joint is the critical failure point because during heating and cooling, stress and strain accumulate in these joints due to mismatches in the CTE of the package and the printed circuit board (PCB), eventually leading to failure.

We further established two-dimensional (2D) and three-dimensional (3D) models and verified their consistency in terms of the mean-time-to-failure (MTTF). To reduce computing time, the 2D model was used to optimize the reliability life by varying the upper and lower pad diameters, the CTE of the glass substrate, and the thickness of the buffer layer. Parametric studies revealed that the reliability life of the optimized FO-WLP and PoP exceeds 1000 cycles, satisfying JEDEC condition G.

2. Materials and Methods

2.1. Shape Prediction of the Reflowed Solder Joint

The geometry profile of the solder joints strongly affects the reliability life of a package. Therefore, before analyzing the reliability of the solder joints, the shape of the solder balls must be accurately described. In this study, Surface Evolver [13–15] software based on the energy method was used to describe and predict the solder joint shape.

When a liquid reaches static equilibrium, its total energy tends to be the lowest and its surface area the smallest. The energy of a liquid mainly comprises surface tension energy, gravitational energy, and external energy. From the total energy (Equation (1)), the restoring force in the direction of gravity can be calculated, and the shape and height of the solder ball can be estimated:

$$\delta E_{total} = T_S \iint_S \left(div\,\vec{h} - \vec{n} \cdot D\vec{h} \cdot \vec{n} \right) dA + \rho g \iint_S \left(div \frac{z}{2} \vec{k} \right) \vec{h} - curl\left(\vec{h} \times \frac{z}{2} \vec{k} \right) \cdot dA - P \iint_S \vec{h} \cdot dA \tag{1}$$

If Equation (1) is differentiated once, the restoring force of the solder ball can be expressed as follows:

$$F_r = \frac{\partial E_{total}}{\partial H} = \frac{\partial E_{surface\ tension} + \partial E_{gravity} + \partial E_{external\ force}}{\partial H} \tag{2}$$

where E_{total} is the total energy related to the height of the solder ball H, T_s is the surface tension of the solder ball, P is the pressure caused by the external force, A is the surface area of a single element on the solder ball surface, z is the height of the solder ball of a single element surface parallel to the direction of gravity, $\vec{k}$ represents the unit vector in the direction of gravity, $\vec{n}$ is the unit vector along the positive direction of the element surface, F_r is the restoring force, ρ is the density of the solder ball, g is the acceleration due to gravity, and V is the volume of the solder ball. Furthermore, $\vec{h}$ is the perturbation equation:

$$\vec{h} = \left[(z_{top} - z) / (z_{top} - z_{base} - H) \right] \vec{k} \tag{3}$$

where z_{top} and z_{base}, respectively, represent the upper and lower boundary conditions of the solder ball (Figure 1). By applying a slight upward or downward interference on the pad, the restoring force of the solder ball in the direction of gravity can be determined. When the restoring force of the solder ball equals the gravity exerted on the solder ball, the molten solder ball achieves static equilibrium. Then, the height and geometry of the solder ball under static equilibrium can be determined.

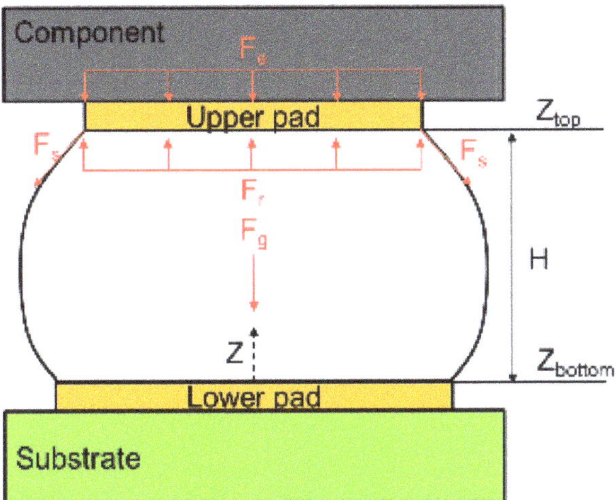

Figure 1. Geometry of the solder joints in the reflow process.

2.2. Life Prediction of the Solder Joints

In electronic packages subjected to accelerated thermal cycling tests, CTE mismatch between the package and the PCB can cause excessive stress and strain to accumulate in the solder joint with the largest distance from the neutral point (DNP); this may cause the first failure of the packaging. The Coffin–Manson strain-based empirical model is widely used to estimate the fatigue life of solder joints, and the empirical equation [16–18] is as follows:

$$N_f = C \left(\varepsilon_{eq}^{in} \right)^{-\eta} \tag{4}$$

where N_f is the mean time to failure (MTTF), and C and η are material constants, usually obtained experimentally. In our model and for SAC305 solder material, these are 0.235 and 1.75, respectively [10–12]. The incremental equivalent inelastic strain in each temperature cycle, ε_{eq}^{in}, is defined as follows:

$$\Delta\varepsilon_{eq.}^{in} = \frac{\sqrt{2}}{3} \left[\left(\Delta\varepsilon_x^{in} - \Delta\varepsilon_y^{in} \right)^2 + \left(\Delta\varepsilon_y^{in} - \Delta\varepsilon_z^{in} \right)^2 + \left(\Delta\varepsilon_z^{in} - \Delta\varepsilon_x^{in} \right)^2 + \frac{3}{2} \left(\Delta\gamma_{xy}^{in\,2} + \Delta\gamma_{yz}^{on\,2} + \Delta\gamma_{zx}^{in\,2} \right) \right]^{\frac{1}{2}} \tag{5}$$

where $\Delta\varepsilon_x^{in}$, $\Delta\varepsilon_y^{in}$, $\Delta\varepsilon_z^{in}$, $\Delta\gamma_{xy}^{in}$, $\Delta\gamma_{yz}^{in}$, and $\Delta\gamma_{zx}^{in}$ are the incremental inelastic strains in the x, y, and z directions and the incremental inelastic shear strain in the xy, yz, and zx directions, respectively.

In this study, we used finite element simulation and the Coffin–Mason empirical model to evaluate the reliability life of the solder joint with the largest DNP.

3. Test Vehicle Structure and Thermal Cycling

3.1. Structure of the Test Vehicle

The target stackable FO-WLP was a cavity-down chip mounted on a glass interposer using TGV and redistribution lines. These chips can be stacked and molded to form PoP-type packaging using through molding via (TMV). The test vehicle [19] was a simplified FO-WLP (a 10 mm × 10 mm × 0.1 mm chip) assembled on a glass substrate of size 14 mm × 14 mm × 0.1 mm using a die attach film (Nitto, EM700). The chip and substrate were covered by a molding compound of size 14 mm × 14 mm × 0.2 mm to form the package. Next, the package was subjected to mechanical debonding (Figure 2). The underside of the glass substrate contained 432 solder joints arranged in a peripheral layout and connected in series. The diameter and pitch of the solder ball were 250 mm and 400 µm, respectively. The test board dimensions were 77 mm × 132 mm, following the

JEDEC (JESD22-B111) design rule. The dimensions of all of the components are listed in Table 1. In the test board, a daisy-chain structure was used to check the electrical resistance of the chained solder joints (Figure 3). The test vehicle was deemed to have failed if its daisy-chain resistance was infinite.

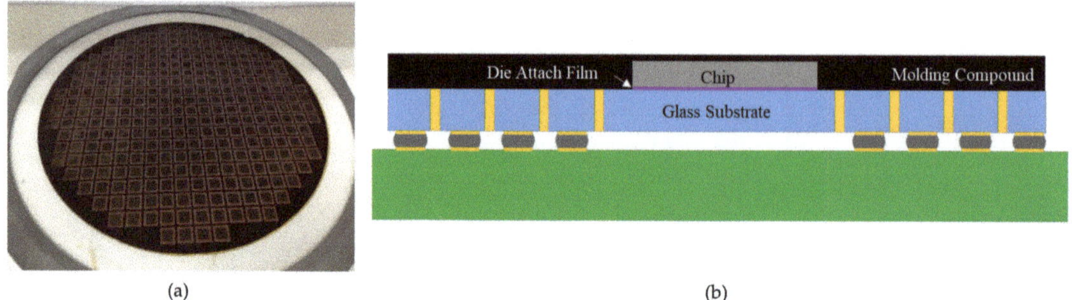

(a)　　　　　　　　　　　　　　　　　　　　(b)

Figure 2. (**a**) Fan-out wafer-level package (FO-WLP) after debonding; (**b**) schematic of the fabricated FO-WLP onboard.

Table 1. Dimensions of the components in a fan-out wafer-level package (FO-WLP).

Component	Size (mm)
Glass substrate	$14 \times 14 \times 0.10$
Chip	$10 \times 10 \times 0.10$
Molding compound	0.19
Die attach film	$10 \times 10 \times 0.01$
Printed circuit board	$77 \times 132 \times 1$
Lower pad	0.24×0.02
Upper pad	0.25×0.002
Stress buffer layer (SBL; polyimide)	0.005
Through glass via	0.025×0.1
Solder ball diameter	0.25
Solder ball pitch	0.4

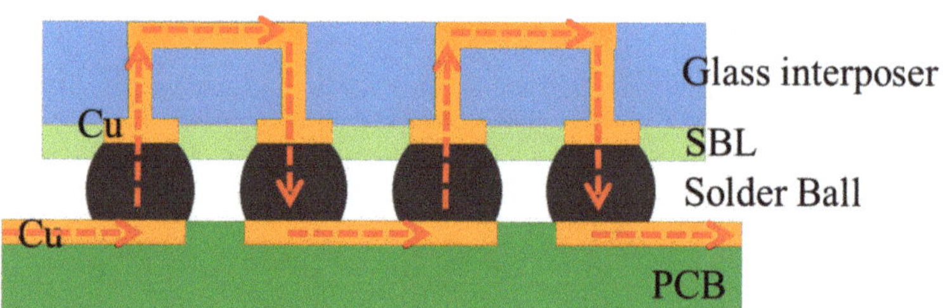

Figure 3. Schematic of the daisy chain. PCB, printed circuit board; SBL, stress buffer layer.

3.2. Thermal Cycling and Weibull Distribution

For reliability testing, we subjected 16 test vehicles to onboard thermal cycling. The test conditions follow the JEDEC standard (condition G): temperature range, −40 to 125 °C; ramp rate, 16.5 °C/min; dwell time, 10 min. The Weibull distribution of the test results (Figure 4) revealed that the mean time to failure was 249 cycles when the cumulative failure percentage equaled 63.2%. Moreover, all 16 samples failed at the top of the outermost solder joint.

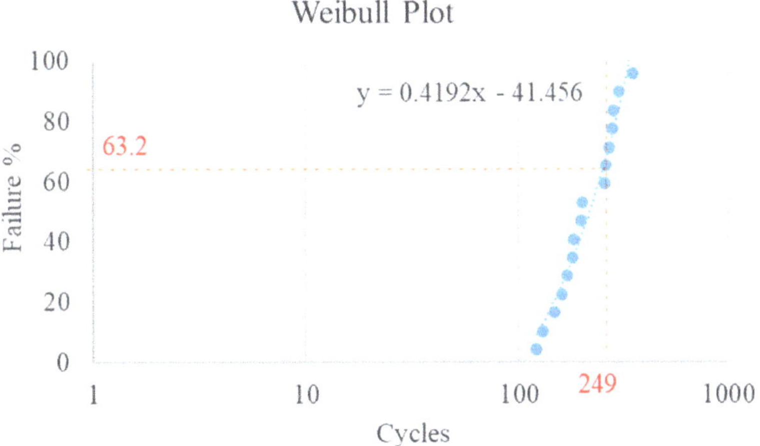

Figure 4. Weibull distribution of the glass interposer fan-out package.

4. Finite Element Analysis of FO-WLP

We established a 2D diagonal semi-symmetric plane strain model and a 3D quarter model to study the thermomechanical behavior of the solder joints in the FO-WLP structure. As the 2D model had fewer nodes than the 3D model, it required less simulation time. However, because plane strain was assumed in the 2D model, it cannot truly represent the actual state of the test vehicle (e.g., the semi-spherical nature of the solder ball). Therefore, a 3D model closer in geometry to the actual vehicle was established to verify the robustness of the 2D model.

After verifying the finite element simulations against the experimental results, we added a TMV component to the FO-WLP structure to make it stackable. The stacked PoP architecture was evaluated by simulation under the JEDEC standard (condition G) testing condition.

4.1. Material Parameters

The test vehicle was composed of an Si chip, a die attach film, a glass substrate, a stress buffer layer (SBL), copper, SAC305 solder balls, a PCB, and a molding compound. We used three types of glass substrates whose CTE ranged from 3.17 to 9.8 ppm/°C and whose modulus ranged from 64 to 74 GPa, with little difference in Poisson's ratio. The material parameters are listed in Table 2. The temperature-dependent characteristics of all of the materials were linear, except for the SAC305 solder balls. Figure 5 shows the stress–strain curve of the solder balls at different temperatures [20].

Table 2. Summary of the material properties of the components in the FO-WLP. CTE, coefficient of thermal expansion.

Material	Young's Modulus (GPa)	Poisson's Ratio	CTE (ppm/°C)
Silicon	150	0.28	2.62
Stress buffer layer	2	0.33	55
Copper	68.9	0.34	16.7
SAC305 solder	Nonlinear/creep	0.4	22.36
Die attach film	1.66	0.26	17
Glass A	73.6	0.23	3.17
Glass B	71.7	0.21	8.37
Glass C	74	0.23	9.8
Printed circuit board	18.2	0.19	16
Molding compound	8.96	0.35	15

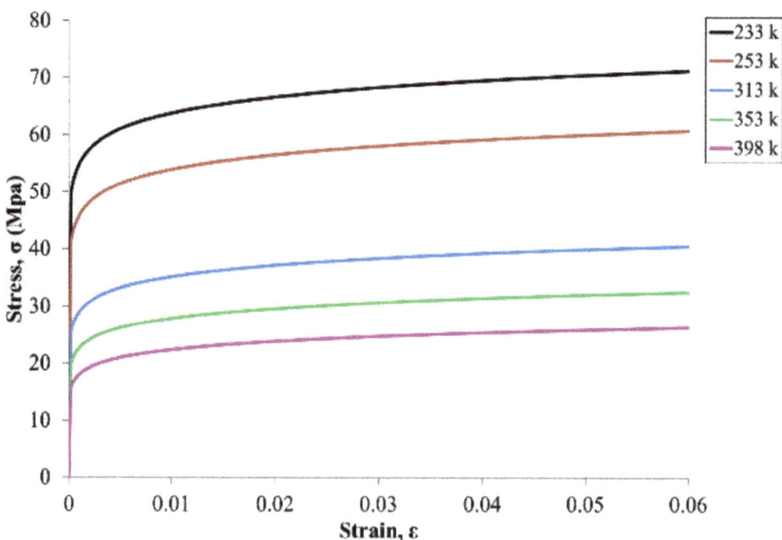

Figure 5. Stress–strain curves of Sn–3Ag–0.5Cu solder balls at different temperatures.

To model the creep behavior [7] of the solder balls, we adopted the Garofalo–Arrhenius hyperbolic sine model (Equation (6)), which has been widely used to simulate the thermal cycle load of solder joints:

$$\frac{d\varepsilon}{dt} = A\left(sinh(B\sigma)^n exp\left(\frac{-Q}{RT}\right)\right) \tag{6}$$

where ε is the strain, σ is the stress, A and B are the material constants, R is the gas constant, T is the absolute temperature, Q is the activation energy, and n is the stress index. Table 3 lists the values of these constants as used in this study [13].

Table 3. Constants in the creep equation [13].

Material	A (1/s)	B (1/MPa)	n	Q (J/mol)
Value	2631	0.0425	4.96	52,400

4.2. 2D Plane Strain Model

For the finite element analysis of the 2D model, we assumed plane strain and used PLANE42 and PLANE182 (ANSYS, 2020R2). Surface Evolver was used to define the shape of the solder joints. Figure 6 depicts the top view of the test vehicle, where the solder joints located in the periphery array format are marked in white. The finite element model was a 2D half model built along the diagonal of the test vehicle (Figure 7). Its position is the yellow line in Figure 6. Figure 8 illustrates the finite element model of the stacked FO-WLP (PoP) structure. The upper and lower stacks of the package were connected through soldering and TMV. The TMV was 0.25 mm in diameter and located atop the solder balls.

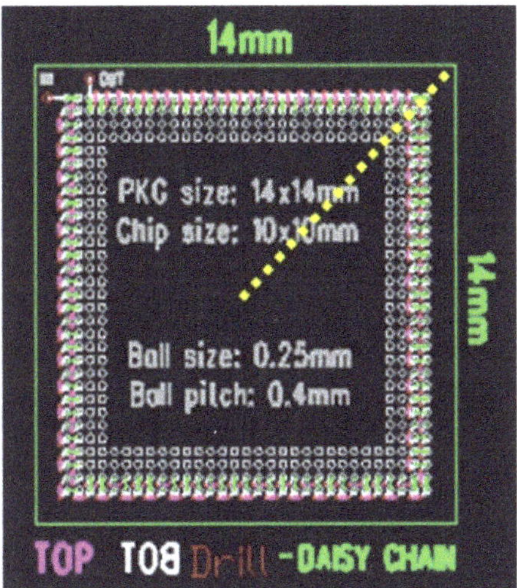

Figure 6. Top view of the test vehicle.

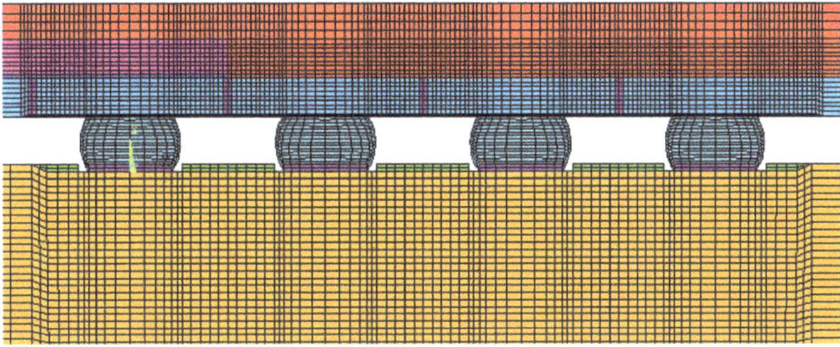

Figure 7. Two-dimensional (2D) finite element model of the FO-WLP.

Figure 8. 2D finite element model of the stacked FO-WLP (i.e., packaging-on-packaging (PoP)).

4.3. 3D Quarter Symmetry Model

Unlike a 3D finite element model, a 2D plane strain model cannot accurately represent semi-spherical-type features. However, a 3D model with a fine mesh contains a large number of elements and necessitates excessive simulation computing time, making parametric study infeasible. Two approaches can be used to overcome these drawbacks. The first is to build a 3D model to verify the 2D plane strain model, which can then be used to evaluate the packaging structure, and the second is to reduce the number of elements in the 3D model. In this work, we used multipoint constraint (MPC) technology to significantly reduce the number of elements. Panels (a) and (c) in Figure 9 illustrate 1/4 models of a simple WLP with full fine meshes, and panels (b) and (d) illustrate their MPC-reduced counterparts. This simple 1/4 WLP model will not take too much simulation time and was selected in this research to verify that the MPC model can obtain stress/strain results similar to the full fine mesh model.

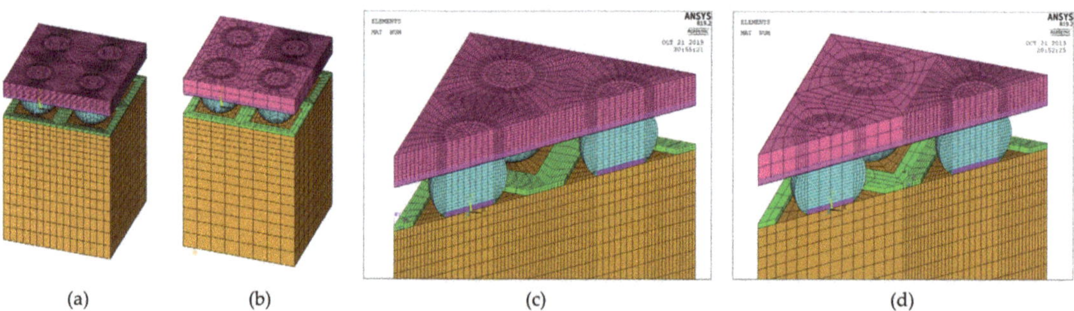

(a) (b) (c) (d)

Figure 9. Three-dimensional (3D) finite element model: (**a**) Full fine-mesh model; (**b**) multipoint constraint (MPC) model; (**c**) cross-section view of the full fine-mesh model; (**d**) cross-section view of the MPC model.

After verifying the MPC model against the full model, the MPC model was used to simulate the test vehicle and the stacked PoP architecture. Figure 10 shows the 3D quarter symmetry finite element model.

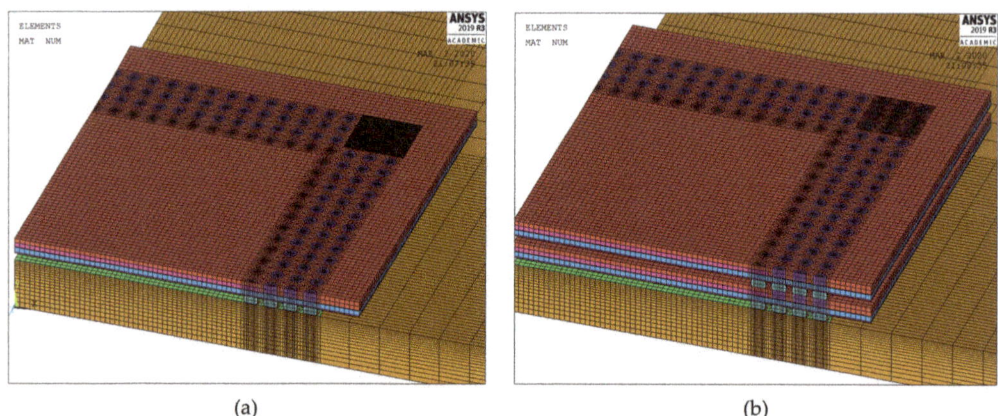

(a) (b)

Figure 10. Three-dimensional MPC-reduced finite element models for the (**a**) FO-WLP and (**b**) FO-WLP PoP.

5. Results and Discussion

5.1. 2D Diagonal Plane Strain Model

After OBTCT, we inspected the failed test vehicle and found that the crack was located atop the solder joint with the largest DNP. Consistent with the experiment wherein we used glass substrate A, the simulation also showed that the maximum equivalent inelastic strain/stress was atop the outermost solder joint (Figure 11). In addition, the 248 life cycles simulated by Equation (4) are in good agreement with the experiment results of 249 mean cycles to failure (Figure 4). Thus, the failure prediction of the 2D model is in good agreement with the experiment data.

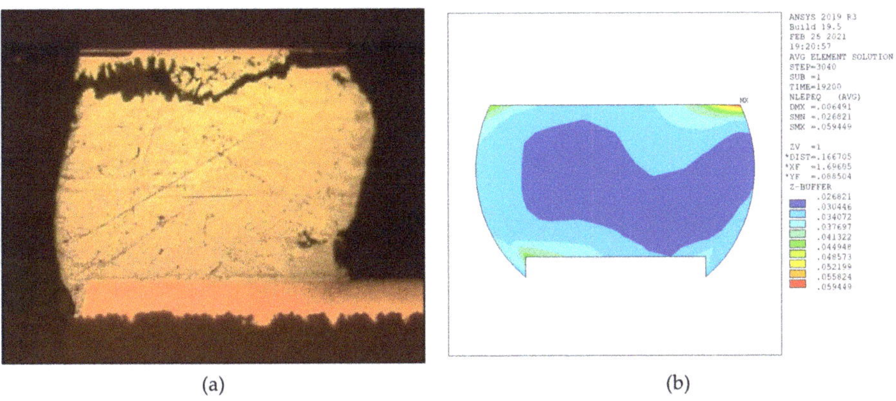

(a) (b)

Figure 11. Failure point of solder joint after onboard thermal cycling testing (OBTCT): (**a**) Cross-section view; (**b**) equivalent inelastic strain distribution.

Next, we conducted a parametric study to optimize the reliability life under different combinations of upper and lower pad diameters. As evident in Figure 12, an upper pad larger than a lower pad resulted in better reliability. The Young's modulus of silicon material is stronger than that of PCB, so a larger upper pad size is required to reduce the equivalent inelastic strain. A lower strain can have better reliability life, and the strain of the solder ball is related to the pad size, contact angle, and standoff height of the solder ball. In the simulation, when the upper pad diameter was 250 μm and the lower pad diameter was 180 μm, the best reliability life cycles could be achieved. That is, the pad on the interposer side should be larger than that on the PCB side. The optimal reliability life of 368 cycles was achieved at an upper pad/lower pad diameters ratio of approximately 1:0.72.

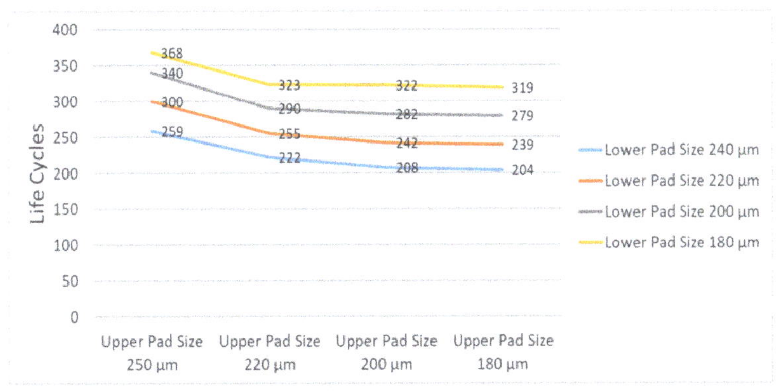

Figure 12. Simulated reliability life under different combinations of pad thicknesses.

To meet condition G of the JEDEC standard, the thermal cycling life of the product must exceed 1000 cycles. To meet this standard with the optimized pad combination (upper pad 250 μm and lower pad 180 μm), we further optimized the SBL thickness and CTE of the glass substrate (Figure 13). Specifically, we simulated SBL thicknesses of 5, 10, and 20 μm and both substrate glasses A and B. Glasses A and B differ little in terms of Poisson's ratio, but differ substantially in CTE (3.17 and 8.37 ppm/°C, respectively). Under the same conditions, the reliability life was higher for glass B than for glass A by 600 cycles on average, meaning that the reliability life is most sensitive to glass CTE. Moreover, a thicker SBL reduced solder stress/strain and thus favorably affected the thermal cycling reliability. Overall, the reliability life improved from 248 cycles to 1432 cycles with the following optimized combination: upper pad diameter, 250 μm; lower pad diameter, 180 μm; SBL thickness, 20 μm; glass CTE, 8.35 ppm/°C.

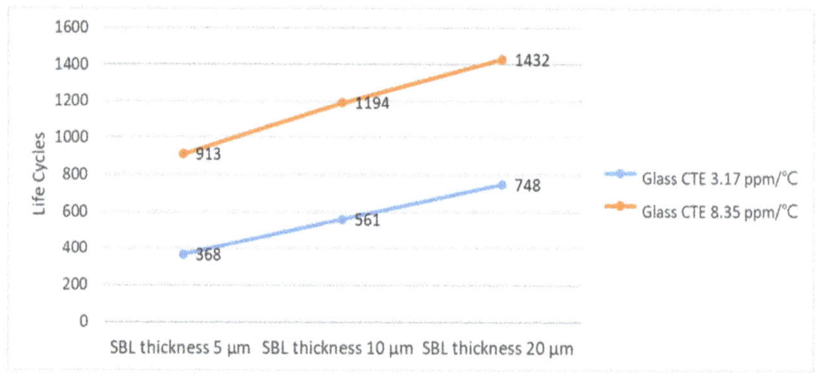

Figure 13. Simulated reliability life for stress buffer layers of different thicknesses.

5.2. 3D Quarter Symmetry Model

The 3D FO-WLP PoP simulation model is much larger than the nonstacked FO-WLP, which makes 2D simulation the only feasible option for the parametric study of a stacked FO-WLP. To verify whether a 2D model can be applied to stacked packaging, a 3D model must first be established. To reduce the computing time, the number of elements and nodes must be decreased. First, we established a small WLP model (four solder balls per quarter model; Figure 9) both as a full fine-mesh model (86,656 elements; Figure 9a) and an MPC model (32,648 elements; Figure 9b). The corresponding computing times were 5 h 43 min and 2 h 13 min. We applied the same thermal loading as in the earlier analyses (−40 to 125 °C). Strain was found to be concentrated in a corner of the ball and the upper edge (Figure 14). After one to three thermal cycles, the strain increment in the two models differed by less than 1% (Figure 15). These simulation results demonstrate the feasibility of applying the MPC approach to simulate FO-WLP PoP with 112 solder balls per quarter model, as shown in Figure 10.

Next, we applied the design optimized in Section 5.1 (upper pad diameter, 250 μm; lower pad diameter, 180 μm; SBL thickness, 20 μm; glass CTE, 8.35 ppm/°C) to the 3D quarter symmetry model. Figure 16 shows the equivalent inelastic strain distribution in the FO-WLP. Strain accumulated in the solder ball with the maximum DNP, and the cycling life predicted by the 3D model differed from the 2D model prediction by only 1.3%.

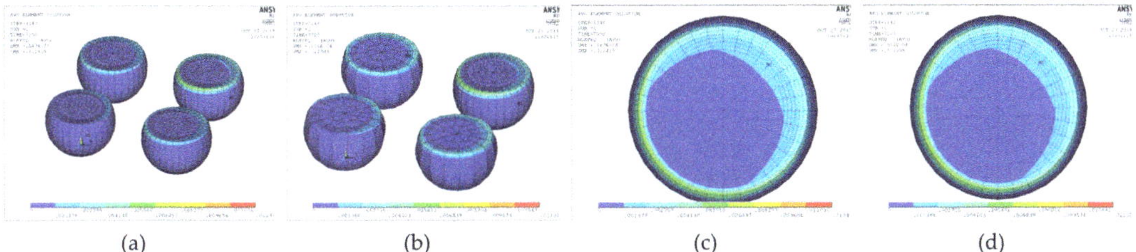

(a)   (b)   (c)   (d)

Figure 14. Equivalent inelastic strain distribution: (**a**) Full fine-mesh model; (**b**) MPC model; (**c**) maximum strain in the full model (top view); (**d**) maximum strain in the MPC model (top view).

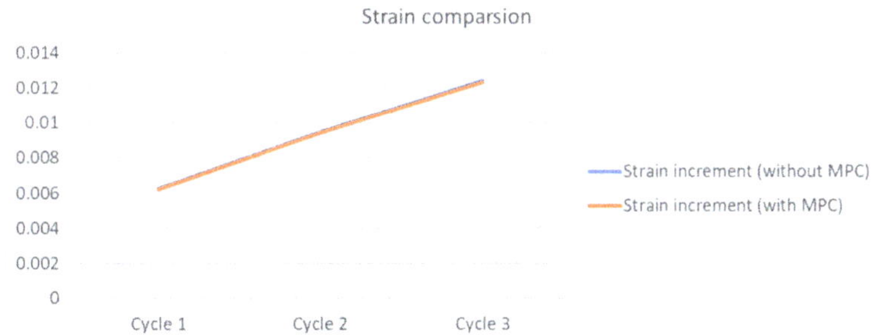

Figure 15. Strain increment in the full and MPC models.

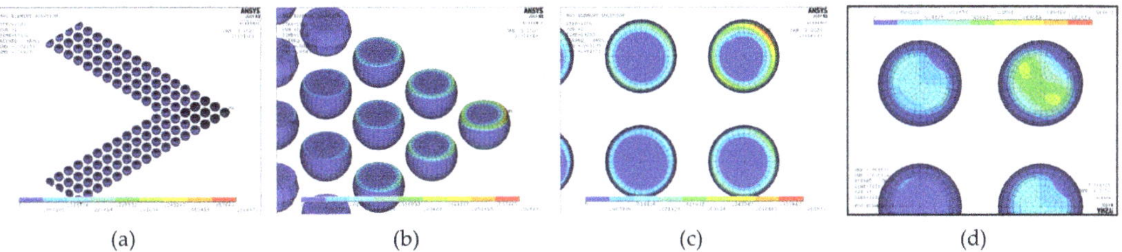

(a)   (b)   (c)   (d)

Figure 16. Equivalent inelastic strain distribution of the FO-WLP: (**a**) All solder joints; (**b**) location of the maximum strain; (**c**) top view of the bottom solder joint with the maximum distance from the neutral point (DNP); (**d**) bottom view of the solder joint with the maximum DNP.

We repeated the above simulation for the FO-WLP PoP structure. Again, we found that strain concentrated in the bottom layer of the solder ball with the maximum DNP, replicating the trend of failure at a corner location. Moreover, the maximum strain was in the upper pad (Figure 17). The upper and lower parts of the solder balls in the bottom layer were connected to the glass substrate and the PCB, respectively. Therefore, these solder balls could withstand greater stress than could solder balls in the upper layer. However, this double-layer structure reduced the cycling life to 974 cycles. In the PoP structure, the percentage of glass in this package increased; therefore, the equivalent inelastic strain due to CTE mismatch also increased, and a higher strain may lead to earlier failure.

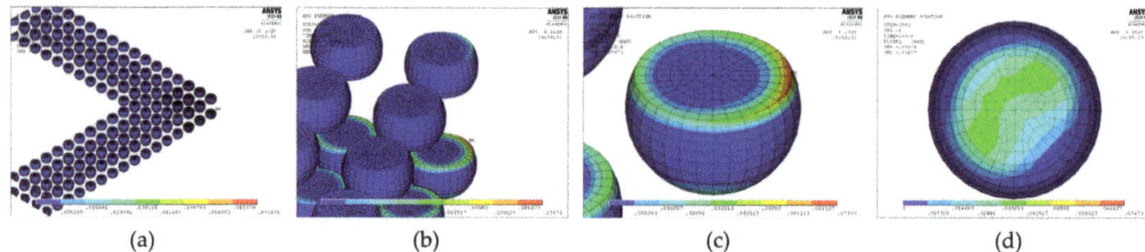

Figure 17. Equivalent inelastic strain distribution in FO-WLP PoP: (**a**) All solder joints; (**b**) location of the maximum strain; (**c**) top view of the solder joint with the maximum DNP; (**d**) bottom view of the solder joint with the maximum DNP.

Herein, the 2D and 3D models exhibited good consistency, with only 5% difference in reliability life. Therefore, we used the verified 2D models in the parametric analysis to optimize the design parameters in order to extend the reliability life of the FO-WLP PoP structure to more than 1000 cycles.

5.3. Parametric Analysis of FO-WLP PoP Structure Using the 2D Plane Strain Finite Element Model

In addition to the already optimized pad combination, three design parameters were considered in our parametric study: SBL thickness, chip thickness, and CTE of the glass substrate. In this parametric analysis, SBL thicknesses of 20, 25, and 30 μm, chip thicknesses of 90, 100, and 150 μm, and glass substrate CTEs of 8.35 and 9.8 ppm/°C were evaluated, yielding 16 combinations (Table 4). SBLs thicker than 30 μm were not evaluated due to the manufacturing difficulty. The results showed that the thicker the SBL, the higher the reliability and the higher the released stress/strain concentration. Furthermore, a lower chip thickness yielded a higher molding compound volume, which in turn increased the CTE of the whole package and thus narrowed the CTE mismatch between the package and PCB. Most importantly, glass CTE was found to have the strongest decreasing effect on the CTE mismatch between the glass substrate and the PCB. In summary, the reliability life of FO-WLP PoP can be increased by optimizing the SBL thickness and glass CTE (Figure 18). Of the 16 simulated combinations, combination 13—SBL thickness, 30 μm; chip thickness, 90 μm; glass substrate CTE, 9.8 ppm/°C—yielded the longest life cycle of 1427 cycles (Table 4).

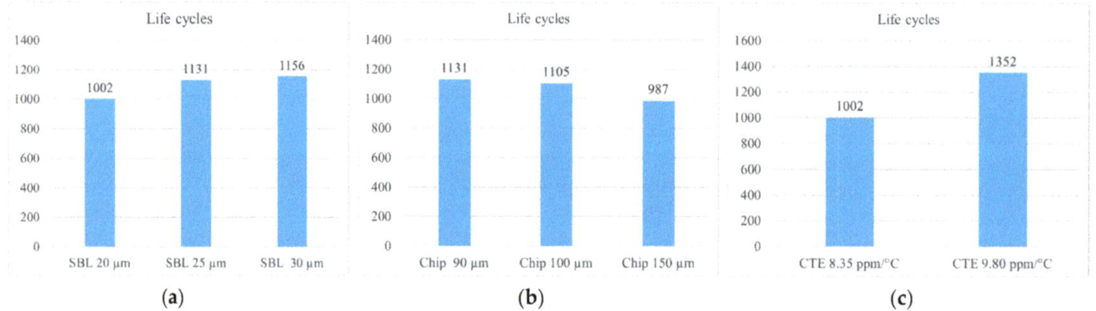

Figure 18. Life predictions of the FO-WLP PoP under different (**a**) SBL thicknesses, (**b**) chip thicknesses, and (**c**) CTEs.

Table 4. Combination of the design parameters in the parametric analysis of the FO-WLP PoP.

Item	SBL Thickness (μm)			Chip Thickness (μm)			CTE (ppm/°C)		Life Prediction
	20	25	30	90	100	150	8.35	9.8	
1	V			V			V		1002
2	V			V				V	1352
3	V				V		V		976
4	V				V			V	1337
5	V					V	V		923
6	V					V		V	1301
7		V		V			V		1131
8		V		V				V	1398
9		V			V		V		1105
10		V			V			V	1241
11		V				V	V		987
12			V	V			V		1156
13			V	V				V	1427
14			V		V		V		1129
15			V		V			V	1410
16			V			V	V		998

6. Conclusions

The simulation of 3D models is infeasible because of the high processing power and time requirements. In this study, to optimize the reliability life of FO-WLP and FO-WLP-type PoP under thermal cycling loading, 2D and 3D MPC technique finite element simulation models were established and experimentally verified to have similar accuracy. The reliability life of the verified 2D models was then optimized through parametric analysis. The CTE of the glass substrate was found to exert the strongest effect on reliability life. In addition, the upper and lower pad diameters and buffer layer thickness significantly affected the reliability life of the FO-WLP and FO-WLP-type PoP. Specifically, the reliability life of the solder joints was highest when the upper pad was larger than the lower pad (1:0.72). Moreover, a thicker buffer layer released a greater stress/strain concentration and thus positively affected the reliability life. However, SBLs thicker than 30 μm were not evaluated in this study due to the manufacturing difficulty.

Author Contributions: P.-H.W. and Y.-W.H. studied the reliability life prediction and structural optimization methods of the PoP architecture. K.-N.C. is a professor who led and discussed the content of this research. All authors have read and agreed to the published version of the manuscript.

Funding: The authors would like to thank the Ministry of Science and Technology (Project MOST 108-2221-E-007-077-MY3) of Taiwan for supporting this research.

Conflicts of Interest: The authors declare no conflict of interest.

References

1. Lau, J.H.; Li, M.; Tian, D.; Fan, N.; Kuah, E.; Kai, W.; Yong, Q. Warpage and Thermal Characterization of Fan-out Wafer-Level Packaging. In Proceedings of the IEEE 67th Electronic Components and Technology Conference, Lake Buena Vista, FL, USA, 30 May–2 June 2017; pp. 595–602.
2. Tseng, C.F.; Liu, C.S.; Wu, C.H.; Yu, D. InFO (Wafer Level Integrated Fan-Out) Technology. In Proceedings of the IEEE 66th Electronic Components and Technology Conference, Las Vegas, NV, USA, 31 May–3 June 2016; pp. 1–6.

3. Tanaka, M.; Kuramochi, S.; Tai, T.; Sato, Y.; Kidera, N. High Frequency Characteristics of Glass Interposer. In Proceedings of the 2020 IEEE 70th Electronic Components and Technology Conference (ECTC), Las Vegas, NV, USA, 26–29 May 2020; pp. 601–610. [CrossRef]

4. Tsai, C.H.; Liu, S.W.; Chiang, K.N. Warpage Analysis of Fan-Out Panel-Level Packaging Using Equivalent CTE. *IEEE Trans. Device Mater. Reliab.* **2019**, *20*, 51–57. [CrossRef]

5. Che, F.X.; Yamamoto, K.; Rao, V.S.; Sekhar, V.N. Panel Warpage of Fan-Out Panel-Level Packaging Using RDL-First Technology. *IEEE Trans. Compon. Packag. Manuf. Technol.* **2019**, *10*, 304–313. [CrossRef]

6. Shah, U.; Liljeholm, J.; Campion, J.; Ebefors, T.; Oberhammer, J. Low-Loss, High-Linearity RF Interposers Enabled by Through Glass Vias. *IEEE Microw. Wirel. Compon. Lett.* **2018**, *28*, 960–962. [CrossRef]

7. Barbera, D.; Chen, H. On the Plastic Strain Accumulation in Notched Bars During High-Temperature Creep Dwell. *J. Mech.* **2020**, *36*, 167–176. [CrossRef]

8. Cheng, H.C.; Wu, Z.D.; Liu, Y.C. Viscoelastic Warpage Modeling of Fan-Out Wafer-Level Packaging During Wafer-Level Mold Cure Process. *IEEE Trans. Compon. Packag. Manuf. Technol.* **2020**, *10*, 1240–1250. [CrossRef]

9. Demir, K.; Armutlulu, A.; Sundaram, V.; Raj, P.M.; Tummala, R.R. Reliability of Copper Through-Package Vias in Bare Glass Interposers. *IEEE Trans. Compon. Packag. Manuf. Technol.* **2017**, *7*, 829–837. [CrossRef]

10. Yang, C.C.; Su, Y.F.; Liang, S.Y.; Chiang, K.N. Simulation of Wire Bonding Process Using Explicit Fem with Ale Remeshing Technology. *J. Mech.* **2020**, *36*, 47–54. [CrossRef]

11. Ramachandran, V.; Chiang, K.N. Feasibility Evaluation of Creep Model for Failure Assessment of Solder Joint Reliability of Wafer-Level Packaging. *IEEE Trans. Device Mater. Reliab* **2017**, *17*, 672–677. [CrossRef]

12. Wu, P.L.; Wang, P.H.; Chiang, K.N. Empirical Solutions and Reliability Assessment of Thermal Induced Creep Failure for Wafer Level Packaging. *IEEE Trans. Device Mater. Reliab.* **2018**, *19*, 126–132. [CrossRef]

13. Brakke, K.A. *Surface Evolver Manual*; Version 2.01; The Geometry Center: Minneapolis, MN, USA, 1996.

14. Li, L.; Yeung, B. Wafer level and flip chip design through solder prediction models and validation. *IEEE Trans. Compon. Packag. Technol.* **2001**, *24*, 650–654. [CrossRef]

15. Yeung, B.H.; Lee, T.-Y.T. Evaluation and optimization of package processing and design through solder joint profile prediction. *IEEE Trans. Electron. Packag. Manuf.* **2003**, *26*, 68–74. [CrossRef]

16. Coffin, L.F. A study of the effects of cyclic thermal stress on a ductile metal. *Transactions ASME* **1954**, *76*, 931–950.

17. Mason, S.S. Behavior of materials under conditions of thermal stress. *Natl. Advis. Comm. Aeronaut. Tech. Note* **1953**, *2933*, 317–350.

18. Chou, P.H.; Chiang, K.; Liang, S.Y. Reliability Assessment of Wafer Level Package using Artificial Neural Network Regression Model. *J. Mech.* **2019**, *35*, 829–837. [CrossRef]

19. Wang, P.H.; Lee, Y.; Lee, C.K.; Chang, H.H.; Chiang, K.N. Solder Joint Reliability Assessment and Pad Size Studies of FO-WLP with Glass Substrate. *IEEE Trans. Device Mater. Reliab.* **2021**, *1*, 96–101. [CrossRef]

20. Lee, H.H.; Kwak, J.B. Realistic Creep Characterization for Sn3.0Ag0.5Cu Solder Joints in Flip Chip BGA Package. *J. Electron. Mater.* **2019**, *48*, 6857–6865. [CrossRef]

 *micromachines*

Article

A Hybrid Fuzzy Decision Model for Evaluating MEMS and IC Integration Technologies

Qian-Yo Lee [1], Ming-Xuan Lee [2] and Yen-Chun Lee [3,*]

[1] Department of Biomechatronic Engineering, National Chiayi University, No. 300, Syuefu Rd.,
Chiayi City 600355, Taiwan; yolee2300@gmail.com
[2] Graduate Institute of Electronics Engineering, National Taiwan University, No. 1, Sec. 4, Roosevelt Rd.,
Taipei City 10617, Taiwan; ken.lee1998@gmail.com
[3] Institute of Management of Technology, National Yang Ming Chiao Tung University, No. 1001, University Rd.,
Hsinchu City 300, Taiwan
* Correspondence: focuslee2@gmail.com; Tel.: +886-3-5712121 (ext. 31981)

Abstract: Integrated devices incorporating MEMS (microelectromechanical systems) with IC (integrated circuit) components have been becoming increasingly important in the era of IoT (Internet of Things). In this study, a hybrid fuzzy MCDM (multi-criteria decision making) model was proposed to effectively evaluate alternative technologies that incorporate MEMS with IC components. This model, composed of the fuzzy AHP (analytic hierarchy process) and fuzzy VIKOR (VIseKriterijumska Optimizacija I Kompromisno Resenje) methods, solves the decision problem of how best to rank MEMS and IC integration technologies in a fuzzy environment. The six important criteria and the major five alternative technologies associated with our research themes were explored through literature review and expert investigations. The priority weights of criteria were derived using fuzzy AHP. After that, fuzzy VIKOR was deployed to rank alternatives. The empirical results show that development schedule and manufacturing capability are the two most important criteria and 3D (three-dimensional) SiP (system-in-package) and monolithic SoC (system-on-chip) are the top two favored technologies. The proposed fuzzy decision model could serve as a reference for the future strategic evaluation and selection of MEMS and IC integration technologies.

Keywords: technology evaluation; MEMS and IC integration; MCDM; fuzzy AHP; fuzzy VIKOR

Citation: Lee, Q.-Y.; Lee, M.-X.; Lee, Y.-C. A Hybrid Fuzzy Decision Model for Evaluating MEMS and IC Integration Technologies. *Micromachines* **2021**, *12*, 276. https://doi.org/10.3390/mi12030276

Academic Editor: Seonho Seok

Received: 2 February 2021
Accepted: 4 March 2021
Published: 7 March 2021

Publisher's Note: MDPI stays neutral with regard to jurisdictional claims in published maps and institutional affiliations.

1. Introduction

The rapid development of IoT (Internet of Things) provides great opportunities for disruptive innovations [1]. It is a remarkable fact that MEMS (microelectromechanical systems) transducers are the key enablers of IoT [2], because IoT requires a variety of input quantities that are sensed from external environments. MEMS devices act as transducers that measure or control physical, optical, or chemical quantities, such as acceleration, radiation, or fluids [3]. A typical configuration of an accelerometer sensor system in an airbag system is shown in Figure 1 [4]. The MEMS device interacts with the signal conditioning ICs to produce the amplified signal that fires the ignitor if the deceleration is high enough.

To enable MEMS sensors to function well, the electrical interfaces with the outside world need to be realized through ICs (integrated circuits) that provide electronic systems with operating intelligence [5]. ICs can definitely provide signal conditioning functions, such as analog-to-digital conversion, amplification, filtering, data processing, and communication between MEMS sensors and the outside world [6].

Until now, several viable technologies for incorporating MEMS with IC components have been either developed or are under development, but an appropriate decision making model for global semiconductor industry remains an open question. In this study, a hybrid fuzzy MCDM (multi-criteria decision making) model was proposed to facilitate the evaluation and selection of MEMS and IC integration technologies. This model, composed of fuzzy AHP (analytic hierarchy process) and fuzzy VIKOR (VIseKriterijumska Optimizacija I Kompromisno Resenje) methods, solves the decision problem of how best to rank MEMS and IC integration technologies in a fuzzy environment.

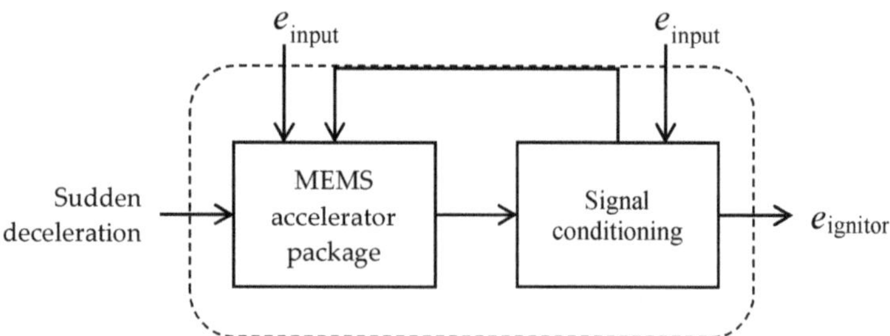

Figure 1. Accelerometer sensor system in an airbag system [4]. MEMS, microelectromechanical systems.

2. Literature Review

In this section, the literature associated with our research themes was reviewed.

2.1. Technology Assessment Using MCDM

In the recent decade, MCDM methods have developed rapidly and have evolved to accommodate various types of applications [7]. For example, Van de Kaa, Rezaei, Kamp, and de Winter [8] applied both crisp AHP and fuzzy AHP methods to a standardization problem for photovoltaic technological systems. Vinodh, Nagaraj, and Girubha [9] used the fuzzy VIKOR method for evaluating rapid prototyping technologies in an agile environment. Liu, You, Lu, and Chen [10] proposed a novel hybrid MCDM model for selection of healthcare waste treatment technologies. Bairagi, Dey, Sarkar, and Sanyal [11] proposed a technique for TOPSIS-based (technique for order preference by similarity to ideal solution) fuzzy MCDM approach for selecting the best robotic system. Lee and Chou [12] explored a technology selection process for evaluating 3DIC (three-dimensional integrated circuit) integration technologies. In addition, Taylan, Alamoudi, Kabli, AlJifri, Ramzi, and Herrera-Viedma [13] integrated fuzzy AHP, fuzzy VIKOR and TOPSIS methods to determine the most eligible energy systems for investment. Salimi, Noori, Bonakdari, Masoompour Samakosh, Sharifi, Hassanvand, Agharazi, and Gharabaghi [14] integrated fuzzy AHP and fuzzy VIKOR methods based on group decision making to examine the role of mass media advertising types on improving the water consumption behavior. In view of the associated literature mentioned above, MCDM methods have demonstrated their importance as optimal candidates in the decision making of emerging technologies.

2.2. Alternative Technologies

Several viable technologies for incorporating MEMS with IC components have been either developed or are under development, and they can basically be divided into two major solutions: (1) hybrid multi-chip solutions and (2) SoC (system-on-chip) solutions [5]. Hybrid integration of MEMS and IC components has been dominated by 2D (two-dimensional) integration approaches in which each of MEMS and IC wafers are fabricated independently [15]. These individual wafers are then separated into discrete chips and finally integrated into MCMs (multi-chip modules) [16]. Another approach for the hybrid integration of MEMS and IC components is SiP (system-in-package), which is also known as vertically stacked MCMs [17,18]. In this case, discrete chips are attached on top of each other and interconnected via wire bonding or flip-chip bonding [18–20]. Moreover, SoP (system-on-package) is another approach in which MEMS and IC chips are integrated with other technologies, enabling highly integrated and miniaturized systems at package levels [21,22].

SoC solutions are characterized by incorporating MEMS with IC components on the same wafers in which chip separation occurs only at or near the end of manufacturing processes [5,23]. They can be categorized into two major approaches: (1) monolithic MEMS and IC integration approaches in which MEMS and IC structures are manufactured altogether on the same wafers [24] and (2) heterogeneous MEMS and IC integration approaches in which MEMS and IC structures are premanufactured on discrete wafers and then merged onto the same substrates via wafer bonding techniques [25]. Monolithic MEMS and IC integration approaches can be further categorized into four techniques [26]: (1) monolithic MEMS and IC integration using MEMS-first processing [27], (2) monolithic MEMS and IC integration using interleaved MEMS and IC processing [28], (3) monolithic MEMS and IC integration using MEMS-last processing via bulk micromachining of IC substrates (also known as CMOS-MEMS (complementary metal-oxide-semiconductor) [29], and (4) monolithic MEMS and IC integration using MEMS-last processing via layer deposition and surface micromachining [30]. Another SoC solution based on heterogeneous MEMS and IC integration approaches can be also categorized into two techniques [31]: (1) heterogeneous MEMS and IC integration via formation during layer transfer and (2) heterogeneous MEMS and IC integration via formation after layer transfer.

3. Research Methods

The methodologies of fuzzy AHP and fuzzy VIKOR methods are illustrated below.

3.1. Fuzzy AHP Method

The AHP method, developed by Saaty [32,33], has been criticized because a decision problem can be structured in a hierarchical manner. However, AHP cannot effectively reflect the ambiguity in human thinking style [34]. To solve this problem, fuzzy AHP was thus proposed [35,36]. The procedure of fuzzy AHP is described in the following steps [12]:

Step 1 Define a problem.
Step 2 Determine important criteria.
Step 3 Establish a hierarchical structure.
Step 4 Determine linguistic variables.
Step 5 Construct fuzzy judgment matrices.

A fuzzy judgment matrix can be defined as follows:

$$\widetilde{A}^k = \left[\widetilde{a}_{ij}\right]^k \tag{1}$$

$$\widetilde{a}_{ij}^k = (1,\ 1,\ 1),\ \forall\, i = j,\ \text{and}\ \widetilde{a}_{ji}^k = 1/\widetilde{a}_{ij}^k,\ \forall\, i,j = 1,\ 2,\ \ldots,\ n$$

where $\widetilde{A}^k$ is a fuzzy judgment matrix evaluated by expert k ($k = 1, 2, \cdots, K$), $\widetilde{a}_{ij}^k$ is fuzzy assessment between criterion i and criterion j evaluated by expert k, and n is the number of criteria at the same level.

Step 6 Check consistency.

If A is consistent, then $\widetilde{A}$ is accordingly consistent [37]. To verify whether A is sufficiently consistent, the maximum eigenvalue $\lambda_{\max}$ can be computed as follows:

$$A \circ W = W' = \lambda_{\max} \circ W \tag{2}$$

where A is a pairwise comparison matrix and W is a weight matrix.

$$\lambda_{\max} = \frac{1}{n}\left(\frac{W'_1}{W_1} + \frac{W'_2}{W_2} + \cdots + \frac{W'_n}{W_n}\right) \tag{3}$$

Saaty [33] proposed a consistency index (*CI*) to check the consistency within pairwise comparison matrices, as well as that of the entire hierarchy. The *CI* is formulated as follows:

$$CI = (\lambda_{\max} - n)/(n - 1) \tag{4}$$

where n is the dimension of matrix A.

The consistency ratio (*CR*) is accordingly defined as follows:

$$CR = CI/RI \tag{5}$$

where RI is the random consistency index.

The pairwise comparison matrix A is considered consistent if the resulting ratio CR is less than 0.1.

Step 7 Integrate experts' opinions.

In order to integrate experts' opinions, a fuzzy synthetic judgment matrix can be obtained using the geometric mean technique [38] to compute fuzzy geometric means of each criterion [39]. Then, fuzzy weights of each criterion can be computed using the arithmetic mean technique as follows:

$$\widetilde{B} = \left[\widetilde{b}_{ij} \right], \, \forall \, i, j \tag{6}$$

$$\widetilde{b}_{ij} = \left(\widetilde{a}_{ij}^1 \otimes \cdots \otimes \widetilde{a}_{ij}^k \otimes \cdots \otimes \widetilde{a}_{ij}^K \right)^{1/K} \tag{7}$$

$$\widetilde{r}_i = \left(\widetilde{b}_{ij}^1 \otimes \widetilde{b}_{ij}^2 \otimes \cdots \otimes \widetilde{b}_{ij}^n \right)^{1/n} \tag{8}$$

$$\widetilde{w}_i = \widetilde{r}_i \otimes \left(\widetilde{r}_1 \oplus \widetilde{r}_2 \oplus \cdots \oplus \widetilde{r}_n \right)^{-1} \tag{9}$$

where $\widetilde{B}$ is a fuzzy synthetic judgment matrix of total K experts, $\widetilde{b}_{ij}$ is a geometric mean of fuzzy assessment of total K experts, $\widetilde{r}_{ij}$ is a geometric mean of a row of the fuzzy synthetic judgment matrix $\widetilde{B}$, and $\widetilde{w}_i$ is a fuzzy weight of the ith criterion.

Step 8 Defuzzify fuzzy weights.

Using the centroid defuzzification technique to locate BNP (best nonfuzzy performance) values [12,40], fuzzy weights of criteria can be defuzzified as crisp values.

$$DF_{\widetilde{w}_i} = BNP_{\widetilde{w}_i} \cong \frac{1}{3} \left(L_{\widetilde{w}_i} + M_{\widetilde{w}_i} + U_{\widetilde{w}_i} \right) \tag{10}$$

where $L_{\widetilde{w}_i}$, $M_{\widetilde{w}_i}$, and $U_{\widetilde{w}_i}$ represent the lower, middle, and upper values of the fuzzy weight of the ith criterion.

3.2. Fuzzy VIKOR Method

The VIKOR method helps decision-makers to determine compromise solutions for a problem and to rank and select from a set of alternatives over conflicting and incommensurable criteria for reaching ideal/aspired levels [41]. In the past few decades, an extension of VIKOR, namely fuzzy VIKOR, has been further combined with fuzzy set theory to determine compromise solutions under the fuzzy environment where both criteria and weights could be fuzzy sets [42,43]. The procedure of fuzzy VIKOR is described in the following steps [44–46].

Step 1 Determine a group of experts.

Let $A_i \, (i = 1, 2, \cdots, m)$ be a be a finite set of m alternatives that are to be evaluated by hte kth expert $(E_k, \, k = 1, 2, \cdots, K)$ with respect to a set of n criteria $(C_j, \, j = 1, 2, \cdots, n)$.

Step 2 Determine linguistic variables.
Step 3 Obtain fuzzy performance rating matrices.

A typical fuzzy VIKOR questionnaire can be expressed in a matrix format as follows:

$$\widetilde{X}_k = \left[\widetilde{x}_{ijk}\right]_{m \times n} \tag{11}$$

where $\widetilde{x}_{ijk}$ is the fuzzy performance rating of alternative A_i with respect to criterion C_j evaluated by expert E_k. $\widetilde{x}_{ijk} = \left(x^l_{ijk}, x^m_{ijk}, x^u_{ijk}\right)$ is a linguistic variable denoted by a TFN (triangular fuzzy number).

Step 4 Construct an aggregated fuzzy performance rating matrix.

An aggregated fuzzy performance rating $\widetilde{D}$ can be constructed with m alternatives and n criteria as follows:

$$\widetilde{D} = \left[\widetilde{D}_{ij}\right]_{m \times n} = \begin{array}{c} \\ A_1 \\ \vdots \\ A_i \\ \vdots \\ A_m \end{array} \begin{array}{ccccc} \widetilde{C}_1 & \cdots & \widetilde{C}_j & \cdots & \widetilde{C}_n \\ \left[\begin{array}{ccccc} \widetilde{x}_{11} & \cdots & \widetilde{x}_{1j} & \cdots & \widetilde{x}_{1n} \\ \vdots & & \vdots & & \vdots \\ \widetilde{x}_{i1} & \cdots & \widetilde{x}_{ij} & \cdots & \widetilde{x}_{in} \\ \vdots & & \vdots & & \vdots \\ \widetilde{x}_{m1} & \cdots & \widetilde{x}_{mj} & \cdots & \widetilde{x}_{mn} \end{array}\right] \end{array} \tag{12}$$

where $\widetilde{x}_{ij} = \left(x^l_{ij}, x^m_{ij}, x^u_{ij}\right)$ is the fuzzy performance rating of ith alternative with respect to jth criterion, and

$$x^l_{ij} = \frac{1}{K}\sum_{k=1}^{K} x^l_{ijk} x^m_{ij} = \frac{1}{K}\sum_{k=1}^{K} x^m_{ijk} x^m_{ij} = \frac{1}{K}\sum_{k=1}^{K} x^m_{ijk} \tag{13}$$

Step 5 Determine the fuzzy best value and fuzzy worst value.

The fuzzy best value and the fuzzy worst value are determined as follows:

$$\widetilde{x}^+_j = \left\{ \left(\max_i \widetilde{x}_{ij} \middle| j \in B\right), \left(\min_i \widetilde{x}_{ij} \middle| j \in C\right) \middle| \forall j = 1, 2, \cdots, n \right\} \tag{14}$$

$$\widetilde{x}^-_j = \left\{ \left(\min_i \widetilde{x}_{ij} \middle| j \in B\right), \left(\max_i \widetilde{x}_{ij} \middle| j \in C\right) \middle| \forall j = 1, 2, \cdots, n \right\} \tag{15}$$

where $\widetilde{x}^+_j$ is the fuzzy positive ideal solution (FPIS) and $\widetilde{x}^-_j$ is the fuzzy negative ideal solution (FNIS) for the jth criterion. B belongs to the benefit criteria and C belongs to the cost criteria.

Step 6 Calculate utility and regret measures.

The VIKOR ranking indicates that the preferred alternative is proximate to the ideal solution, starting from the L^p-metric used as an aggregating function in a compromise programming method as follows:

$$L^p_i = \left\{ \sum_{j=1}^{n} \left| \widetilde{w}_j \frac{\left(\widetilde{x}^+_j - \widetilde{x}_{ij}\right)}{\left(\widetilde{x}^+_j - \widetilde{x}^-_j\right)} \right|^p \right\}^{1/p}, 1 \leq p \leq \infty; i = 1, 2, \cdots, m \tag{16}$$

In the VIKOR method, $L^{p=1}_i$ as $\widetilde{S}_i$ and $L^{p=\infty}_i$ as $\widetilde{R}_i$ are used to formulate the ranking measure as follows:

$$\widetilde{S}_i = \sum_{j=1}^{n} \left| \widetilde{w}_j \widetilde{r}_{ij} \right| = \sum_{j=1}^{n} \left| \widetilde{w}_j \frac{\left(\widetilde{x}^+_j - \widetilde{x}_{ij}\right)}{\left(\widetilde{x}^+_j - \widetilde{x}^-_j\right)} \right|, i = 1, 2, \cdots, m \tag{17}$$

$$\widetilde{R}_i = \max_j \left| \widetilde{w}_j \widetilde{r}_{ij} \right| = \max_j \left| \widetilde{w}_j \frac{\left(\widetilde{x}_j^+ - \widetilde{x}_{ij} \right)}{\left(\widetilde{x}_j^+ - \widetilde{x}_j^- \right)} \right| , i = 1, 2, \cdots, m \tag{18}$$

where $\widetilde{S}_i$ and $\widetilde{R}_i$ represent the utility measure and the regret measure, respectively; $\widetilde{S}_i$ is shown as the average gap for achieving the aspired level; $\widetilde{R}_i$ is shown as the maximal gap for improving the priority; $\widetilde{S}_i$ refers to the separation measure of A_i from the positive-ideal solution; $\widetilde{R}_i$ is the separation measure of A_i from the negative-ideal solution; and $\widetilde{w}_j$ are the fuzzy weights of criteria.

Step 7 Compute index value.

To obtain the ranking results, the index value $\widetilde{Q}_i$ is computed as follows:

$$\widetilde{Q}_i = v \frac{\left(\widetilde{S}_i - \widetilde{S}^+ \right)}{\left(\widetilde{S}^- - \widetilde{S}^+ \right)} + (1 - v) \frac{\left(\widetilde{R}_i - \widetilde{R}^+ \right)}{\left(\widetilde{R}^- - \widetilde{R}^+ \right)} , i = 1, 2, \cdots, m \tag{19}$$

where

$$\widetilde{S}^+ = \min_i \widetilde{S}_i \ ; \ \widetilde{S}^- = \max_i \widetilde{S}_i, \ \widetilde{R}^+ = \min_i \widetilde{R}_i \ ; \ \widetilde{R}^- = \max_i \widetilde{R}_i \tag{20}$$

The indices $\widetilde{S}^+$ and $\widetilde{R}^+$ are related to a maximum group utility (majority rule) and a minimum individual regret of an opponent strategy, respectively. The parameter $v \in [0, 1]$ is defined as the weight for the decision-making strategy of maximum group utility, whereas $(1 - v)$ is defined as the weight for the decision-making strategy of minimum individual regret.

Step 10 Defuzzify TFNs

Using the centroid defuzzification technique to locate BNP values [12,40], the TFNs $\widetilde{S}_i = \left(S_i^l, S_i^m, S_i^u \right)$, $\widetilde{R}_i = \left(R_i^l, R_i^m, R_i^u \right)$, and $\widetilde{Q}_i = \left(Q_i^l, Q_i^m, Q_i^u \right)$ can be defuzzified as the crisp values S_i, R_i, and Q_i, respectively.

$$\begin{aligned} S_i &\cong \tfrac{1}{3} \left(S_i^l + S_i^m + S_i^u \right) \\ R_i &\cong \tfrac{1}{3} \left(R_i^l + R_i^m + R_i^u \right) \\ Q_i &\cong \tfrac{1}{3} \left(Q_i^l + Q_i^m + Q_i^u \right) \end{aligned} \tag{21}$$

Step 11 Rank alternatives.

An index Q_i represents the separation measure of alternative A_i from a positive-ideal solution. The smaller the value of Q_i, the better the alternative.

Step 12 Derive a compromise solution.

A solution with a minimum Q_i value in the ranking list is considered the optimal compromise one, if the following two conditions are satisfied:

Condition 1 Acceptable advantage

An alternative $A^{(1)}$ has an acceptable advantage if $\left(\widetilde{Q}\left(A^{(2)} \right) - \widetilde{Q}\left(A^{(1)} \right) \right) / \left(\widetilde{Q}\left(A^{(n)} \right) - \widetilde{Q}\left(A^{(1)} \right) \right) \geq 1/(n-1)$, where $A^{(1)}$ is the best ranked alternative and $A^{(2)}$ is the alternative with second position in the ranking list by the measure $\widetilde{Q}$. n is the number of alternatives.

Condition 2 Acceptable stability

An alternative $A^{(1)}$ must also be the best ranked by S or/and R. The compromise solution is stable within a decision-making process, which could be with "voting by majority rule" (when $v > 0.5$ is needed), with "consensus" (when $v = 0.5$), or with "veto" (when $v < 0.5$).

4. Numerical Analysis

In this section, an empirical work is definitely conducted and illustrated step-by-step according to the proposed hybrid fuzzy decision model.

Step 1 Establish the hierarchical model.

This study explored the six important criteria and concluded on the major five alternative technologies associated with our research themes through literature review and expert investigations. The hierarchical model was thus established as shown in Figure 2.

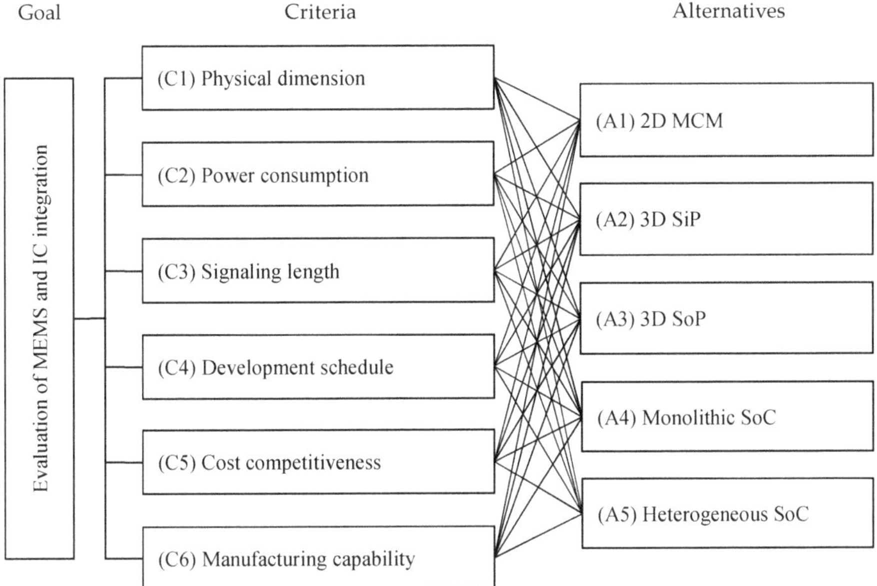

Figure 2. The hierarchical model for evaluating MEMS and IC (integrated circuit) integration technologies. MCM, multi-chip module; SiP, system-in-package; SoP, system-on-package; SoC, system-on-chip.

Step 2 Derive the fuzzy preference weights of the criteria.

(1)   Design a questionnaire for the data collection.

A typical AHP questionnaire used a nine-point rating scale (see Table 1) to represent the relative importance of each criterion.

Table 1. Linguistic variables and corresponding TFNs (triangular fuzzy numbers).

TFNs	Linguistic Variables	Scale of TFNs
$\tilde{9}$	Extremely more important	$(8, 9, 10)$
$\tilde{8}$	Intermediate	$(7, 8, 9)$
$\tilde{7}$	Very strongly more important	$(6, 7, 8)$
$\tilde{6}$	Intermediate	$(5, 6, 7)$
$\tilde{5}$	Strongly more important	$(4, 5, 6)$
$\tilde{4}$	Intermediate	$(3, 4, 5)$
$\tilde{3}$	Moderately more important	$(2, 3, 4)$
$\tilde{2}$	Intermediate	$(1, 2, 3)$
$\tilde{1}$	Equally important	$(1, 1, 1)$

(2)   Generate the fuzzy judgment matrices.

The fuzzy judgment matrices for the criteria from twelve experts were generated by Equation (1). Table 2 shows an individual fuzzy judgment matrix of expert 1.

Table 2. Individual fuzzy judgment matrix of expert 1.

Criteria	C1	C2	C3	C4	C5	C6
C1	(1, 1, 1)	(1, 2, 3)	(1, 2, 3)	(1, 1, 1)	(1, 2, 3)	(1, 1, 1)
C2	(1/3, 1/2, 1)	(1, 1, 1)	(1, 1, 1)	(1/3, 1/2, 1)	(1, 1, 1)	(1/3, 1/2, 1)
C3	(1/3, 1/2, 1)	(1, 1, 1)	(1, 1, 1)	(1/4, 1/3, 1/2)	(1/3, 1/2, 1)	(1/3, 1/2, 1)
C4	(1, 1, 1)	(1, 2, 3)	(2, 3, 4)	(1, 1, 1)	(1, 2, 3)	(1, 1, 1)
C5	(1/3, 1/2, 1)	(1, 1, 1)	(1, 2, 3)	(1/3, 1/2, 1)	(1, 1, 1)	(1/3, 1/2, 1)
C6	(1, 1, 1)	(1, 2, 3)	(1, 2, 3)	(1, 1, 1)	(1, 2, 3)	(1, 1, 1)

$\lambda_{max} = 6.056$; $CI = 0.022$; $CR = 0.018 \leq 0.1$

(3) Check the consistency.

Using Equations (4) and (5), the CI value is 0.022 and the CR value is 0.018 (less than 0.1), indicating the consistency of the collected data in the questionnaires and the robustness of fuzzy judgment matrices.

(4) Integrate the experts' opinions.

Using Equations (6) and (7) to compute the fuzzy geometric means of each criterion, the fuzzy synthetic judgment matrix was thus obtained as shown in Table 3.

Table 3. The fuzzy synthetic judgment matrix.

Criteria	C1	C2	C3	C4	C5	C6
C1	(1, 1, 1)	(1.26, 1.93, 2.51)	(1.26, 2.29, 3.3)	(0.69, 0.79, 1)	(1, 1.33, 1.58)	(0.69, 0.79, 1)
C2	(0.4, 0.52, 0.79)	(1, 1, 1)	(1.26, 1.72, 2.09)	(0.4, 0.52, 0.79)	(0.76, 0.84, 1)	(0.3, 0.44, 0.79)
C3	(0.3, 0.44, 0.79)	(0.48, 0.58, 0.79)	(1, 1, 1)	(0.27, 0.37, 0.59)	(0.4, 0.52, 0.79)	(0.4, 0.52, 0.79)
C4	(1, 1.26, 1.44)	(1.26, 1.93, 2.51)	(1.68, 2.71, 3.72)	(1, 1, 1)	(1.26, 1.93, 2.51)	(1, 1.26, 1.44)
C5	(0.63, 0.75, 1)	(1, 1.19, 1.32)	(1.26, 1.93, 2.51)	(0.4, 0.52, 0.79)	(1, 1, 1)	(0.4, 0.52, 0.79)
C6	(1, 1.26, 1.44)	(1.26, 2.29, 3.3)	(1.26, 1.93, 2.51)	(0.69, 0.79, 1)	(1.26, 1.93, 2.51)	(1, 1, 1)

After that, using Equations (8) and (9), the fuzzy preference weight of each criterion was obtained as shown in Table 4.

Table 4. The fuzzy preference weights of the criteria.

Fuzzy Weights	L	M	U
$\widetilde{w}_1$	0.118	0.194	0.312
$\widetilde{w}_2$	0.074	0.116	0.205
$\widetilde{w}_3$	0.053	0.084	0.160
$\widetilde{w}_4$	0.145	0.247	0.388
$\widetilde{w}_5$	0.087	0.137	0.229
$\widetilde{w}_6$	0.130	0.222	0.358

(5) Defuzzify the fuzzy weights.

Equation (10) was used to compute the BNP value of the fuzzy preference weight of each criterion. Table 5 shows the empirical results in which C4 (development schedule) and C6 (manufacturing capability) are the two most important criteria.

Table 5. Best nonfuzzy performance (BNP) values of the criteria.

Criteria	BNP Values	Rank
C1 (Physical dimension)	0.208	3
C2 (Power consumption)	0.131	5
C3 (Signaling length)	0.099	6
C4 (Development schedule)	0.260	1
C5 (Cost competitiveness)	0.151	4
C6 (Manufacturing capability)	0.237	2

Step 3 Rate the alternatives.

(1) Design a questionnaire for the data collection.
(2) Generate the fuzzy performance rating matrices.

Using Equation (11), the fuzzy performance rating matrices for the alternatives with respect to the criteria were generated (see Table 6).

Table 6. Individual fuzzy performance rating matrix of expert 1.

Criteria	A1	A2	A3	A4	A5
C1	(2, 3, 4)	(7, 8, 9)	(8, 9, 10)	(7, 8, 9)	(8, 9, 10)
C2	(3, 4, 5)	(7, 8, 9)	(7, 8, 9)	(6, 7, 8)	(6, 7, 8)
C3	(2, 3, 4)	(5, 6, 7)	(5, 6, 7)	(6, 7, 8)	(8, 9, 10)
C4	(8, 9, 10)	(7, 8, 9)	(3, 4, 5)	(5, 6, 7)	(2, 3, 4)
C5	(7, 8, 9)	(6, 7, 8)	(4, 5, 6)	(5, 6, 7)	(5, 6, 7)
C6	(7, 8, 9)	(7, 8, 9)	(3, 4, 5)	(6, 7, 8)	(2, 3, 4)

(3) Integrate the experts' opinions.

Using Equations (12) and (13), the fuzzy synthetic performance rating matrix for the alternatives with respect to the criteria was thus obtained as shown in Table 7.

Table 7. The fuzzy synthetic performance rating matrix.

Criteria	A1	A2	A3	A4	A5
C1	(2.42, 3.42, 4.42)	(6.67, 7.67, 8.67)	(6.75, 7.75, 8.75)	(7.17, 8.17, 9.17)	(7.25, 8.25, 9.25)
C2	(1.75, 2.75, 3.75)	(6.67, 7.67, 8.67)	(6.67, 7.67, 8.67)	(6.17, 7.17, 8.17)	(5.67, 6.67, 7.67)
C3	(1.75, 2.75, 3.75)	(4.83, 5.83, 6.83)	(5.33, 6.33, 7.33)	(6.33, 7.33, 8.33)	(7.67, 8.67, 9.67)
C4	(7.58, 8.58, 9.58)	(7.17, 8.17, 9.17)	(3.67, 4.67, 5.67)	(5.83, 6.83, 7.83)	(2.08, 3.08, 4.08)
C5	(7.33, 8.33, 9.33)	(6.83, 7.83, 8.83)	(5.33, 6.33, 7.33)	(6.33, 7.33, 8.33)	(5.83, 6.83, 7.83)
C6	(7.33, 8.33, 9.33)	(7.25, 8.25, 9.25)	(3.83, 4.83, 5.83)	(5.92, 6.92, 7.92)	(2.25, 3.25, 4.25)

(4) Determine the fuzzy best values and fuzzy worst values.

Using Equations (14) and (15), the FPIS and NPIS reference points for each criterion were determined as shown in Table 8.

Table 8. The FPIS (fuzzy positive ideal solution) and FNIS (fuzzy negative ideal solution).

Criteria	FPIS $\widetilde{x}_j^+$	FNIS $\widetilde{x}_j^-$
C1	(7.25, 8.25, 9.25)	(2.42, 3.42, 4.42)
C2	(6.67, 7.67, 8.67)	(1.75, 2.75, 3.75)
C3	(7.67, 8.67, 9.67)	(1.75, 2.75, 3.75)
C4	(7.58, 8.58, 9.58)	(2.08, 3.08, 4.08)
C5	(7.33, 8.33, 9.33)	(5.33, 6.33, 7.33)
C6	(7.33, 8.33, 9.33)	(2.25, 3.25, 4.25)

(5) Obtain the weighted fuzzy synthetic normalized rating matrix.

Using the equation $\left|\widetilde{w}_j \widetilde{r}_{ij}\right| = \left|\widetilde{w}_j \dfrac{\left(\widetilde{x}_j^+ - \widetilde{x}_{ij}\right)}{\left(\widetilde{x}_j^+ - \widetilde{x}_j^-\right)}\right|$, $i = 1, 2, \cdots, m$, the weighted fuzzy synthetic normalized rating matrix was obtained as shown in Table 9.

Table 9. The weighted fuzzy synthetic normalized rating matrix.

Criteria	A1	A2	A3	A4	A5
C1	(0.12, 0.19, 0.31)	(0.01, 0.02, 0.04)	(0.01, 0.02, 0.03)	(0, 0, 0.01)	(0, 0, 0)
C2	(0.07, 0.12, 0.2)	(0, 0, 0)	(0, 0, 0)	(0.01, 0.01, 0.02)	(0.01, 0.02, 0.04)
C3	(0.05, 0.08, 0.16)	(0.03, 0.04, 0.08)	(0.02, 0.03, 0.06)	(0.01, 0.02, 0.04)	(0, 0, 0)
C4	(0, 0, 0)	(0.01, 0.02, 0.03)	(0.1, 0.18, 0.28)	(0.05, 0.08, 0.12)	(0.14, 0.25, 0.39)
C5	(0, 0, 0)	(0.02, 0.03, 0.06)	(0.09, 0.14, 0.23)	(0.04, 0.07, 0.11)	(0.07, 0.1, 0.17)
C6	(0, 0, 0)	(0, 0, 0.01)	(0.09, 0.15, 0.25)	(0.04, 0.06, 0.1)	(0.13, 0.22, 0.36)

(6)   Rank the alternatives.

The separation values $\widetilde{S}_i$ and $\widetilde{R}_i$ were calculated through Equations (17) and (18), respectively, and the index value $\widetilde{Q}_i$ was computed based on Equations (19) and (20). The BNP values of $\widetilde{Q}_i$ were then obtained through Equation (21). The defuzzified crisp value Q_i represents the separation measure of alternative A_i from a positive-ideal solution. Table 10 shows that A2 (3D SiP) and A4 (monolithic SoC) are the top two favored technologies.

Table 10. Performance ratings of the alternatives. MCM, multi-chip module; SiP, system-in-package; SoP, system-on-package; SoC, system-on-chip; BNP, best nonfuzzy performance.

Alternatives	$\widetilde{S}_i$	$\widetilde{R}_i$	$\widetilde{Q}_i$	BNP of $\widetilde{Q}_i$	Rank
A1 (2D MCM)	(0.24, 0.39, 0.68)	(0.12, 0.19, 0.31)	(0.26, 0.66, 1.82)	0.91	3
A2 (3D SiP)	(0.07, 0.12, 0.21)	(0.03, 0.04, 0.08)	(0, 0, 0)	0.00	1
A3 (3D SoP)	(0.31, 0.52, 0.85)	(0.10, 0.18, 0.28)	(0.28, 0.75, 1.98)	1.00	4
A4 (Monolithic SoC)	(0.15, 0.24, 0.40)	(0.05, 0.08, 0.12)	(0.08, 0.22, 0.54)	0.28	2
A5 (Heterogeneous SoC)	(0.36, 0.60, 0.96)	(0.14, 0.25, 0.39)	(0.38, 1.00, 2.64)	1.34	5

5. Discussion

The two most important criteria were reviewed in Section 5.1. While 3D SiP and monolithic SoC were selected as the two most favored technologies, the rationale for them is also discussed in Section 5.2.

5.1. Rationale for the Two Most Important Criteria

In view of the fuzzy AHP results shown in Table 5, the two most important criteria are development schedule (0.260) and manufacturing capability (0.237). The development schedule indicates the roadmap of technology development through the expense of R&D (research and development) resources. Semiconductor firms would gain more competitive advantages if the development schedule of new technologies can be further reduced. Manufacturing capability refers to the technical and physical limitations of semiconductor firms. Higher manufacturing capability always leads to higher manufacturing efficiency and yield. Hence, the development schedule and manufacturing capability are the two key factors that should be considered first while evaluating the optimal alternative technology for incorporating MEMS with IC components.

5.2. Rationale for the Top Two Preferable Technologies

In view of the fuzzy VIKOR results shown in Table 10, 3D SiP is of primary interest to semiconductor firms among the five alternatives. The key advantages of 3D SiP technology are its higher integration densities, shorter signaling lengths, and smaller package footprints in comparison with 2D MCM. This method yields very compact packages (i.e., benefiting by physical dimensions and signaling length) [19,47] and has been employed in a number of commercial products (i.e., benefiting by development schedule and manufacturing capability) [48,49]. 3D SoP is another technology that enables a highly integrated and miniaturized system at the package level, but it went a step beyond 3D SiP by integrating thin-film components on a package substrate (i.e., suffering from development schedule and manufacturing capability) [50].

The second-ranked technology is monolithic SoC. The CMOS-MEMS technique provides advantages such as that it can be implemented using existing IC infrastructure and MEMS components can be formed in completed wafers using cost-effective processing steps (i.e., benefiting by development schedule and cost competitiveness) [5]. The heterogeneous SoC integration approach can be also supported by an existing foundry structure, but it often requires accurate substrate-to-substrate or wafer-to-wafer alignment during bonding, reliable electrical interconnections, and/or even a greater number of manufacturing steps (i.e., suffering from development schedule and manufacturing capability).

While comparing CMOS-MEMS with 3D SiP, 3D SiP has greater performance ratings than CMOS-MEMS in terms of the major two criteria—development schedule and manufacturing capability. On the other hand, CMOS-MEMS has greater performance ratings than 3D SiP in terms of the major two criteria—physical dimension and signaling length.

6. Conclusions

MEMS sensors are now prevalent in the era of IoT. The global semiconductor industry expects a higher integration of mechanical structures with electronics that can be manufactured by CMOS technologies as usual. A variety of microfabrication and integration approaches have been attempted, but each has distinguishing features. This study successfully proposed a hybrid fuzzy MCDM model that effectively facilitates the evaluation and selection of MEMS and IC integration technologies in a fuzzy environment. The six important criteria and the major five alternative technologies associated with our research themes were first explored through literature review and expert investigations. The priority weights of criteria were then derived using fuzzy AHP, and the two most important criteria are development schedule and manufacturing capability. After that, fuzzy VIKOR was deployed to rate the alternatives, and 3D SiP and monolithic SoC are the top two favored technologies. The proposed fuzzy decision model could serve as a reference for the future strategic evaluation and selection of MEMS and IC integration technologies.

Author Contributions: Conceptualization, Q.-Y.L., M.-X.L. and Y.-C.L.; Methodology, Q.-Y.L., M.-X.L. and Y.-C.L.; Software, Y.-C.L.; Validation: Q.-Y.L., M.-X.L. and Y.-C.L.; Formal analysis, Q.-Y.L. and M.-X.L.; Investigation, Q.-Y.L. and M.-X.L.; Resources: Q.-Y.L., M.-X.L. and Y.-C.L.; Data curation, Q.-Y.L., M.-X.L. and Y.-C.L.; Writing—original draft preparation, Q.-Y.L., M.-X.L. and Y.-C.L.; Writing—review and editing, Q.-Y.L., M.-X.L. and Y.-C.L.; Visualization, Q.-Y.L. and M.-X.L.; Supervision, M.-X.L. and Y.-C.L.; Project administration, Y.-C.L.; Funding acquisition, Q.-Y.L., M.-X.L. and Y.-C.L. All authors have read and agreed to the published version of the manuscript.

Funding: This research received no external funding.

Conflicts of Interest: The authors declare no conflict of interest.

References

1. Lee, I.; Lee, K. The Internet of Things (IoT): Applications, investments, and challenges for enterprises. *Bus. Horiz.* **2015**, *58*, 431–440. [CrossRef]
2. Li, S.; Xu, L.D.; Zhao, S. The internet of things: A survey. *Inform. Syst. Front.* **2015**, *17*, 243–259. [CrossRef]

3.   Song, Z.; Du, Y.; Liu, M.; Yang, S.; Wu, D.; Wang, Z. Three-dimensional integration of suspended single-crystalline silicon MEMS arrays with CMOS. In Proceedings of the 28th IEEE International Conference on Micro Electro Mechanical Systems (MEMS), Estoril, Portugal, 18–22 January 2015; pp. 304–307.
4.   Adams, T.M.; Layton, R.A. *Introductory MEMS: Fabrication and Applications*; Springer Science & Business Media: Berlin/Heidelberg, Germany, 2009.
5.   Fischer, A.C.; Forsberg, F.; Lapisa, M.; Bleiker, S.J.; Stemme, G.; Roxhed, N.; Niklaus, F. Integrating MEMS and ICs. *Microsyst. Nanoeng.* **2015**, *1*, 1–16. [CrossRef]
6.   Vavassori, P.; Pancaldi, M.; Perez-Roldan, M.J.; Chuvilin, A.; Berger, A. Remote magnetomechanical nanoactuation. *Small* **2016**, *12*, 1013–1023. [CrossRef]
7.   Lee, Y.-C.; Chung, P.-H.; Shyu, J.Z. Performance evaluation of medical device manufacturers using a hybrid fuzzy MCDM. *J. Sci. Ind. Res.* **2017**, *76*, 28–31.
8.   van de Kaa, G.; Rezaei, J.; Kamp, L.; de Winter, A. Photovoltaic technology selection: A fuzzy MCDM approach. *Renew. Sustain. Energ. Rev.* **2014**, *32*, 662–670. [CrossRef]
9.   Vinodh, S.; Nagaraj, S.; Girubha, J. Application of fuzzy VIKOR for selection of rapid prototyping technologies in an agile environment. *Rapid Prototyp. J.* **2014**, *20*, 523–532. [CrossRef]
10. Liu, H.-C.; You, J.-X.; Lu, C.; Chen, Y.-Z. Evaluating health-care waste treatment technologies using a hybrid multi-criteria decision making model. *Renew. Sustain. Energ. Rev.* **2015**, *41*, 932–942. [CrossRef]
11. Bairagi, B.; Dey, B.; Sarkar, B.; Sanyal, S. Selection of robotic systems in fuzzy multi criteria decision-making environment. *Int. J. Comput. Syst. Eng.* **2015**, *2*, 32–42. [CrossRef]
12. Lee, Y.-C.; Chou, C.J. Technology evaluation and selection of 3DIC integration using a three-stage fuzzy MCDM. *Sustainability* **2016**, *8*, 114. [CrossRef]
13. Taylan, O.; Alamoudi, R.; Kabli, M.; AlJifri, A.; Ramzi, F.; Herrera-Viedma, E. Assessment of energy systems using extended fuzzy AHP, fuzzy VIKOR, and TOPSIS approaches to manage non-cooperative opinions. *Sustainability* **2020**, *12*, 2745. [CrossRef]
14. Salimi, A.H.; Noori, A.; Bonakdari, H.; Samakosh, J.M.; Sharifi, E.; Hassanvand, M.; Agharazi, M.; Gharabaghi, B. Exploring the role of advertising types on improving the water consumption behavior: An application of integrated fuzzy AHP and fuzzy VIKOR method. *Sustainability* **2020**, *12*, 1232. [CrossRef]
15. Yang, H.S.; Bakir, M.S. 3D integration of CMOS and MEMS using mechanically flexible interconnects (MFI) and through silicon vias (TSV). In Proceedings of the 60th Electronic Components and Technology Conference, Las Vegas, NV, USA, 1–4 June 2010; pp. 822–828.
16. Fischer, A.C.; Korvink, J.G.; Roxhed, N.; Stemme, G.; Wallrabe, U.; Niklaus, F. Unconventional applications of wire bonding create opportunities for microsystem integration. *J. Micromech. Microeng.* **2013**, *23*, 083001. [CrossRef]
17. Lau, J.H. Design and process of 3D MEMS system-in-package (SiP). *J. Microelectron. Electron. Packag.* **2010**, *7*, 10–11. [CrossRef]
18. Kumar, A.; Verma, G.; Nath, V.; Choudhury, S. IC Packaging: 3D IC Technology and Methods. In Proceedings of the International Conference on Nano-electronics, Circuits & Communication Systems, Jharkhand, India, 11–12 November 2017; pp. 303–317.
19. Xu, G.; Yan, P.; Chen, X.; Ning, W.; Luo, L.; Jiao, J. Wafer-level chip-to-wafer (C2W) integration of high-sensitivity MEMS and ICs. In Proceedings of the 12th International Conference on Electronic Packaging Technology and High Density Packaging (ICEPT-HDP), Shanghai, China, 8–11 August 2011; pp. 1–5.
20. Marenco, N.; Reinert, W.; Warnat, S.; Lange, P.; Gruenzig, S.; Hillmann, G.; Kostner, H.; Bock, G.; Guadagnuolo, S.; Conte, A. Vacuum encapsulation of resonant MEMS sensors by direct chip-to-wafer stacking on ASIC. In Proceedings of the 10th Electronics Packaging Technology Conference, Singapore, 9–12 December 2008; pp. 773–777.
21. Lau, J.H. Critical issues of TSV and 3D IC integration. *J. Microelectron. Electron. Packag.* **2010**, *7*, 35–43. [CrossRef]
22. Li, C.; Tian, K.-M.; Huang, H. Structure design of the MEMS package based on QFN and SOP. *Adv. Mater. Res. Switz.* **2015**, *5*, 18.
23. Mita, Y.; Lebrasseur, E.; Okamoto, Y.; Marty, F.; Setoguchi, R.; Yamada, K.; Mori, I.; Morishita, S.; Imai, Y.; Hosaka, K. Opportunities of CMOS-MEMS integration through LSI foundry and open facility. *Jpn. J. Appl. Phys.* **2017**, *56*, 06GA03. [CrossRef]
24. Chaehoi, A.; O'Connell, D.; Weiland, D.; Adamson, R.; Bruckshaw, S.; Ray, S.; Begbie, M.; Bruce, J. Monolithic CMOS MEMS technology development: A piezoresistive-sensors case study. In Proceedings of the NSTI-Nanotech, Anaheim, CA, USA, 21–25 June 2010; pp. 284–287.
25. Topart, P.; Picard, F.; Ilias, S.; Alain, C.; Chevalier, C.; Fisette, B.; Paultre, J.E.; Généreux, F.; Legros, M.; Lepage, J.-F. Heterogeneous MEMS device assembly and integration, SPIE MOEMS-MEMS. In Proceedings of the SPIE 8975, Reliability, Packaging, Testing, and Characterization of MOEMS/MEMS, Nanodevices, and Nanomaterials XIII, San Francisco, CA, USA, 3–4 February 2014; p. 89750E-13.
26. Bahr, B.; Marathe, R.; Weinstein, D. Theory and design of phononic crystals for unreleased CMOS-MEMS resonant body transistors. *J. Microelectromech. Syst.* **2015**, *24*, 1520–1533. [CrossRef]
27. Knese, K.; Armbruster, S.; Weber, H.; Fischer, M.; Benzel, H.; Metz, M.; Seidel, H. Novel technology for capacitive pressure sensors with monocrystalline silicon membranes. In Proceedings of the 22nd International Conference on Micro Electro Mechanical Systems, Sorrento, Italy, 25–29 January 2009; pp. 697–700.
28. Smith, J.; Montague, S.; Sniegowski, J.; Murray, J.; McWhorter, P. Embedded micromechanical devices for the monolithic integration of MEMS with CMOS. In Proceedings of the IEDM'95, Washington, DC, USA, 10–13 December 1995; pp. 609–612.

29. Ding, X.; Czarnocki, W.; Schuster, J.; Roeckner, B. DSP-based CMOS monolithic pressure sensor for high volume manufacturing. In Proceedings of the Dig Internation Conference on Solid-State Sensors and Actuators, Sendai, Japan, 7–10 June 1999; pp. 362–365.
30. Fedder, G.K.; Howe, R.T.; Liu, T.-J.K.; Quevy, E.P. Technologies for cofabricating MEMS and electronics. *Proc. IEEE* **2008**, *96*, 306–322. [CrossRef]
31. Koyanagi, M. Recent progress in 3D integration technology. *IEICE Electron. Expr.* **2015**, *12*, 1–17. [CrossRef]
32. Saaty, T.L. Axiomatic foundation of the analytic hierarchy process. *Manag. Sci.* **1986**, *32*, 841–855. [CrossRef]
33. Saaty, T.L. How to make a decision: The analytic hierarchy process. *Eur. J. Oper. Res.* **1990**, *48*, 9–26. [CrossRef]
34. Tzeng, G.-H.; Huang, J.-J. *Multiple Attribute Decision Making: Methods and Applications*; Taylor & Francis: Milton Park, UK, 2011.
35. Lin, G.T.R.; Lee, Y.-C. Evaluation and decision making in Taiwan semiconductor industry through silicon via technology. *J. Sci. Ind. Res.* **2014**, *73*, 456–460.
36. Lee, Y.-C.; Lin, G.T.R.; Hsi, P.-H.; Lim, S.S. Evaluating the commercial potential of original technologies in universities. *J. Sci. Ind. Res.* **2016**, *75*, 463–465.
37. Csutora, R.; Buckley, J.J. Fuzzy hierarchical analysis: The Lambda-Max method. *Fuzzy Set. Syst.* **2001**, *120*, 181–195. [CrossRef]
38. Buckley, J.J. Fuzzy hierarchical analysis. *Fuzzy Set. Syst.* **1985**, *17*, 233–247. [CrossRef]
39. Hsieh, T.-Y.; Lu, S.-T.; Tzeng, G.-H. Fuzzy MCDM approach for planning and design tenders selection in public office buildings. *Int. J. Proj. Manag.* **2004**, *22*, 573–584. [CrossRef]
40. Opricovic, S.; Tzeng, G.-H. Defuzzification within a multicriteria decision model. *Int. J. Uncertain. Fuzz.* **2003**, *11*, 635–652. [CrossRef]
41. Opricovic, S.; Tzeng, G.-H. Extended VIKOR method in comparison with outranking methods. *Eur. J. Oper. Res.* **2007**, *178*, 514–529. [CrossRef]
42. Hu, S.-K.; Lu, M.-T.; Tzeng, G.-H. Exploring smart phone improvements based on a hybrid MCDM model. *Expert Syst. Appl.* **2014**, *41*, 4401–4413. [CrossRef]
43. Mohammady, P.; Amid, A. Integrated fuzzy AHP and fuzzy VIKOR model for supplier selection in an agile and modular virtual enterprise. *Fuzzy Inf. Eng.* **2011**, *3*, 411–431. [CrossRef]
44. Opricovic, S.; Tzeng, G.-H. Compromise solution by MCDM methods: A comparative analysis of VIKOR and TOPSIS. *Eur. J. Oper. Res.* **2004**, *156*, 445–455. [CrossRef]
45. Opricovic, S. A fuzzy compromise solution for multicriteria problems. *Int. J. Uncertain. Fuzz.* **2007**, *15*, 363–380. [CrossRef]
46. Alguliyev, R.M.; Aliguliyev, R.M.; Mahmudova, R.S. Multicriteria personnel selection by the modified fuzzy VIKOR method. *Sci. World J.* **2015**, *2015*, 1–16. [CrossRef] [PubMed]
47. Choi, W.K.; Premachandran, C.; Xie, L.; Ong, S.C.; He, J.H.; Yap, G.J.; Yu, A. A novel die to wafer (D2W) collective bonding method for MEMS and electronics heterogeneous 3D integration. In Proceedings of the 60th Electronic Components and Technology Conference (ECTC), Las Vegas, NV, USA, 1–4 June 2010; pp. 829–833.
48. Vigna, B.; Lasalandra, E.; Ungaretti, T. Motion MEMS and sensors, today and tomorrow. In *Nyquist AD Converters. Sensor Interfaces, and Robustness*; Springer: Berlin/Heidelberg, Germany, 2013; pp. 117–127.
49. Oouchi, A. Plastic molded package technology for MEMS sensor evolution of MEMS sensor package. In Proceedings of the International Conference on Electronics Packaging (ICEP), Toyama, Japan, 23–25 April 2014; pp. 371–375.
50. Tummala, R. SoC vs. MCM vs SiP vs. SoP. 2006. Available online: https://sst.semiconductor-digest.com/2006/07/soc-vs-mcm-vs-sip-vs-sop/ (accessed on 23 July 2020).

micromachines

Article

A RF Redundant TSV Interconnection for High Resistance Si Interposer

Mengcheng Wang [1], Shenglin Ma [1,*], Yufeng Jin [2], Wei Wang [3], Jing Chen [3], Liulin Hu [4] and Shuwei He [4]

1. Department of Mechanical & Electrical Engineering, Xiamen University, Xiamen 361102, China; 19920181152234@stu.xmu.edu.cn
2. School of Electronic and Computer Engineering, Shenzhen Graduate School of Peking University, Shenzhen 518055, China; yfjin@pku.edu.cn
3. National Key Laboratory of Science and Technology on Micro/Nano Fabrication, School of Electronic Engineering and Computer Science, Peking University, Beijing 100871, China; w.wang@pku.edu.cn (W.W.); j.chen@pku.edu.cn (J.C.)
4. Chengdu Ganide Technology Co., Ltd, Chengdu 610000, China; huliul@sina.com (L.H.); heshuwei_3@163.com (S.H.)
* Correspondence: mashenglin@xmu.edu.cn

Abstract: Through Silicon Via (TSV) technology is capable meeting effective, compact, high density, high integration, and high-performance requirements. In high-frequency applications, with the rapid development of 5G and millimeter-wave radar, the TSV interposer will become a competitive choice for radio frequency system-in-package (RF SIP) substrates. This paper presents a redundant TSV interconnect design for high resistivity Si interposers for millimeter-wave applications. To verify its feasibility, a set of test structures capable of working at millimeter waves are designed, which are composed of three pieces of CPW (coplanar waveguide) lines connected by single TSV, dual redundant TSV, and quad redundant TSV interconnects. First, HFSS software is used for modeling and simulation, then, a modified equivalent circuit model is established to analysis the effect of the redundant TSVs on the high-frequency transmission performance to solidify the HFSS based simulation. At the same time, a failure simulation was carried out and results prove that redundant TSV can still work normally at 44 GHz frequency when failure occurs. Using the developed TSV process, the sample is then fabricated and tested. Using L-2L de-embedding method to extract S-parameters of the TSV interconnection. The insertion loss of dual and quad redundant TSVs are 0.19 dB and 0.46 dB at 40 GHz, respectively.

Keywords: millimeter-wave; redundant TSV; equivalent circuit model; S-parameters extraction

Citation: Wang, M.; Ma, S.; Jin, Y.; Wang, W.; Chen, J.; Hu, L.; He, S. A RF Redundant TSV Interconnection for High Resistance Si Interposer. *Micromachines* **2021**, *12*, 169. https://doi.org/10.3390/mi12020169

Received: 22 January 2021
Accepted: 5 February 2021
Published: 8 February 2021

1. Introduction

With the development of 5G communication technology and millimeter-wave radar, system-in-package (SIP) for high-frequency devices has become a popular research subject in industrial and academic fields. This is due to the technical benefits of small size, light weight, high integration, high density, and system performance improvement [1]. Traditionally, radio frequency (RF) SIP solution-based microwave printed circuit boards or high-performance ceramic substrates, such as HTCC (high temperature co-fired ceramics) and LTCC (low temperature co-fired ceramics), have faced challenges in terms of their precision of critical dimension and minimum size of redistribution lines and pitch. Due to the precise wiring capacity and the low mismach in material coefficient of thermal expansion (CTE), research works have been done to explore the feasibility as well as the technical advantage of TSV technology for RF application [2–6]. It has been found that the RF property of TSV becomes the key issue in this field as the natural property of Si as semiconductor, which is characterized in term of S-parameters. S-parameters are network parameters based on the relationship between incident wave and reflected wave. S_{11} named the return loss represents the reflection coefficient of incident port while S_{21} named insertion loss represents the transmission coefficient from incident port to the destination. To improve the S_{21} of TSV interconnection, optimization methods in materials, structural, and process flow are proposed. For example [7,8] using a high-resistance silicon substrate, the measured insertion loss of a single TSV is 0.35 dB at 20 GHz. References [9,10] designed the coaxial TSV structure containing two layers of conductors, the measured S_{21} of a TSV is −0.48 dB at 10 GHz. By optimizing the important electroplating process in the TSV manufacturing process [11–13], TSV can achieve bottom-up Cu filling, and a single TSV will have low DC resistance of 36.7 mΩ to ensure low RF loss. To our knowledge, the best test result is demonstrated by [14], which has an insertion loss of 0.53 dB at 75 GHz for a single TSV.

Various types of defects have been found in the manufacturing process of TSV. These include discontinuities and voids in the metal inside TSV caused by a poor sputtering seed layer or plating failure [15], pinholes and cracks of TSV oxidation caused by impurities in the insulating materials or deposition methods. The discontinuity of metal in the hole causes the signal channel to open and reflect most of the transmitted signal. Pinholes in the insulator around the TSV will cause a leakage current between the TSV and the substrate, resulting in a resistive short circuit [16]. Voids will cause the resistance of the interconnect to change, resulting in increased signal loss. These defects affect the signal transmission from the input to the receiver in different ways.

To address TSV failure, designs for a redundant TSV have been proposed. Samsung proposed a TSV redundant architecture with a switching method for 3D DDR3 DRAM (Samsung, Seoul, Korea) products [17]. Hsieh proposed a method for repairing shifted TSV [18]. Reference [19] proposed a redundant architecture based on routers. The current research on redundant TSVs is mainly focused on the logic 3D IC (integrated circuit) application, while there is little research regarding TSV's RF application [20,21]. In this field, dense TSVs are not required for RF transmission which is favorable for redundant design. However, unlike the redundant design of TSV in logic IC, the participation of redundant TSV may change the characteristic impedance of RF TSV, and finally cause the degradation in RF insertion loss or electromagnetic compatibility issue. Therefore, special attention should be paid to RF redundant TSV design to guarantee that it can maintain an equivalent RF property with single RF TSV design whatever a defect occurs, which is the key point.

Therefore, a dual redundant and four redundant TSV interconnection designs are proposed for a high-resistivity Si interposer in this paper. The high-frequency performance is analyzed using a 3D field solver and a modified equivalent circuit model and compared with a single TSV interconnection. S-parameters are simulated when the redundant TSV has a via failure. Based on the proposed redundant TSV design, through the typical TSV process, samples are manufactured and tested. S-parameters of TSV interconnection are obtained by de-embedding. In view of the process factors that cause the measured radio frequency performance to decline, analysis and optimization simulation are carried out and an agreement is obtained. Finally, the high-frequency performance of the redundant TSV structure proposed in this paper is compared with the published single TSV to show its technological advantage.

2. Structural Design

Figure 1 shows the proposed redundant RF TSV. Figure 1b,c are dual and quad redundant RF TSVs on high resistivity Si substrate, respectively, while Figure 1a is a traditional single RF TSV as a reference. For ease of use in further test and S-parameters extraction, RF TSVs on high resistivity Si substrate are connected by coplanar waveguide (CPW) lines, which have a designed impedance of 50 Ω and dimensions are summarized in Table 1.

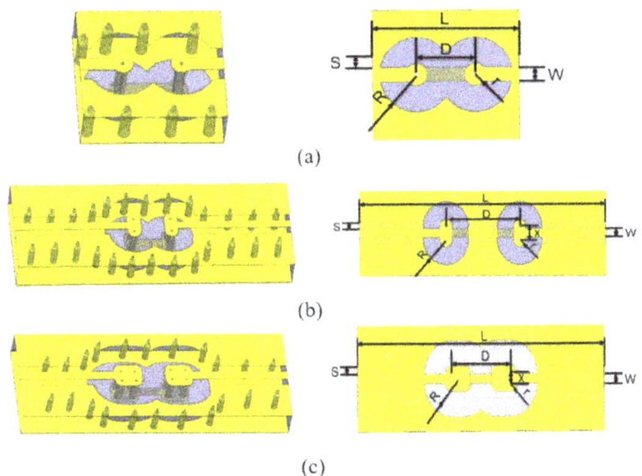

Figure 1. Test structure HFSS model: (a) single TSV; (b) dual redundant TSV; (c) quad redundant TSV.

Table 1. Parameters of the test structure (μm).

Table 1	L	S	W	D	R	r	x
Single TSV	1000	70	100	400	250	75	-
Dual redundant TSV	3000	70	100	400	250	75	160
Quad redundant TSV	3000	70	100	640	250	75	120

Figure 2 shows simulated S-parameters with HFSS model. It can be seen that at a frequency of 40 GHz, the insertion losses of a single TSV interconnect test structure, dual redundant TSV interconnect test structure, and quad redundant TSV interconnect test structure design are 0.197 dB, 0.538 dB, and 0.998 dB, respectively. The S_{11} parameter is less than -15 dB. The results show that it has a good high-frequency transmission performance. At the same time, as the number of redundant TSVs increases, the S_{21} parameter shows that the insertion loss value gradually degrades, and the S_{11} parameter shows that the resonance frequency gradually decreases.

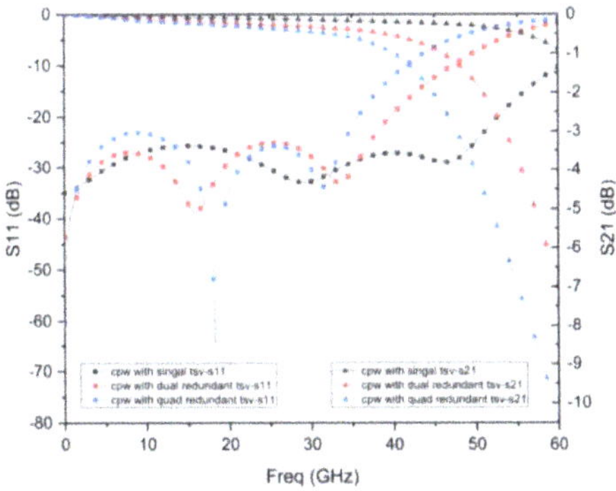

Figure 2. Test structure simulation result.

To better understand the effect of the redundant TSVs on the high-frequency transmission performance, a modified lumped circuit model is established for the dual redundant TSV test structure in this paper, as shown in Figure 3. The parameters of the symbols in the model are listed in Table 2. The values of each lumped element can be calculated by applying the dimensions [22–25] and material properties [26–28] in the following equations.

The diameter, width, length, thickness, height, and spacing is symbolized d, w, l, t, h, and p, respectively.

$$\delta_{Cu} = \frac{1}{\sqrt{\pi f \mu_0 \sigma_{Cu}}} \tag{1}$$

$$R_{RDL\_DC} = \frac{l_{RDL}}{\sigma_{Cu} w_{RDL} t_{RDL}} \tag{2}$$

$$R_{RDL\_AC} = \frac{l_{RDL}}{\sigma_{Cu} w_{RDL} \delta_{Cu}} \tag{3}$$

$$R_{RDL} = \sqrt{R_{RDL\_DC}^2 + R_{RDL\_AC}^2} \tag{4}$$

$$L_{RDL} = \frac{\mu_0 l_{RDL}}{2\pi} \left(\ln\left(\frac{2 l_{RDL}}{w_{RDL} + t_{RDL}} \right) + \frac{1}{2} + \frac{w_{RDL} + t_{RDL}}{3 l_{RDL}} \right) \tag{5}$$

$$C_{RDLinSub} = \frac{\varepsilon_0 \varepsilon_{si} l_{RDL} w_{RDL}}{h_{si}} \tag{6}$$

$$G_{RDLinSub} = \frac{\sigma_{si} l_{RDL} w_{RDL}}{h_{si}} \tag{7}$$

$$R_{TSV\_DC} = \frac{h_{TSV}}{\sigma_{Cu} \pi (d_{TSV}/2)^2} \tag{8}$$

$$R_{TSV\_AC} = \frac{h_{TSV}}{\sigma_{Cu} \pi ((d_{TSV}/2)^2 - (d_{TSV}/2 - \delta_{Cu})^2)} \tag{9}$$

$$R_{TSV} = \sqrt{R_{TSV\_DC}^2 + R_{TSV\_AC}^2} \tag{10}$$

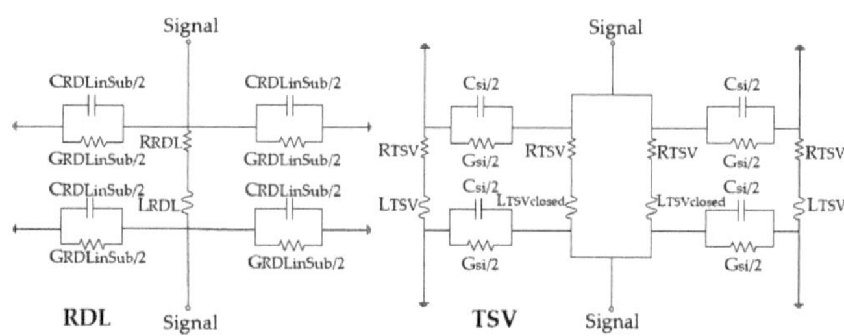

Figure 3. Equivalent circuit models of dual redundant TSV test structure.

Table 2. Symbols and parameters in the equivalent circuit model for TSV and RDL.

Symbol	Parameter
R_{RDL}/R_{TSV}	Resistance of RDL/TSV
L_{RDL}	Self-Inductance of RDL
$L_{TSVclosed}$	Inductance of RDL under the influence of proximity effect
$C_{RDLinSub}/C_{sub}$	Capacitance between RDL/TSV and substrate
$G_{RDLinSub}/G_{sub}$	Conductance between RDLs/TSVs in silicon substrate
PF	Proximity effect correction factor
d_{Cu}	Skin depth

When the alternating current in the same direction flows in the redundant TSV copper column, the alternating magnetic field generated by each current will generate eddy currents on adjacent TSVs, resulting in uneven current distribution in the TSV copper column and the proximity effect. The electric field distribution of the redundant TSV design proposed in this paper is shown in Figure 4. It is obvious that the internal electric field of TSV is concentrated in the edge area. Therefore, the equivalent resistance and inductance of TSV are changed, which needs to be considered when calculating the equivalent circuit.

$$PF = \frac{p_{TSV}/d_{TSV}}{\sqrt{(p_{TSV}/d_{TSV})^2 - 1}} \tag{11}$$

$$R_{TSV_{closed}} = PF \cdot R_{TSV\_AC} \tag{12}$$

$$L_{TSV} = \frac{\mu_0 h_{TSV}}{2\pi}\left[\ln\left(\frac{h_{TSV}}{d_{TSV}/2} + \sqrt{(\frac{h_{TSV}}{d_{TSV}/2})^2 + 1}\right) + \frac{d_{TSV}/2}{h_{TSV}} - \sqrt{(\frac{d_{TSV}/2}{h_{TSV}})^2 + 1}\right] + \frac{R_{TSV\_AC}}{2\pi f} \tag{13}$$

$$M_{TSV} = \frac{\mu_0 h_{TSV}}{2\pi}\left[\ln\left(\frac{h_{TSV}}{p_{TSV}} + \sqrt{(\frac{h_{TSV}}{p_{TSV}})^2 + 1}\right) + \frac{p_{TSV}}{h_{TSV}} - \sqrt{(\frac{p_{TSV}}{h_{TSV}})^2 + 1}\right] \tag{14}$$

$$L_{TSV_{closed}} = L_{TSV} + M_{TSV} \tag{15}$$

$$Csub = \frac{\pi \varepsilon_0 \varepsilon_{si} h_{TSV}}{\cosh^{-1}\left(\frac{p_{TSV}}{d_{TSV}}\right)} \tag{16}$$

$$Gsub = \frac{\pi \sigma_{si} h_{TSV}}{\cosh^{-1}\left(\frac{p_{TSV}}{d_{TSV}}\right)} \tag{17}$$

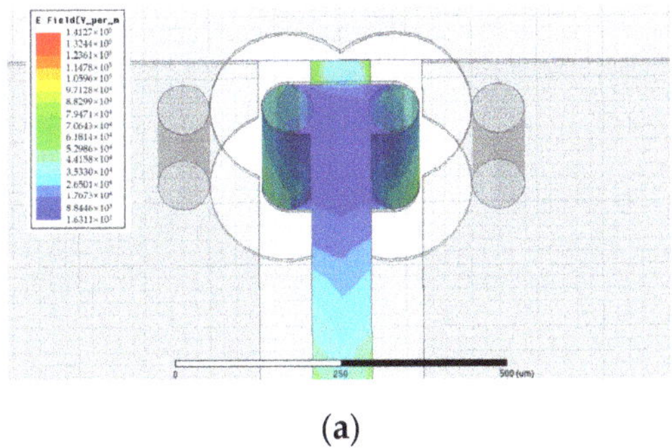

(**a**)

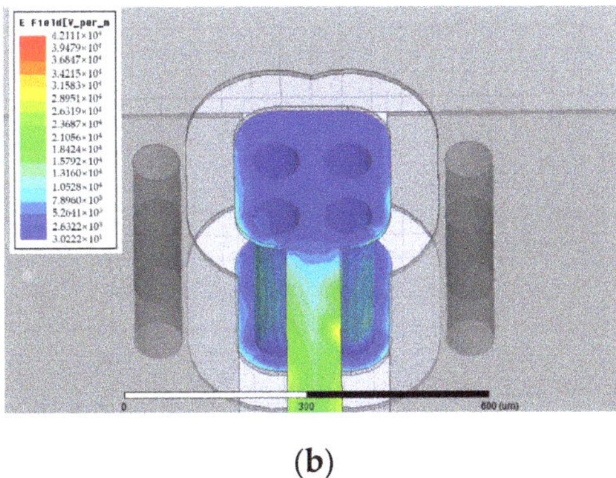

(**b**)

Figure 4. Electric field diagram: (**a**) dual redundant TSV; (**b**) quad redundant TSV.

The simulation results with HFSS and the established equivalent circuit model in Figure 3 are shown in Figure 5. The similarity of the curvature of the amplitude curves of S_{11} and S_{21} indicates that the establishment of the equivalent circuit model is correct. It can be seen from the equivalent circuit model that, at high frequencies, the main factors that affect the S-parameters are inductance and resistance. As the redundant structure contains multiple RF TSVs, the proximity effect between TSVs will increase, and the overall inductance and resistance will increase due to the additional RDL required to connect multiple TSVs. Therefore, an increase in the number of redundant TSVs will cause the resonance frequency to appear in the lower frequency range, and the high-frequency loss will degrade significantly.

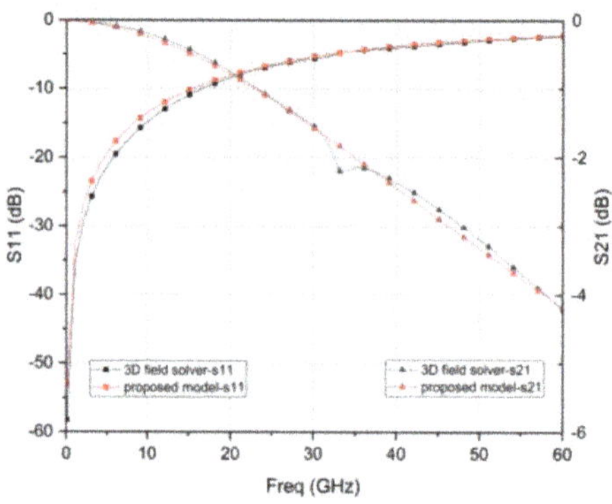

Figure 5. S-parameters results of dual redundant TSV structure obtained by 3D FEM solver and equivalent circuit model.

In order to verify the feasibility of the RF redundant TSV scheme, Figures 6 and 7 shows that the S-parameters of dual and quad redundant RF TSV test structure when failure occurs. It can be found that it has a higher resonance frequency and smaller insertion loss as the number of failed TSV increases. This result is basically consistent with the above analysis. Furthermore, it can achieve reliable function at 0–44 GHz regardless of what failure occurs.

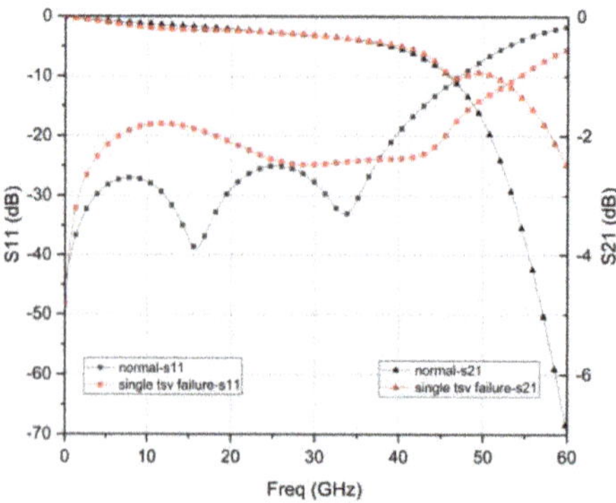

Figure 6. Dual redundant TSV failure simulation results.

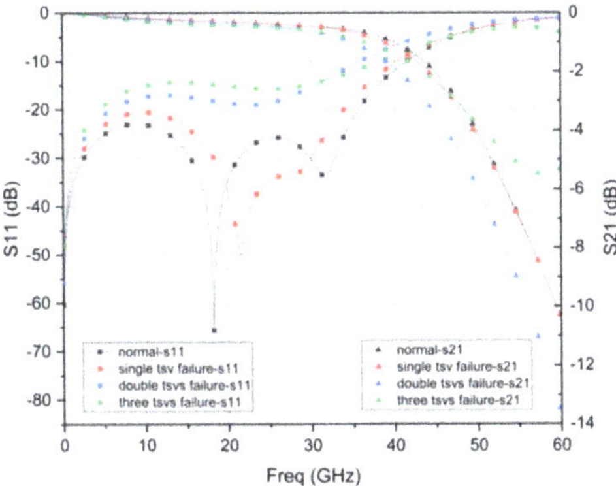

Figure 7. Quad redundant TSV failure simulation results.

3. Fabrication

Figure 8 shows the main steps for fabricating the redundant RF TSV sample. They include the following: (a) First, a 300 μm high-resistance silicon wafer is cleaned in acetone and isopropanol. (b) The fabrication of the large backside TSV is completed by photolithography and deep reactive ion etching. (c) The fabrication of the small front side TSV is completed by back engraving and deep reactive ion etching. (d) After standard cleaning, a high-temperature thermal oxygen process is used to form a dense 100 nm SiO2 insulating layer on the surface of the high resistance silicon wafer and the sidewall of the TSV. (e) The double-sided sputtered adhesion layer Ti and seed layer Cu is fabricated. (f) Double-sided lithography and thickening of the surface local copper layer and TSV hole copper layer is conducted by electroplating in a copper sulfate solution. (g) Copper plating area mask protection is carried out. Removing the excess Cu seed layer and Ti adhesion layer by wet etching. (h) Electroless nickel-gold plating is con-ducted on the Cu layer. Figure 9 provides a physical diagram of the completed production. As shown in Figure 10, the metal filling in the TSV hole is good under X-ray detection. To extract the S-parameters of the TSV interconnect structure, a transmission line is also manufactured at the same wafer.

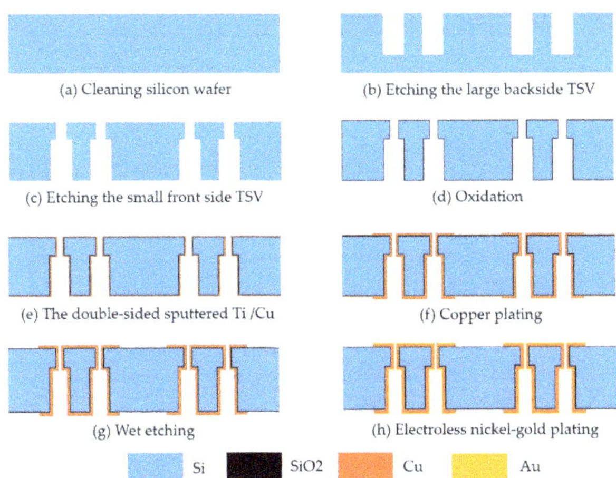

Figure 8. Process design.

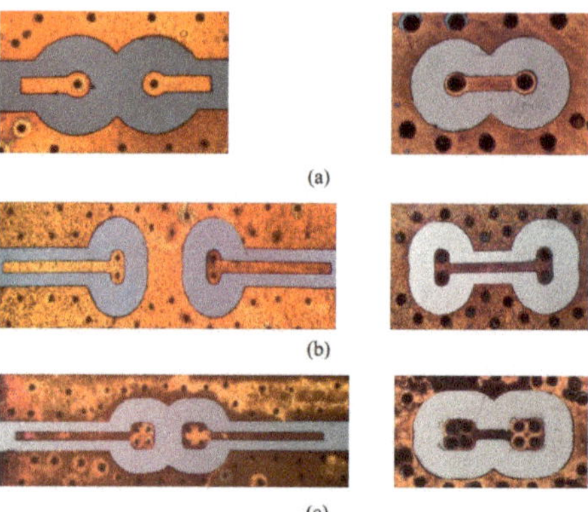

Figure 9. Test structure physical photo: (**a**) single TSV; (**b**) dual redundant TSV; (**c**) quad redundant TSV.

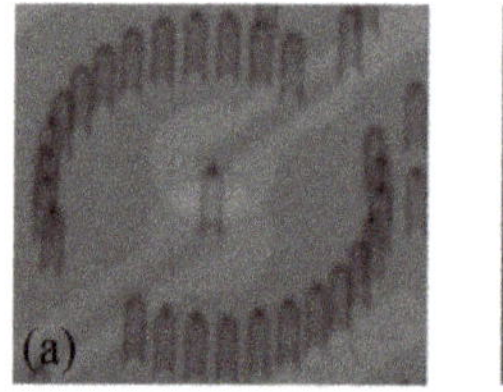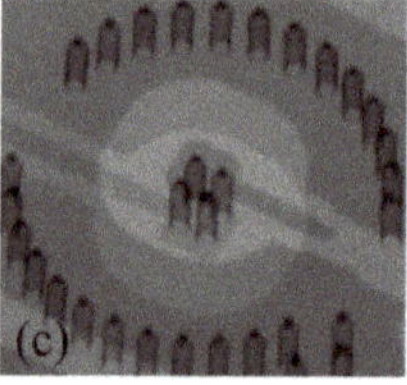

Figure 10. Test structure X-ray detection map: (**a**) single TSV; (**b**) dual redundant TSV; (**c**) quad redundant TSV.

4. RF Test and Analysis

The redundant RF TSV samples were tested using a GSG probe in a semi-automatic probe station, which was connected with an AV3629 high-performance microwave integrated vector network analyzer. Before the test, the measurement system was firstly calibrated using the classic SOLT calibration method, including short circuit, open circuit, load, and straight through four standard structures, to correct the system error, stripping probe and cable parasitic parameters [29]. The measured insertion losses at 40 GHz for a single TSV interconnect test structure, dual redundant TSV interconnect test structure, and quad redundant TSV interconnect test structure are 0.721 dB, 1.18 dB, and 1.635 dB, respectively.

Compared with the simulation results in Section 2, the maximum deviations of the insertion loss of the simulation and testing of a single TSV, dual redundant TSV and four redundant TSV test structures are 0.53 dB, 0.84 dB, and 0.95 dB in the range of 0–40 GHz, which may be caused by the use of the ideal Cu layer in the simulation. However, the fabricated Cu layer has some differences with ideal Cu layer, such as surface roughness and resistivity. Figure 11 is a captured photo of the Cu layer by a profiler during the process, which shows the roughness is approximately 60–70 nm and some local regions reach about 150 nm due to oxidation. Table 3 summarizes the tested resistivity, which has an average resistivity of 12.79 $\mu\Omega\cdot$cm. Because surface roughness of the copper generates parasitic inductance, the surface impedance will change and results in conductor loss [30]. Especially when the skin depth corresponding to the operating frequency is less than or equal to the surface roughness, the effect of surface roughness will become very significant [31,32]. Additionally, the resistivity of the conductor also affects the conductor loss. To testify this point, using the monitored data, the simulation is optimized and repeated in HFSS model, and the results are compared with the test results shown in Figure 12. It can be seen that the deviation is relatively reduced, and the higher the frequency, the better the fit. This proves that the roughness and resistivity of the conductor have an impact on high-frequency performance. It can also be seen in the figure that the gap between the measured and simulated results of the four-redundancy is significantly larger compared to a single TSV. Since the resistivity and roughness of the conductor in the hole cannot be measured, the parameters of the conductor on the plane can only be used instead. As the number of RF TSV holes increases, the error accumulation is greater.

Table 3. Resistivity test results.

Points	Resistivity/$\mu\Omega\cdot$cm
1	4.56
2	12.44
3	13.95
4	12.43
5	18.24
6	12.36
Average value	12.79

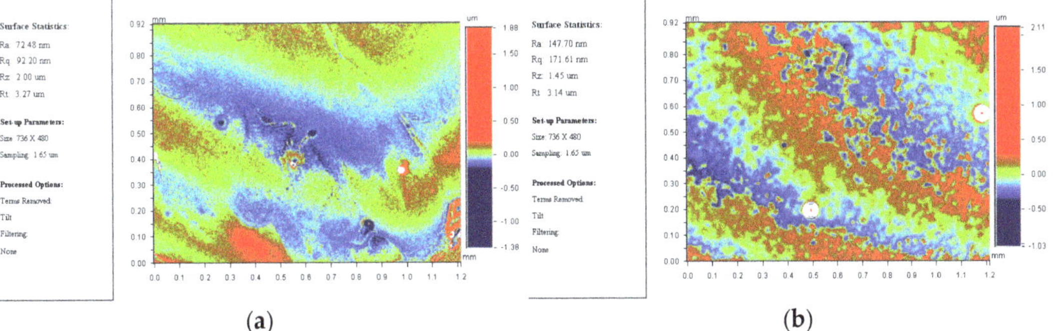

(a)   **(b)**

Figure 11. Surface roughness test results: (**a**) small roughness area; (**b**)large roughness area.

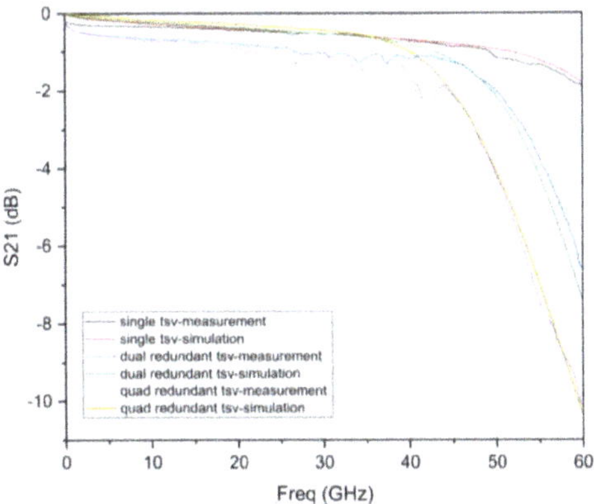

Figure 12. Test structure S_{21} parameter measurement results and optimized simulation results of HFSS.

5. S-Parameter Extraction

To obtain the precise value of insertion loss contributed by RF redundant TSVs, de-embedding was conducted. According to the relevant theory of microwave network parameters, conversion into ABCD parameters with cascade characteristics for RF redundant TSV sample structures was carried out via parameter transformation and matrix operation [33]. The 1000 μm CPW test structure is viewed as four 250 μm CPW connections. A single RF TSV interconnect and redundant RF TSV interconnect test structure is viewed as three CPW and two TSV interconnect structures, as shown in Figure 13. To simplify the description, J1, J2, J3, and J4 are used to represent the CPW, single TSV interconnect, dual redundant TSV interconnect, and quad redundant TSV interconnect test structure. L1 represents 250 μm CPW and S-TSV represents a single TSV mutual connected structure, D-TSV represents a dual redundant TSV interconnect structure, and Q-TSV represents a quad redundant TSV interconnect structure.

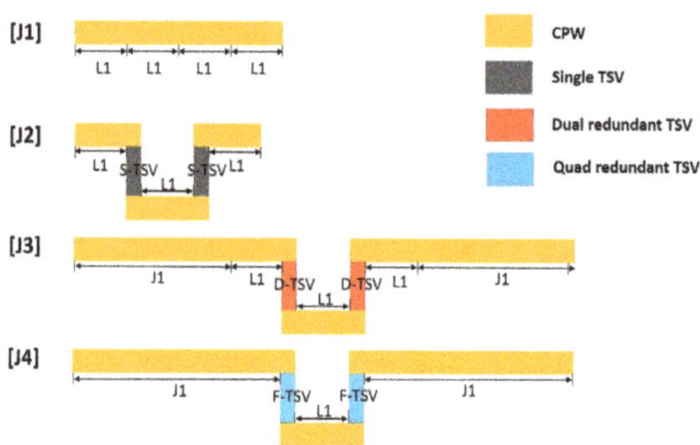

Figure 13. Schematic diagram of test structure de-embedding.

The ABCD parameters corresponding to the four unit structures are represented by square brackets "[]" and the tag name. The ABCD parameters of the unit structure are multiplied by the unit structure to represent the ABCD parameters of the three test structures J1, J2, J3, and J4 as

$$[J1] = [L1][L1][L1][L1] \tag{18}$$

$$[J2] = [L1][S - TSV][L1][S - TSV][L1] \tag{19}$$

$$[J3] = [J1][L1][D - TSV][L1][D - TSV][L1][J1] \tag{20}$$

$$[J4] = [J1][Q - TSV][L1][Q - TSV][J1] \tag{21}$$

The number of frequency points of the high-frequency measurement is marked as N, and the size of the ABCD parameters matrix of all the above test structures and unit structures is $2 \times 2 \times N$. In the calculation, an N-step loop is set to perform a 2×2 matrix operation. The de-embedding process solves a single TSV interconnect S-TSV, a dual redundant TSV interconnect D-TSV, and a quad-redundant TSV interconnect Q-TSV using the above four matrix equations. Through the operations of square root and inversion matrix, the ABCD parameters matrix of each unit structure is obtained as

$$[L1] = [J1]^{\frac{1}{4}} \tag{22}$$

$$[S - TSV] = [L1]^{-1}\left([J2][L1]^{-1}\right)^{\frac{1}{2}} \tag{23}$$

$$[D - TSV] = [L1]^{-1}\left([J1]^{-1}[J3][J1]^{-1}[L1]^{-1}\right)^{\frac{1}{2}} \tag{24}$$

$$[L2] = [L1][L1][L1] \tag{25}$$

$$[Q - TSV] = [L1]^{-1}\left([L2]^{-1}[J4][J1]^{-1}\right)^{\frac{1}{2}} \tag{26}$$

where [L2] is an intermediate variable for simplifying expressions. According to the transformation relationship between ABCD parameters and S-parameters, the ABCD parameters of each unit structure is converted into S-parameters, and the transmission characteristics of the three TSV interconnections can be compared. Microwave network parameter conversion and other operations in the de-embedding process are performed in MATLAB software(MATLAB, R2020a, MathWorks, Natick, Mass, USA).

Using the above de-embedding process, the S-parameters of the two kinds of RF TSV design are obtained, as shown in Figure 14. It can be observed that the S_{21} values of the three TSV interconnects are close when the frequency is less than 40 GHz, and the gap is in the range of 0.25 dB. When the frequency is greater than 40 GHz, the S_{21} value of the dual redundant and quad-redundant TSV interconnects degrade rapidly versus frequency, and the insertion loss of the quad-redundant TSV interconnect significantly increases. The test results of the TSV interconnection and the simulation results of the equivalent lumped components models are compared. The following points can be seen from Figure 14: (a) The simulated and measured insertion loss values gradually degrade as frequency increases. (b) Both simulation and measurement results show a similar trend that the insertion loss value degrades as the number of RF Redundant TSV increase. (c) The S-parameter curve of the TSV interconnection extracted from the RF measurement results shows some fluctuation, bringing out maximum deviation values of 0.07 dB, 0.17 dB, and 0.12 dB in the 0–40 GHz range between the simulation and measured results for a single TSV, dual redundant TSV, and four redundant TSV respectively. These fluctuations should be ascribed to the surface roughness or random discontinuities in the deposited Cur layer on the sidewall of TSV, of which, the later one is especially hard to discern or characterize to our knowledge. This is the reason why the current simulation based equivalent lumped component circuit model only considers the change of conductor resistivity as well. However, even with the deviations due to the fluctuations, those founding is sufficient to draw a conclusion that it has a competing RF property with single RF TSV for Redundant RF design, which is the most important to this research and highlighted in the Table 4 as well.

At present, there is no report of RF redundant TSV interconnection sample measurement result, although few papers proposed RF redundant RF TSV design. Table 4 compares redundant RF TSVs proposed in this paper with single TSV products presented in recent years. It can be seen from the table that when the frequency is less than 40 GHz, the RF Redundant TSV interconnection sample fabricated in this work is in the same level with traditional single RF TSV in term of RF insertion loss, but it has a better capacity to resist failure risk of RF TSVs.

Table 4. Comparison RF TSV for high frequency applications.

Ref.	Substrate Material	Type of Vias	Transmission Loss of One Transition (dB)		Via Size (μm)	Via Length (μm)
			10 GHz	**40 GHz**		
[34]	Glass	Single TGV	0.03	0.22	Φ55	366
[35]	LCP	Single Via	0.071	0.12	Φ55	51
[36]	Si(HR)	Single TSV	0.05	—	Φ100	300
[37]	Si(HR)	Single TSV	0.04	—	Φ8 & Φ90	25 & 280
[38]	Si(HR)	Single TSV	1.6	—	Φ40	120
[39]	Si(HR)	Single TSV	0.37	—	Φ10	100
This Work	Si(HR)	Single TSV	0.11	0.22	Φ40 & Φ80	50 & 250
		Dual redundant TSV	0.14	0.19		
		Quad redundant TSV	0.2	0.46		

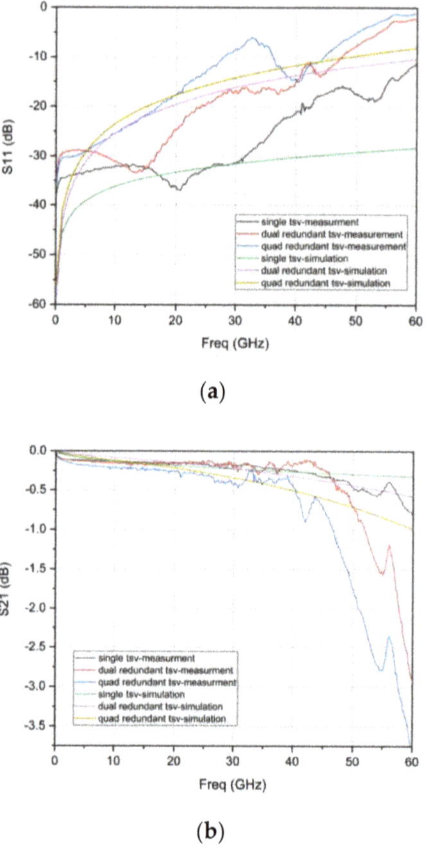

(a)

(b)

Figure 14. TSV interconnects S_{21} parameter measurement results and simulation results of the equivalent lumped components models: (a) S_{11}; (b) S_{21}.

6. Conclusions

This paper presented a RF redundant TSV design for a high resistance Si interposer as a package substrate. To verify the feasibility of the scheme, two test structures for connecting CPW transmission lines through redundant TSVs were designed to be able to work in 0–40 GHz. Modeling and simulation were carried out using HFSS, conclusion can be drawn from the obtained S-parameters that the more the number of RF redundant TSV, the lower the resonance frequency, and the greater the insertion loss. This was also solidified with the results obtained with the established modified equivalent circuit model for RF redundant TSV interconnection. The simulation also shows that the designed RF redundant TSV interconnection is capable to work in the range of 0–40 GHz without unacceptable RF property degradation when failure occurs. RF redundant TSV test vehicles were fabricated and tested, while an improved simulation factoring in nonideal factors such as surface roughness and resistivity is also taken. The result of the test shows an agreement with simulation. The tested insertion loss of the single TSV, dual redundant TSV and quad redundant TSV after de-embedding is 0.22, 0.19, and 0.46 dB at 40 GHz, respectively, which is close to the reported single TSV design. However, redundant TSV offers a better capacity to resist failure risk of TSV.

Author Contributions: Conceptualization, S.M. and Y.J.; Methodology, S.M.; Software, S.H. and M.W.; Validation, L.H., S.M., and Y.J.; Formal Analysis, S.M.; Investigation, L.H. and S.M.; Resources, J.C. and W.W.; Data Curation, S.H. and M.W.; Writing—original draft preparation, M.W.; Writing–Review & Editing, S.M.; Visualization, M.W.; Supervision, S.M.; Project Administration, L.H., S.M., and Y.J.; Funding Acquisition, L.H., S.M., and Y.J. All authors have read and agreed to the published version of the manuscript.

Funding: This research was funded by the National Natural Science Foundation of China, no. U1613215. This research was also funded by TSV 3D Integrated Micro/Nano system lab, ZDSYS20180-2061805105.

Institutional Review Board Statement: Not applicable.

Informed Consent Statement: Not applicable.

Data Availability Statement: The data presented in this study are available on request from the corresponding author. The data are not publicly available due to the experimental data needs to be further researched in the future.

Conflicts of Interest: The authors declare no conflict of interest.

References

1. Talebbeydokhti, P.; Dalmia, S.; Thai, T.; Sover, R.; Tal, S. Ultra large area SIPs and integrated mmW antenna array module for 5G mmWave outdoor applications. In Proceedings of the 2019 IEEE 69th Electronic Components and Technology Conference (ECTC), Las Vegas, NV, USA, 29–30 May 2019; pp. 294–299. [CrossRef]
2. Hu, S.; Wang, L.; Xiong, Y.-Z.; Lim, T.G.; Zhang, B.; Shi, J.; Yuan, X. TSV Technology for millimeter-wave and terahertz design and applications. *IEEE Trans. Compon. Packag. Manuf. Technol.* **2010**, *1*, 260–267. [CrossRef]
3. Yan, J.; Ma, S.; Jin, Y.; Wang, W.; Chen, J.; Luo, R.; Cai, H.; Li, J.; Xia, Y.; Hu, L.; et al. Fabrication and RF property evaluation of high-resistivity Si interposer for 2.5-D/3-D heterogeneous integration of RF devices. *IEEE Trans. Compon. Packag. Manuf. Technol.* **2018**, *8*, 2012–2020. [CrossRef]
4. El Bouayadi, O.; Dussopt, L.; Lamy, Y.; Dehos, C.; Ferrandon, C.; Siligaris, A.; Soulier, B.; Simon, G.; Vincent, P. Silicon interposer: A versatile platform towards full-3D integration of wireless systems at millimeter-wave frequencies. In Proceedings of the 2019 IEEE 69th Electronic Components and Technology Conference (ECTC), Las Vegas, NV, USA, 29–30 May 2019; pp. 973–980.
5. Lamy, Y.; Dussopt, L.; El Bouayadi, O.; Ferrandon, C.; Siligaris, A.; Dehos, C.; Vincent, P. A compact 3D silicon interposer package with integrated antenna for 60GHz wireless applications. In Proceedings of the 2013 IEEE International 3D Systems Integration Conference (3DIC), Francisco, CA, USA, 2–4 October 2013.
6. Thadesar, P.A.; Bakir, M.S. Fabrication and characterization of polymer-enhanced TSVs, inductors, and antennas for mixed-signal silicon interposer platforms. *IEEE Trans. Compon. Packag. Manuf. Technol.* **2016**, *6*, 455–463. [CrossRef]
7. Sekhar, V.N.; Toh, J.S.; Cheng, J.; Sharma, J.; Fernando, S.; Bangtao, C.; Fernando, S. Wafer level packaging of RF MEMS devices using TSV interposer technology. In Proceedings of the 2012 IEEE 14th Electronics Packaging Technology Conference (EPTC), Singapore, 5–7 December 2012.
8. Tenailleau, J.-R.; Brunet, A.; Borel, S.; Voiron, F.; Bunel, C. TSV development, characterization and modeling for 2.5-D interposer applications. In Proceedings of the 2013 IEEE 63rd Electronic Components and Technology Conference, Las Vegas, NV, USA, 28–31 May 2013.
9. Xu, Z.; Lu, J.Q. Three-dimensional coaxial through-silicon-via (TSV) design. *IEEE Electron Device Lett.* **2012**, *10*, 1441–1443. [CrossRef]
10. Yu, P.; Lin, H.; He, Z.; Song, C.; Cai, J.; Wang, Q.; Wang, Z. Coaxial through-silicon-vias using low-κ SiO_2 insulator. In Proceedings of the 2020 IEEE 70th Electronic Components and Technology Conference (ECTC), Orlando, FL, USA, 3–30 June 2020.
11. Hwang, G.; Kalaiselvan, R.; Sam, M.I.E.; Hsiang-Yao, H. Study on through Silicon Via (TSV) filling failures on various electroplating conditions. In Proceedings of the 2019 IEEE 21st Electronics Packaging Technology Conference (EPTC), Singapore, 4–6 December 2019.
12. Hollman, R.; Dimov, O.; Malik, S.; Hichri, H.; Arendt, M. Ultra fine RDL structure fabrication using alternative patterning and bottom-up plating processes. In Proceedings of the 2018 IEEE 68th Electronic Components and Technology Conference (ECTC), San Diego, CA, USA, 29 May–1 June 2018.
13. Murugesan, M.; Fukushima, T.; Mori, K.; Nakamura, A.; Lee, Y.; Motoyoshi, M.; Bea, J.; Watariguchi, S.; Koyanagi, M. Fully-filled, highly-reliable fine-pitch interposers with TSV aspect ratio >10 for future 3D-LSI/IC packaging. In Proceedings of the 2019 IEEE 69th Electronic Components and Technology Conference (ECTC), Las Vegas, NV, USA, 25 May–2 June 2019.
14. Bleiker, S.J.; Fischer, A.C.; Shah, U.; Somjit, N.; Haraldsson, T.; Roxhed, N.; Oberhammer, J.; Stemme, G.; Niklaus, F. High-aspect-ratio through silicon vias for high-frequency application fabricated by magnetic assembly of gold-coatednickel wires. *IEEE Trans. Compon. Packag. Manuf. Technol.* **2015**, *5*, 21–27. [CrossRef]

15. Liu, X.; Chen, Q.; Dixit, P.; Chatterjee, R.; Tummala, R.R.; Sitaraman, S.K. Failure mechanisms and optimum design for electroplated copper Through-Silicon Vias (TSV). In Proceedings of the 2009 59th Electronic Components and Technology Conference, San Diego, CA, USA, 26–29 May 2009.
16. Shen, J.; Chen, P.; Su, L.; Shi, T.; Tang, Z.; Liao, G. X-ray inspection of TSV defects with self-organizing map network and Otsu algorithm. *Microelectron. Reliab.* **2016**, *67*, 129–134. [CrossRef]
17. Kang, U.; Chung, H.J.; Heo, S.; Park, D.H.; Lee, H.; Kim, J.H.; Ahn, S.H.; Cha, S.H.; Ahn, J.; Kwon, D.; et al. 8Gb 3D DDR3 DRAM using through-silicon-via technology. *IEEE J. Solid State Circuits* **2009**, *45*, 130–131.
18. Hsieh, A.-C.; Hwang, T. TSV Redundancy: Architecture and design issues in 3-D IC. *IEEE Trans. Very Large Scale Integr. Syst.* **2011**, *20*, 711–722. [CrossRef]
19. Jiang, L.; Xu, Q.; Eklow, B. On effective TSV repair for 3D-stacked ICs. In Proceedings of the Design, Automation and Test in Europe Conference and Exhibition (DATE), Dresden, Germany, 12–16 March 2012.
20. Ma, S.; Chai, Y.; Yan, J.; Cai, H.; Hu, L.; He, S.; Wang, W.; Chen, J.; Jin, Y. A 2.5D integrated L band receiver based on high resistivity Si interposer. In Proceedings of the 2018 IEEE International Conference on Integrated Circuits, Technologies and Applications (ICTA), Beijing, China, 21–23 November 2018.
21. Rahimi, A.; Somarajan, P.; Yu, Q. Modeling and characterization of through-silicon-vias (TSVs) in radio frequency regime in an active interposer technology. In Proceedings of the 2020 IEEE 70th Electronic Components and Technology Conference (ECTC), Lake Buena Vista, FL, USA, 26–29 May 2020.
22. Lu, K.-C.; Horng, T.-S. Wideband and scalable equivalent-circuit model for differential through silicon vias with measurement verification. In Proceedings of the 2013 IEEE 63rd Electronic Components and Technology Conference, Las Vegas, NV, USA, 28–31 May 2013.
23. Jung, D.H.; Kim, Y.; Kim, J.J.; Kim, H.; Choi, S.; Song, Y.-H.; Bae, H.-C.; Choi, K.-S.; Piersanti, S.; De Paulis, F.; et al. Through silicon via (TSV) defect modeling, measurement, and analysis. *IEEE Trans. Compon. Packag. Manuf. Technol.* **2016**, *7*, 138–152. [CrossRef]
24. Leferink, F.B.J. Inductance calculations; methods and equations. In Proceedings of the International Symposium on Electromagnetic Compatibility, Atlanta, GA, USA, 14–18 August 1995; IEEE ISEMC. 1995; pp. 16–22.
25. Kim, J.; Pak, J.S.; Cho, J.; Song, E.; Cho, J.; Kim, H.; Song, T.; Lee, J.; Lee, H.; Park, K.; et al. High-frequency scalable electrical model and analysis of a through silicon Via (TSV). *IEEE Trans. Compon. Packag. Manuf. Technol.* **2011**, *1*, 181–195.
26. Hassan, M.D.; Farque, M.R.I. Left-handed metamaterial using Z-shaped SRR for multiband application by azimuthal angular rotationsm. *Mater. Res. Express.* **2017**, *4*, 045801. [CrossRef]
27. Hassan, M.D.; Faruque, M.R.I.; Islam, S.S.; Islam, M.T. New compact double-negative miniaturized metamaterial for wideband operation. *Materials* **2016**, *9*, 830. [CrossRef] [PubMed]
28. Hasan, M.; Faruque, M.R.I.; Islam, M.T. Compact left-handed meta-atom for S-, C- and Ku-band application. *Appl. Sci.* **2017**, *7*, 1071. [CrossRef]
29. Ye, R.; Xu, J. SOLT calibration method and its application to radio-frequency measurement. *Chin. J. Electron Devices* **2006**, *29*, 179.
30. Chen, C.D.; Tzuang, C.K.; Peng, S.T. Full-wave analysis of a lossy rectangular waveguide containing rough inner surfaces. *IEEE Microw. Guided Wave Lett.* **1992**, *2*, 180–181. [CrossRef]
31. Palasantzas, G. Influence of self-affine and mound roughness on the surface impedance and skin depth of conductive materials. *J. Phys. Chem. Solids* **2004**, *65*, 1271–1275. [CrossRef]
32. Scogna, A.C.; Schauer, M. Performance analysis of strip line surface roughness models. In Proceedings of the International Symposium on Electromagnetic Compatibility, Detroit, MI, USA, 18–22 August 2008.
33. Noyan, K.; Aksun, M.I. *Modern Microwave Circuits*; Artech House: Boston, MA, USA, 2008; pp. 58–68.
34. Khan, W.T.; Tong, J.; Sitaraman, S.; Sundaram, V.; Tummala, R.; Papapolymerou, J. Characterization of electrical properties of glass and transmission lines on thin glass up to 50 GHz. In Proceedings of the 2015 IEEE 65th Electronic Components and Technology Conference (ECTC), San Diego, CA, USA, 26–29 May 2015.
35. Chung, D.J.; Bhattacharya, S.K.; Papapolymerou, J. Low loss multilayer transitions using via technology on LCP from DC to 40 GHz. In Proceedings of the 2009 59th Electronic Components and Technology Conference, San Diego, CA, USA, 26–29 May 2009.
36. Chen, B.; Sekhar, V.N.; Jin, C.; Lim, Y.Y.; Toh, J.S.; Fernando, S.; Sharma, J. Low-loss broadband package platform with surface passivation and TSV for wafer-level packaging of RF-MEMS devices. *IEEE Trans. Compon. Packag. Manuf. Technol.* **2013**, *3*, 1443–1452. [CrossRef]
37. Ebefors, T.; Fredlund, J.; Perttu, D.; Van Dijk, R.; Cifola, L.; Kaunisto, M.; Rantakari, P.; Vaha-Heikkila, T. The development and evaluation of RF TSV for 3D IPD applications. In Proceedings of the 2013 IEEE International 3D Systems Integration Conference (3DIC), Francisco, CA, USA, 2–4 October 2013.
38. Lorival, J.E.; Calmon, F.; Sun, F.; Frantz, F.; Plossu, C.; Le Berre, M.; O'Connor, I.; Valorge, O.; Charbonnier, J.; Henry, D.; et al. An efficient and simple compact modeking approach for 3-D interconnects with IC's stack global electrical context consideration. *Microelectron. J.* **2015**, *46*, 153–165. [CrossRef]
39. Kim, N.; Wu, D.; Kim, D.; Rahman, A.; Wu, P. Interposer design optimization for high frequency signal transmission in passive and active interposer using through silicon via (TSV). In Proceedings of the 2011 IEEE 61st Electronic Components and Technology Conference (ECTC), Lake Buena Vista, FL, USA, 31 May–3 June 2011.

 *micromachines*

Article

Analysis of Frequency Drift of Silicon MEMS Resonator with Temperature

Bo Jiang, Shenhu Huang, Jing Zhang and Yan Su *

School of Mechanical Engineering, Nanjing University of Science and Technology, Nanjing 210094, China; bjiang@njust.edu.cn (B.J.); huge@njust.edu.cn (S.H.); zhangjing@njust.edu.cn (J.Z.)
* Correspondence: suyan@njust.edu.cn

Abstract: High-quality-factor Micro-Electro-Mechanical System (MEMS) resonators have been widely used in sensors and actuators to obtain great mechanical sensitivity. The frequency drift of resonator with temperature is a problem encountered practically. The paper focuses on the resonator frequency distribution law in the temperature range of—40 to 60 °C. The four-layer models were established to analyze thermal stress caused by temperature due to the mismatch of thermal expansion coefficients. The temperature variation leads to the transformation of stress, which leads to the shift of resonance frequency. The paper analyzes the influence of hard and soft adhesive package on the temperature coefficient of frequency. The resonant accelerometer was employed for the frequency measurements in the paper. In experiments, three types of adhesive dispensing patterns were implemented. The results are consistent with the simulation well. The optimal packaging method achieves -24.1 ppm/°C to -30.2 ppm/°C temperature coefficient of the resonator in the whole temperature range, close to the intrinsic property of silicon (-31 ppm).

Keywords: MEMS resonator; temperature coefficient; thermal stress

1. Introduction

Silicon capacitive resonator, as a primary motion unit, has been widely used in Micro-Electro-Mechanical System (MEMS) accelerometers [1,2], gyroscopes [3,4], pressure sensors [5], and microphones [6]. The capacitive resonator is advantageous to realize displacement amplification under the same driving force, which is suitable for applying MEMS sensors. Silicon material, with high Young's modulus in the micron scale, is an ideal elastic material. However, the silicon resonator frequency changes with temperature due to the material softening of silicon [7,8]. The temperature coefficient of frequency characterizes the thermal frequency stability of resonators. Different ambient temperature leads to different temperature drift coefficient, affecting the device performance. Many works have been done to improve the temperature performance of the resonator [9,10]. An electronically temperature compensated oscillator based on capacitive silicon micromechanical resonators was implemented to overcome the temperature issues [11]. The oscillator exhibits a frequency drift of 39 ppm over 100 °C as compared to uncompensated frequency drift of 2830 ppm over the same range. Some other works focus on presenting novel structures enlarging the tuning frequency range for temperature drift compensation [12–14]. Among them, compensation with different materials or additional static-electrical stiffness was mentioned. The composite resonator with a linear temperature coefficient of frequency was fabricated utilizing silicon and silicon dioxide's opposing temperature coefficients of Young's modulus [15]. The temperature-dependent static-electrical forces were applied to reduce the temperature coefficients of polysilicon micromechanical resonators [16].

All these efforts significantly improve the temperature coefficient of the silicon resonators. However, the frequency and temperature characteristics of the resonator are also affected by the packaging method. Packaging processes bring different materials, leading to intrinsic stress with different temperatures. The Leadless Ceramic Chip (LCC) package

Citation: Jiang, B.; Huang, S.; Zhang, J.; Su, Y. Analysis of Frequency Drift of Silicon MEMS Resonator with Temperature. *Micromachines* **2021**, *12*, 26. https://doi.org/10.3390/mi12010026

Received: 8 December 2020
Accepted: 28 December 2020
Published: 29 December 2020

Publisher's Note: MDPI stays neutral with regard to jurisdictional claims in published maps and institutional affiliations.

chip is made of ceramic materials. The silicon dies, containing a structural layer and a substrate layer, is bonded to the shell by adhesive resistance, as illustrated in Figure 1. The adhesive is generally cured at high temperatures. Its thermal expansion coefficient (TEC) is quite different from that of silicon and ceramic materials [17]. The resonator's stress is produced and varied during cooling to operating temperature, ranging from $-40\,^\circ$C to $60\,^\circ$C. Internal stress changes the resonator's natural frequency, leading to the vibration frequency varies with temperature.

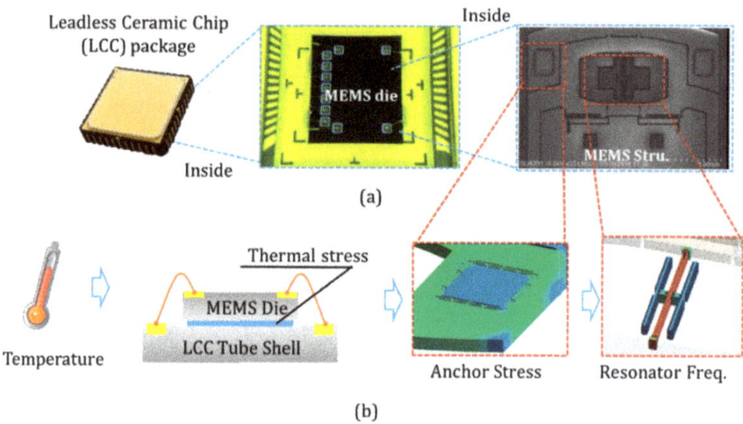

Figure 1. (**a**) Leadless Ceramic Chip (LCC) package with MEMS chip; (**b**) The mechanism of resonator frequency shift due to temperature.

The temperature coefficient of the resonator comes from two parts. The one is the material intrinsic frequency variations with the temperature. The other one is the change of supporting environment, which served as boundary conditions in the mechanical model. The variations of pre-stress lead to the frequency drift due to temperature. The thermal stress during the adhesion process is introduced to study the frequency drift law of the resonator. The four-layer model was established, consists of a ceramic shell, adhesive resist, silicon handle wafer, and silicon device layer. The resonator frequency variations with temperature under different support architectures were calculated and analyzed. The resonant beam accelerometer was used in the experiments to verify the theoretical model. Different adhesive resists were implemented in the packaging of the accelerometer chip and ceramic shell. We also changed the bonding processes to find the optimized packaging method with the minimum temperature coefficient. The results indicate that the optimal packaging method achieves the -24.1 ppm/$^\circ$C to -30.2 ppm/$^\circ$C temperature coefficient of the resonator, close to the intrinsic value of silicon (-31 ppm/K).

2. Materials and Methods

2.1. Resonator Frequency and Intrinsic Stress

The resonant accelerometer is a typical MEMS device that consists of a proof mass, two resonant beams with capacitive combs, flexural beams, a frame, and anchors [18,19]. The resonator has two modes with the movement in the same direction or the opposite direction. The mode I illustrated in Figure 2 is symmetric motion, where two resonant beams move in the same direction. Mode II is the antisymmetric mode, and the two vibrant beams exhibit opposite movements. The symmetric motion is the operating mode of the accelerometer. The capacitive combs provide electrostatic driving forces and displacement detection as well as suspension mass of the resonance beam. The proof mass generates displacement along the axial direction when external acceleration is input. The displacement is transferred through levers and comes into being compressive and tensile stress of resonate beams to change the frequency. So, the accelerometer has differential resonators that effectively resist common-mode interference. It is an ideal candidate for the temperature characteristics of the resonator.

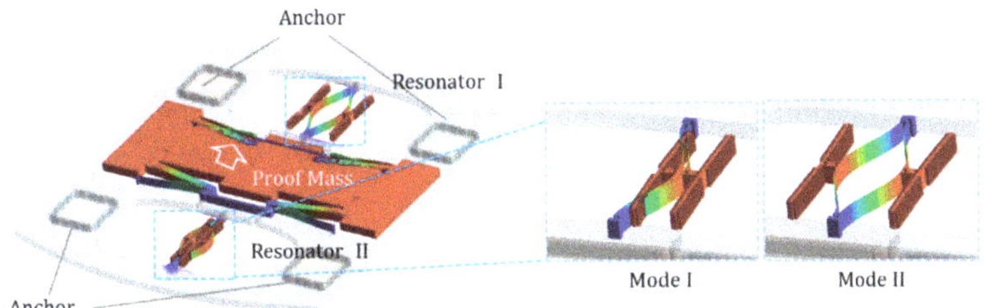

Figure 2. Operating principle of resonant accelerometer with two resonators.

Figure 3 indicates the relationship between pre-stress and resonate frequencies through the finite element analysis. The symmetric mode frequency is about 400 Hz lower than the antisymmetric modes, located in the first order. Resonator frequency varies with internal stress, which decreases under compressive stress and increases under tensile stress. Strains occur in the structure due to the difference in thermal expansion coefficients of materials, resulting in stress, which leads to the change of resonator frequency [10]. Thermomechanical stresses induced by the packaging assembly processes are complicated, and different packaging causes various temperature coefficients [20].

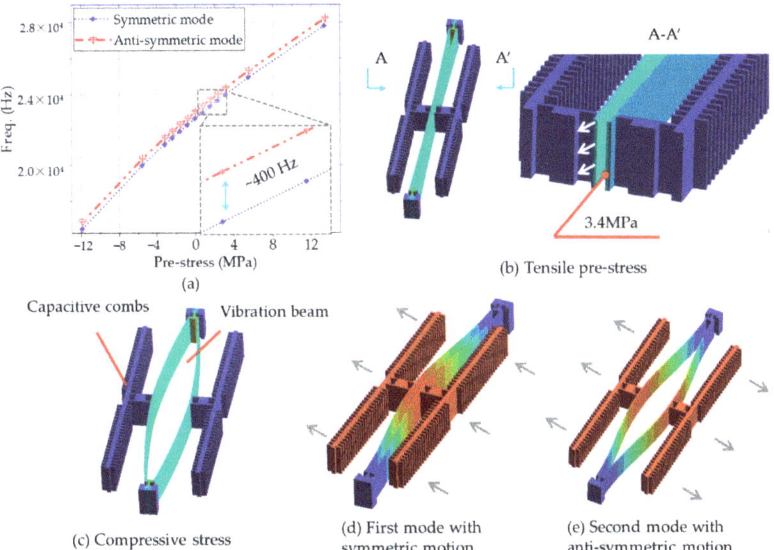

Figure 3. (**a**) The relationship between pre–stress and resonate frequencies; (**b**) Strain and stress distributions on the cross–section of the resonant beam under tensile stress; (**c**) Strain and stress under compressive stress; (**d**,**e**) Symmetric and antisymmetric motion of resonance beam.

2.2. The Model of the Resonator Temperature Characteristic

Four-layer structure models were established to address the resonator temperature characteristics due to thermal expansion coefficient mismatch [21]. There are ceramic shell, adhesive layer, silicon handle layer, and silicon device layer. As shown in Figure 4, the anchors are mounted in the axial direction of the resonant beam. The resonator produces a bending moment at operating temperature due to temperature variation.

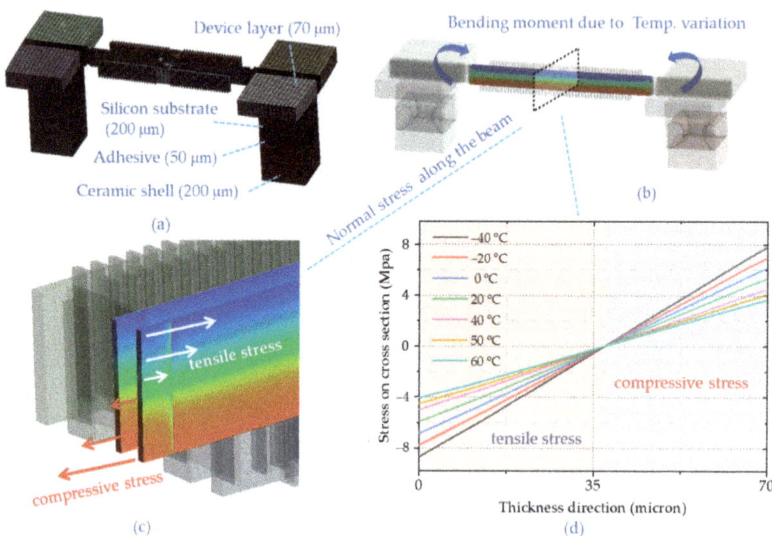

Figure 4. (**a**) The double-clamped resonator model for temperature characteristic; (**b**) Bending moment deformation caused by thermal stress; (**c**) Distribution of axial normal stress on cross-section profile; (**d**) The normal stress distribution along the longitudinal direction of the beam with different temperature.

The normal stress distributions on cross-section are according to the bending moment's characteristics. The tensile stress is above the neutral axis, and the compressive stress is below the neutral axis. The stress changes linearly along the longitudinal direction of the beam, from tensile stress to compressive stress. The bending moment decreases with the rising of temperature, as illustrated in Figure 4d. The axial normal stress of the resonance beam varies depending on the fastening architectures. The resonator is mounted by a frame, where anchors are located on both sides of the resonance beam (Figure 5). Bending moments are generated between two anchors due to the thermal stress. The normal stress distribution along the longitudinal direction of the beam is compressive stress with decreasing amplitude.

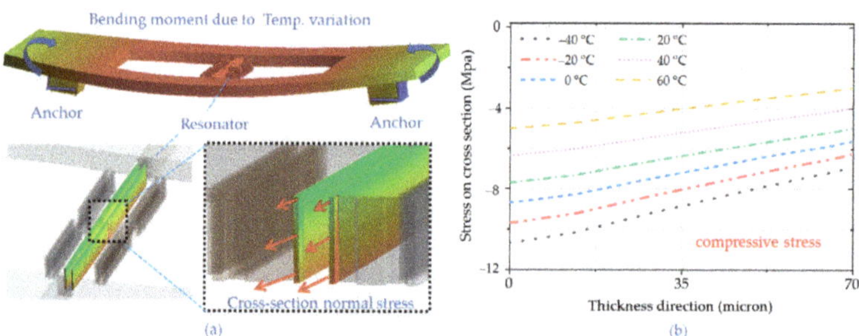

Figure 5. (**a**) Resonator model mounted by frame and the axial normal stress distributions; (**b**) The normal stress distribution along the longitudinal direction of the beam with different temperatures.

This paper selected two adhesive types to determine the influence of different adhesive properties on packaging stress and frequency characteristics. The hard adhesive is the material with a higher Young's modulus. As listed in Table 1, the ABP 84-3JT adhesive from LocTite was used in the experiments and calculations. The modulus is 2.95 GPa below25 °C and 0.589 GPa above 150 °C. The soft adhesive employed in the paper is a silicone adhesive with a lower Young's modulus, which is 5 MPa in solid-state. TEC among materials varies greatly, where TEC of hard adhesive is about 300 ppm/K linearly. The TEC of silicon is 2 ppm/K, nearly two orders of magnitude lower than adhesive.

Table 1. The material parameters applied in the calculation.

Material	Density (kg/m^3)	Young's Modulus (MPa)	Poisson's Ratio	Thermal Conductivity (W/m·K)	Coefficient of Thermal Expansion (ppm/K)
Silicon	2330	1.69×10^{11}	0.27	130	2
Ceramics	3970	3.1×10^{11}	0.28	14	7.1
Hard Adhesive	1500	$2.95 \times 10^9 / 5.89 \times 10^8$ [1]	0.3	0.6	63/117 [2]
Soft Adhesive	1100	5×10^6	0.3	0.8	300

[1] The Modulus at 25 °C is 2950 MPa and 589 MPa at 150 °C. [2] The data below glass transition temperature (56 °C)/above transition temperature.

The stress distribution analyzed above shows that the resonator temperature characteristic with different adhesive in two models is illustrated in Figure 6. The temperature coefficient is expressed as,

$$\alpha = \frac{f_{i+1} - f_i}{(T_{i+1} - T_i)f_i} \tag{1}$$

where f_i is the resonator frequency at the temperature T_i. In the double-clamped resonator model, the frequency decreases with the rising temperature for both hard or soft adhesive. The temperature coefficient utilizing hard adhesive is from -23 ppm to -174 ppm, which is larger than that with soft adhesive (about -40 ppm within the temperature range). The situation is different for the frame-mounted model. The temperature coefficient utilizing hard adhesive is positive, while that of soft adhesive is negative. The compressive stress is dominant and released with the increase of temperature, resulting in the rising frequency.

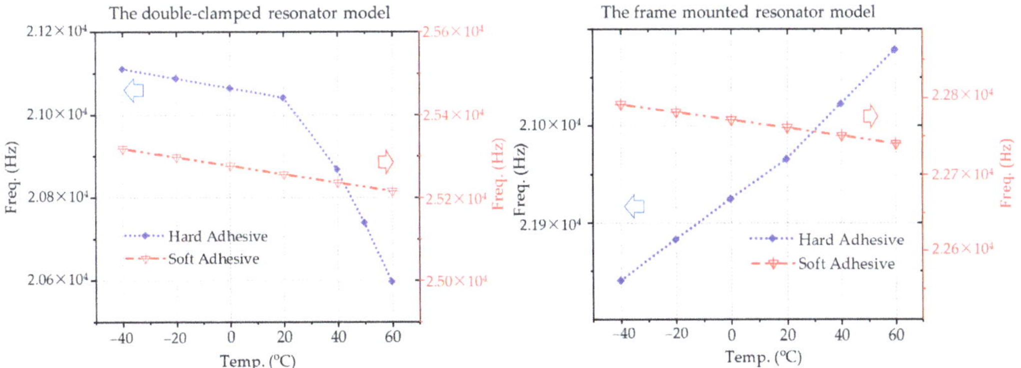

Figure 6. The resonator temperature characteristic with different adhesive in two models.

2.3. The Modeling and Experiments of the Resonant Accelerometer

The finite element model of the Resonant accelerometer was established to study the stress and temperature characteristics of the resonator. Considering that the structure is symmetrical, half of the accelerometer is modeled to reduce the calculation amount. The frame sags to the center, and the stress distribution of the resonance beam is asymmetrically bonded with hard adhesive at $-40\,°C$, as illustrated in Figure 7. The stress decreases from 26.5 kPa to 21 kPa from top to bottom in cross-section A, and the value is maintained at 24 kPa in cross-section B. Both of them are tensile stress. The stress inside the resonate beam determines the resonate frequency. The structure's frequency temperature characteristic is obtained by calculating the models of different temperature points from $-40\,°C$ to $60\,°C$.

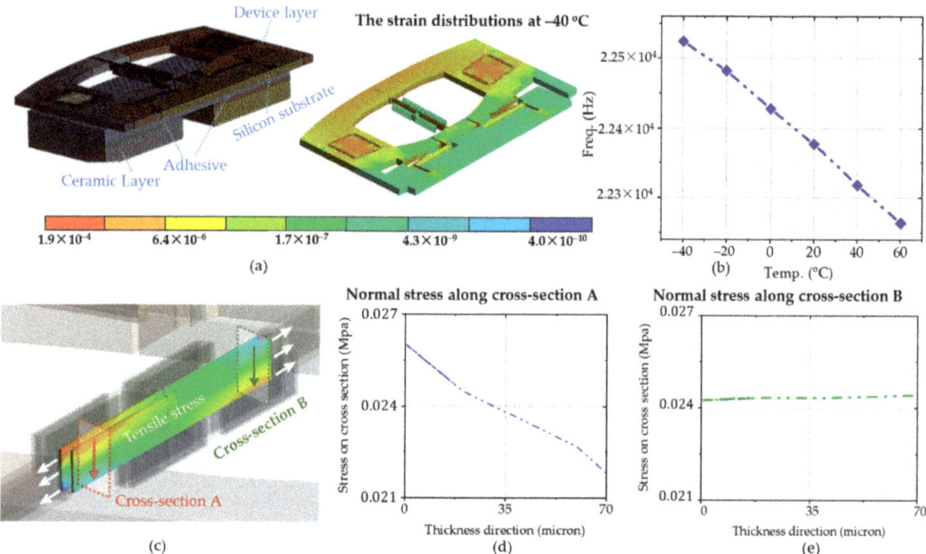

Figure 7. (**a**) The finite element model of the Resonant accelerometer and the train distributions under $-40\,°C$ with hard adhesive; (**b**) The relationship between temperature and symmetric motion resonate frequency; (**c**) The resonance beam exhibits tensile stress in operating temperature; (**d**,**e**) Normal stress along with cross-section A & B.

Three types of adhesive dispensing methods were taken place in the experiments to optimize the bonding method, as shown in Figure 8. The fastening center and eight radiation directions are confirmed by the double-cross pattern shown in the first dispensing type to achieve good shear strength. The second type fixes the four corners of the chip. A center dispensing point constructs the third pattern. After packaging with different types, the temperature experiments were carried out every $10\,°C$ in the range of $-40\,°C$ to $60\,°C$. Each temperature point was kept for 2 h to reach the equilibrium state, and then the resonate frequencies were tested.

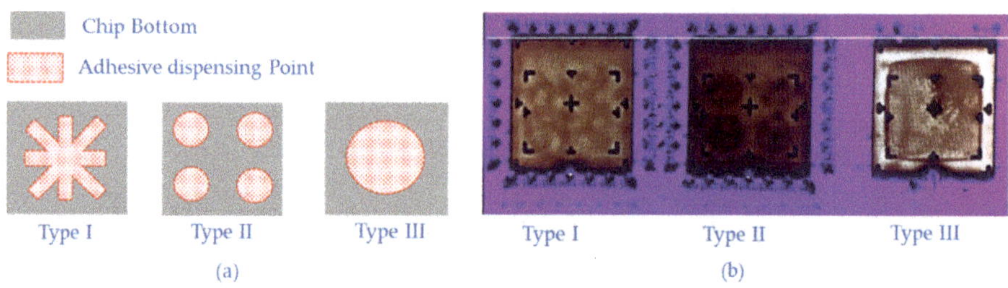

Figure 8. (**a**) Three types of adhesive dispensing method; (**b**) The photography of chips bottom with different adhesive dispensing ways.

3. Results

Temperature characteristic tests of resonant accelerometers with different bonding methods and adhesives were taken place every 10 °C as listed in Tables 2 and 3. The frequency decreases with the rising of the ambient temperature. The frequency of resonators with hard adhesive is higher than that with a soft one. The frequency temperature coefficient is obtained by the frequency difference ratio to temperature variation multiplied by frequency value. The type III fixed pattern samples' amplitude is slightly larger than that of Type I and II.

Table 2. The temperature characteristic test results of the resonator with soft adhesive.

Temp. (°C)	Freq. (Hz)			Freq. Diff. (Hz)			Freq. Temp. Coef. (ppm/°C)		
	Type I	Type II	Type III	Type I	Type II	Type III	Type I	Type II	Type III
−40	24,891.2	24,835.4	25,148.4	-	-	-	-	-	-
−30	24,885.2	24,829.3	25,141.5	−6.0	−6.1	−6.9	−24.1	−24.5	−27.6
−20	24,878.9	24,822.8	25,134.3	−6.2	−6.4	−7.2	−25.1	−25.9	−28.7
−10	24,872.4	24,816.0	25,126.7	−6.5	−6.9	−7.6	−26.3	−27.7	−30.3
0	24,865.7	24,809.0	25,118.7	−6.6	−6.9	−8.0	−26.7	−28.0	−31.7
10	24,859.0	24,801.9	25,110.5	−6.8	−7.2	−8.2	−27.2	−28.9	−32.7
20	24,851.8	24,794.5	25,101.9	−7.2	−7.4	−8.6	−28.8	−29.7	−34.2
30	24,844.5	24,787.0	25,093.1	−7.3	−7.5	−8.8	−29.4	−30.2	−35.1
40	24,837.1	24,779.4	25,084.1	−7.4	−7.7	−9.0	−29.9	−30.9	−35.8
50	24,829.6	24,771.6	25,075.3	−7.5	−7.7	−8.8	−30.2	−31.2	−35.3
60	24,822.2	24,763.9	25,066.8	−7.4	−7.7	−8.5	−29.8	−31.1	−33.8

Table 3. The temperature characteristic test results of the resonator with hard adhesive (ABP 84-3JT).

Temp. (°C)	Freq. (Hz)			Freq. Diff. (Hz)			Freq. Temp. Coef. (ppm/°C)		
	Type I	Type II	Type III	Type I	Type II	Type III	Type I	Type II	Type III
−40	25,523.0	25,584.3	25,169.6	−	−	−	−	−	−
−30	25,478.5	25,534.2	25,137.3	−44.5	−50.2	−32.3	−174.6	−196.5	−128.4
−20	25,438.3	25,482.0	25,105.9	−40.2	−52.2	−31.4	−158.2	−204.7	−125.2
−10	25,402.2	25,433.0	25,076.8	−36.1	−49.0	−29.1	−142.1	−192.8	−115.9
0	25,374.5	25,390.2	25,051.3	−27.7	−42.8	−25.6	−109.2	−168.5	−102.0
10	25,355.8	25,360.6	25,030.9	−18.7	−29.6	−20.3	−73.6	−116.8	−81.3
20	25,343.1	25,336.2	25,015.4	−12.7	−24.4	−15.5	−50.2	−96.2	−62.0
30	25,333.3	25,316.4	25,003.2	−9.8	−19.9	−12.2	−38.6	−78.4	−48.8
40	25,324.8	25,302.7	24,993.1	−8.5	−13.7	−10.1	−33.5	−54.1	−40.3
50	25,316.8	25,289.7	24,985.3	−8.0	−12.9	−7.9	−31.5	−51.1	−31.5
60	25,309.5	25,276.7	24,978.1	−7.3	−13.0	−7.1	−29.0	−51.6	−28.5

The frequency decreases with the increase of temperature for all samples, which means that the temperature coefficient is negative. Figure 9 indicates the comparison in various situations. The samples' temperature coefficient with soft adhesive, which has a smaller Young's modulus, is lower than the samples with hard adhesive in the whole temperature range. The temperature coefficients of the three packaging methods have little difference with soft adhesive bonding. The high-temperature data is slightly higher than that of low temperature, but the difference is within 6 ppm. The samples with Type I dispensing method achieve −24.1 ppm/°C to −30.2 ppm/°C, which is also the best one among three packaging ways. The performance of the samples with type II adhesive dispensing process is close to that of with Type I, where the temperature coefficient is from −24.5 ppm/°C to −31.2 ppm/°C. The samples with type III packaging process have a slightly higher temperature coefficient in −27.6 ppm/°C to −35.8 ppm/°C.

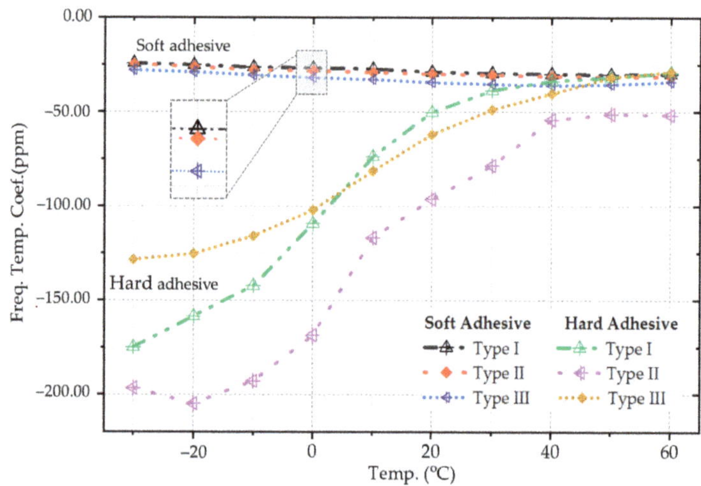

Figure 9. The frequency temperature coefficient in the test with hard & soft adhesive in three dispensing methods.

The situation is different with hard adhesive. The general law is that the temperature coefficient is more extensive at low temperatures and decreases gradually with the increase of temperature. The chips with four corners bonding are worst, which means sizeable normal stress occurs in the resonant beams. In this case, the chip is fully constrained, and the strain caused by the mismatch of TEC is transferred to the resonant beams, resulting in a larger frequency gradient. The central fixed chip achieves better performance. The temperature coefficient is about -128.4 ppm/°C at -30 °C, and in the high-temperature region, the value is -28.5 ppm/°C. The chips with a double-cross bonding pattern (Type I) has better performance above 10 °C. The resonator frequency arises faster with the reduction of temperature below 0 °C. Both packaging methods release thermal stress to decrease the temperature coefficient.

The comparison of experimental and simulation results was illustrated in Figure 10. The frequency at each temperature point is normalized due to silicon's Young's modulus between the simulation and the real value. The frequency variation of the resonator can be studied more clearly by the relative value. All of the data were implemented utilizing Type II dispensing pattern. During the packaging process with soft adhesive, SemiCosil 989(8)/1 K silicone adhesive was selected. The frequency varies by 0.275% in the temperature range of 100 °C, consistent with experimental results. In terms of hard adhesive packaging, ABP 84-3JT and EPO-H65 were used in the experiments. The frequency gradient at a lower temperature is more extensive than at a high temperature for ABP 84-3JT. The Young's Modulus of the adhesive (shown in Figure 10c) significantly affects the frequency variation trend. Compared with the experimental results, the simulation value exhibits better consistency below 0 °C. A slight nonlinearity was observed above 40 °C, but the inflection point appears at 0 °C in the experiment. The experimental results are in better agreement with the simulation value with EPO-H65, Young's modulus of which is 916,396 psi (equals 6.318 GPa), higher than that of ABP 84-3JT. In comparing two hard adhesives, the frequency variation is more extensive for EPO-H65. The temperature coefficient is about -200 ppm/°C in the whole temperature range. The samples with ABP 84-3JT exhibit a variable temperature coefficient, which is higher below 0 °C.

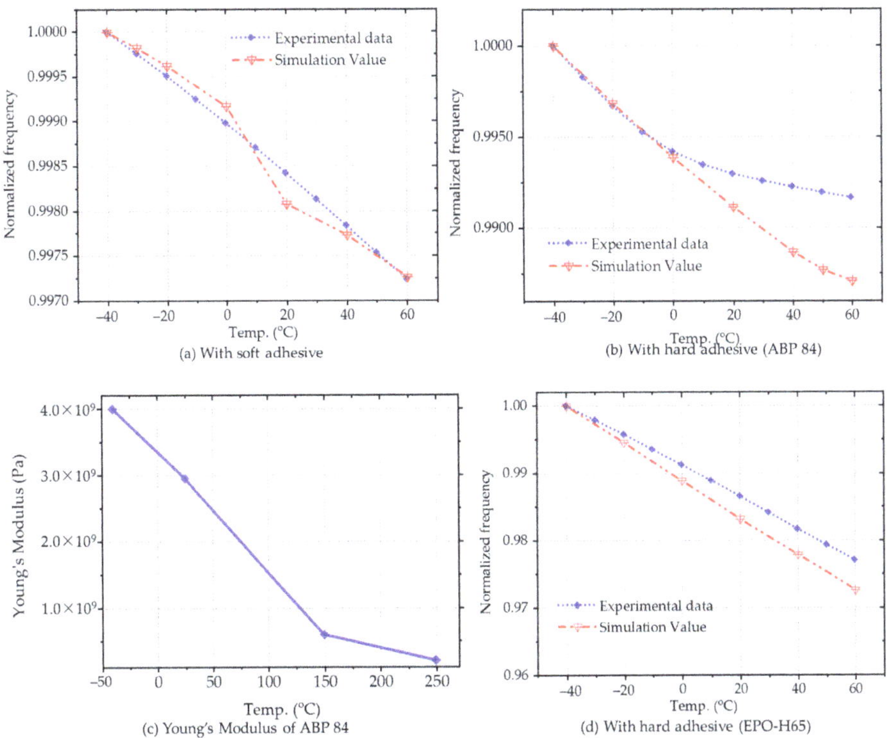

Figure 10. Comparing temperature characteristics between experimental and simulation results, (**a**) the data with soft adhesive; (**b**) the data with hard adhesive ABP 84-3JT; (**c**) Young's Modulus of ABP 84-3JT applied in the simulation; (**d**) the data with hard adhesive EPO-H65.

4. Conclusions

The paper investigates the frequency drift laws caused by the thermal stress in the packaging process. The four-layer model was established to discover the stress distributions in different conditions. The frequency decreases with the rising temperature for both hard or soft adhesive within the double-clamped model. The temperature coefficient utilizing hard adhesive is positive, while that of soft adhesive is negative. The resonant accelerometer was employed for the frequency measurements. Temperature characteristic tests with different bonding methods and adhesives were taken place every 10 °C. Three types of adhesive dispensing methods mean various chip constraints. The results indicate that the frequency varies about -150 ppm/°C at low temperatures. The amplitude of the frequency gradient decreases as the rising temperature with the hard adhesive. The temperature coefficient employing soft adhesive, which has a smaller Young's modulus, is lower in the whole temperature range. The optimal packaging method achieves -24.1 ppm/°C to -30.2 ppm/°C temperature coefficient of the resonator in the whole temperature range, close to the intrinsic property of silicon (-31 ppm/°C).

Author Contributions: Conceptualization, B.J. and Y.S.; Formal analysis, B.J., S.H. and J.Z.; Investigation, J.Z.; Supervision, Y.S.; Writing—original draft, B.J. All authors have read and agreed to the published version of the manuscript.

Funding: This research was funded by National Natural Science Foundation (61805121, 62001223, and 61971466), Jiangsu Natural Science Foundation (BK20170850, BK20190471), China.

Conflicts of Interest: The authors declare no conflict of interest.

References

1. Comi, C.; Corigliano, A.; Langfelder, G.; Longoni, A.; Tocchio, A.; Simoni, B. A Resonant Microaccelerometer with High Sensitivity Operating in an Oscillating Circuit. *J. Microelectromech. Syst.* **2010**, *19*, 1140–1152. [CrossRef]
2. Pinto, D.; Mercier, D.; Kharrat, C.; Colinet, E.; Nguyen, V.; Reig, B.; Hentz, S. A Small and High Sensitivity Resonant Accelerometer. *Procedia Chem.* **2009**, *1*, 536–539. [CrossRef]
3. Fan, B.; Guo, S.; Cheng, M.; Yu, L.; Zhou, M.; Hu, W.; Zheng, F.; Bu, F.; Xu, D. Frequency Symmetry Comparison of Cobweb-Like Disk Resonator Gyroscope With Ring-Like Disk Resonator Gyroscope. *IEEE Electron Device Lett.* **2019**, *40*, 1515–1518. [CrossRef]
4. Cao, H.; Liu, Y.; Kou, Z.; Zhang, Y.; Liu, J. Design, Fabrication and Experiment of Double U-Beam MEMS Vibration Ring Gyroscope. *Micromachines* **2019**, *10*, 186. [CrossRef] [PubMed]
5. Fragiacomo, G.; Reck, K.; Lorenzen, L.; Thomsen, E.V. Novel designs for application specific MEMS pressure sensors. *Sensors* **2010**, *10*, 9541–9563. [CrossRef] [PubMed]
6. Elko, G.W.; Pardo, F.; López, D.; Bishop, D.; Gammel, P. Capacitive MEMS microphones. *Bell Labs Tech. J.* **2010**, *10*, 187–198. [CrossRef]
7. Melamud, R.; Kim, B.; Chandorkar, S.A.; Hopcroft, M.A.; Agarwal, M.; Jha, C.M.; Kenny, T.W. Temperature-compensated high-stability silicon resonators. *Appl. Phys. Lett.* **2007**, *90*, 637–701. [CrossRef]
8. Melamud, R.; Hopcroft, M.; Jha, C.; Kim, B.; Kenny, T.W. Effects of stress on the temperature coefficient of frequency in double clamped resonators. In Proceedings of the The 13th International Conference on Solid-State Sensors, Actuators and Microsystems, TRANSDUCERS '05, Seoul, Korea, 5–9 June 2005.
9. Takigawa, R.; Tomimatsu, T.; Higurashi, E.; Asano, T. Residual Stress in Lithium Niobate Film Layer of LNOI/Si Hybrid Wafer Fabricated Using Low-Temperature Bonding Method. *Micromachines* **2019**, *10*, 136. [CrossRef] [PubMed]
10. Kuo, J.T.W.; Yu, L.; Meng, E. Micromachined Thermal Flow Sensors—A Review. *Micromachines* **2012**, *3*, 550–573. [CrossRef]
11. Sundaresan, K.; Ho, G.K.; Pourkamali, S.; Ayazi, F. Electronically Temperature Compensated Silicon Bulk Acoustic Resonator Reference Oscillators. *IEEE J. Solid State Circuits* **2007**, *42*, 1425–1434. [CrossRef]
12. Toan, N.V.; Toda, M.; Kawai, Y.; Ono, T. A capacitive silicon resonator with a movable electrode structure for gap width reduction. *J. Micromech. Microeng.* **2014**, *24*, 216–251. [CrossRef]
13. Bhatia, V.; Campbell, D.K.; Sherr, D.; D'Alberto, T.; Zabaronick, N.; Ten Eyck, G.A.; Murphy, K.A.; Claus, R.O. Temperature-insensitive and strain-insensitive long-period grating sensors for smart structures. *Opt. Eng.* **1997**, *36*, 1872–1876. [CrossRef]
14. Lee, J.I.; Jeong, B.; Park, S.; Eun, Y.; Kim, J. Micromachined Resonant Frequency Tuning Unit for Torsional Resonator. *Micromachines* **2017**, *8*, 342. [CrossRef] [PubMed]
15. Melamud, R.; Chandorkar, S.A.; Kim, B.; Lee, H.K.; Salvia, J.C.; Bahl, G.; Hopcroft, M.A.; Kenny, T.W. Temperature-Insensitive Composite Micromechanical Resonators. *J. Microelectromech. Syst.* **2009**, *18*, 1409–1419. [CrossRef]
16. Hsu, W.-T.; Nguyen, C.-C. Stiffness-compensated temperature-insensitive micromechanical resonators. In Proceedings of the Technical Digest, MEMS 2002 IEEE International Conference, Fifteenth IEEE International Conference on Micro Electro Mechanical Systems, Las Vegas, NV, USA, 24–24 January 2002; pp. 731–734.
17. Yim, W.M.; Paff, R.J. Thermal expansion of AlN, sapphire, and silicon. *J. Appl. Phys.* **1974**, *45*, 1456–1457. [CrossRef]
18. Jing, Z.; Yan, S.; Qin, S.; Qiu, A.P. Microelectromechanical Resonant Accelerometer Designed with a High Sensitivity. *Sensors* **2015**, *15*, 30293–30310.
19. Krishnamoorthy, U.; Iii, R.H.O.; Bogart, G.R.; Baker, M.S.; Carr, D.W.; Swiler, T.P.; Clews, P.J. In-plane MEMS-based nano-g accelerometer with sub-wavelength optical resonant sensor. *Sens. Actuators A Phys.* **2008**, *145–146*, 283–290. [CrossRef]
20. Zhang, X.; Park, S.; Judy, M.W. Accurate Assessment of Packaging Stress Effects on MEMS Sensors by Measurement and Sensor–Package Interaction Simulations. *J. Microelectromech. Syst.* **2007**, *16*, 639–649. [CrossRef]
21. Peterson, I.M.; Tien, T.Y. Effect of the Grain Boundary Thermal Expansion Coefficient on the Fracture Toughness in Silicon Nitride. *J. Am. Ceram. Soc.* **2010**, *78*, 2345–2352. [CrossRef]

 *micromachines*

Article

Characterization and Analysis of Metal Adhesion to Parylene Polymer Substrate Using Scotch Tape Test for Peripheral Neural Probe

Seonho Seok [1,†], HyungDal Park [2,†] and Jinseok Kim [2,*]

[1] Center for Nanoscience and Nanotechnology (C2N), University-Paris-Saclay, Orsay 91400, France; seonho.seok@u-psud.fr

[2] Center for Bionics, Korea Institute of Science and Technology (KIST), Seongbuk-gu, Seoul 02792, Korea; hyungdal@kist.re.kr

* Correspondence: jinseok@kist.re.kr

† Both authors are equally-contributed to this manuscript.

Received: 3 June 2020; Accepted: 19 June 2020; Published: 22 June 2020

Abstract: This paper presents measurement and FEM (Finite Element Method) analysis of metal adhesion force to a parylene substrate for implantable neural probe. A test device composed of 300 nm-thick gold and 30 nm-thick titanium metal electrodes on top of parylene substrate was prepared. The metal electrodes suffer from delamination during wet metal patterning process; thus, CF_4 plasma treatment was applied to the parylene substrate before metal deposition. The two thin film metal layers were deposited by e-beam evaporation process. Metal electrodes had 200 μm in width, 300 μm spacing between the metal lines, and 5 mm length as the neural probe. Adhesion force of the metal lines to parylene substrate was measured with scotch tape test. Angle between the scotch tape and the test device substrate changed from 60° to 90° during characterization. Force exerted the scotch tape was recorded as the function of displacement of the scotch tape. It was found that a peak was created in measured force-displacement curve due to metal delamination. Metal adhesion was estimated 1.3 J/m^2 by referring to the force peak and metal width at the force-displacement curve. Besides, the scotch tape test was simulated to comprehend delamination behavior of the test through FEM modeling.

Keywords: adhesion; thin film metal; parylene; neural probe; scotch tape test; FEM

1. Introduction

Many research efforts have been made to develop and improve of the prosthetic hands and arms for the amputees, and, in recent years, much progress has been observed in the development of life-like robotic hands and the means of controlling them with greater degree of freedom. For this purpose, micro-electro-mechanical systems (MEMS) technologies have been used to fabricate neural interface probes [1–6]. However, existing MEMS-based neural electrodes would have a limitation on the neural interface due to its material characteristics. Thus, flexible neural electrodes have been recently proposed to minimize mechanical mismatch between the electrode and tissue after the electrode's implantation for a stable long-term recording and stimulation. To this sense, peripheral neural interface (PNI) devices have appeared to retrieve and send neural signals directly from and to the residual or existing peripheral nerves in this field [7]. Recently, thin film flexible polymeric devices are being used for measuring nerve impulse from the central or peripheral nerve systems [8–13]. Such flexible polymeric devices tend to be designed several μm in thickness and a few mm in length due to its nature of interfacing with neurons in human body. Consequently, metal electrodes on such a thin and long polymeric substrate have constraints to be at best several hundred nanometers. Thus, metallization technology frequently used is PVD (Physical Vapor Deposition), such as evaporation or sputtering. The PVD metal layers are formed into long and thin metal lines through wet-etch or lift-off technique. Although thin-film polymeric devices are flexible and biocompatible, they are prone to delamination and carry concerns about their mechanical robustness [14]. A nanopillar array created by plasma etching could be used to enhance adhesion among different materials in the parylene-metal-parylene system [15]. ALD (Atomic layer deposition) Al2O3 combined with the silane adhesion promoter A-174 would increase adhesion force between two parylene films [14]. Parylene material has been shown that mechanical properties can be maintained after stored in PBS (Phosphate Buffered Saline) solution up to 12 months [16]. Therefore, mechanical or chemical treatments of the interface between polymer substrate and metal film is frequently required [17]. However, concrete characterization of metal adhesion to parylene substrate dedicated to neural probe is still insufficient. Scotch tape test has been frequently used to estimate adhesion force of metal layers to polymer substrates [18–20]. Tape is often used to overcome the adhesion challenges in what's commonly known as the "tape test," which is a variation of ASTM (American Society for Testing and Materials) D-3359 [21].

In this paper, metal adhesion to parylene substrate has been characterized by using scotch-tape test and it has been analyzed based on Finite Element Method (FEM) modeling. Sample preparation for adhesion test is presented in Section 2. Scotch tape test results are explained in Section 3. FEM modeling and simulation of the scotch tape test is depicted in Section 4. Finally, conclusion will be made in Section 5.

2. Test Sample Preparation for Metal Parylene Adhesion Test

Figure 1 shows the schematic of test pattern; it has length of 50 mm and width of 5 mm. It consists of 6 metal lines; each metal line has 200 μm in width and is spaced at 300 μm as is the design of multi-channel neural probe. Test pattern fabrication process is shown in Figure 2; (a) A parylene (parylene-C) layer having 5 μm in thickness was deposited in a 4-inch Si wafer using by commercial parylene coater system (VPC-500, Paco Engineering, Incheon, Korea). The monomer was deposited on the surface of the silicon wafer at a vapor phase condition with 0.8 μm/min, and deposition temperature was 20 °C. (b) Before metal layers deposition, the parylene surface was etched with CF_4 gas (process conditions; (25 mTorr, 20 sccm, 1.5 min)) without O_2 in order to increase surface roughness reducing the hydrophobicity, as shown in Figure 3. Note that O_2 was not used for surface modification since the parylene is susceptible to be surface oxidation causing degradation of mechanical properties of the polymer neural probe. Moreover, the nanopillar structures on parylene substrate was efficiently built with CF_4 gas rather than O_2 gas, as shown in Figure 3b,c. Besides, there have been two major methods to improve interfacial adhesion between parylene and metal layer: chemical treatment or mechanical surface modification with RIE (Reactive Ion Etching) etch (Plasma-therm 790 MF, Plasma-Therm, Saint Petersburg, FL, USA). Concerning the test sample used in our study, parylene was already deposited on a silicon substrate; thus, chemical treatment may have changed the properties of parylene substrate itself, as well as the interface on which the metal was deposited. Furthermore, RIE etch have shown better performance compared with conventional A-174 saline chemical treatment [15]. Therefore, we modified the parylene surface with CF_4 plasma etch to make a nanopillared surface, increasing the interfacial energy. The effectiveness of the nanopillared parylene surface was confirmed during metal patterning step. Titanium (30 nm) and gold (300 nm) were sequentially deposited with using by E-beam evaporator (ei5, ULVAC, Methuen MA, USA) on the parylene substrate without rupture of vacuum. (c) Photoresist (AZ GXR 601 (46cp), Merck, Kenilworth, NJ, USA) was patterned as a metal etch mask. The process conditions are summarized in Table 1. (d) Gold was first etched with Au etchant (Gold ETCH TFA, Transene Company, Danvers, MA, USA) for 5 min, and then titanium is etched with BOE (Buffered Oxide Etchant) solution (Buffered oxide etch 6:1, VWR International, Radnor, PA, USA) for 10 s. After metal patterning, PR mask was removed with acetone. Etch time was 60 s.

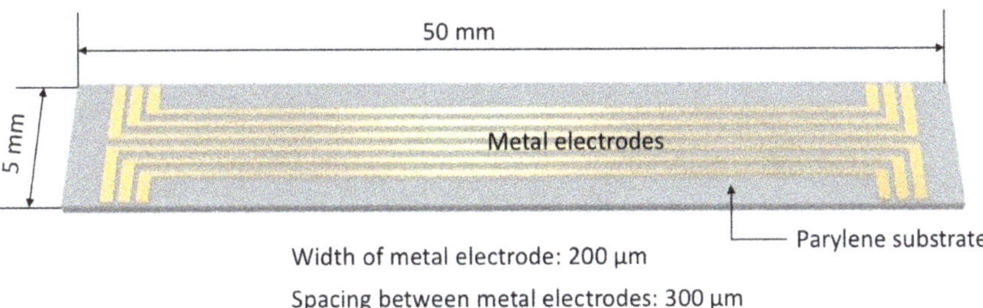

Figure 1. Design of test pattern.

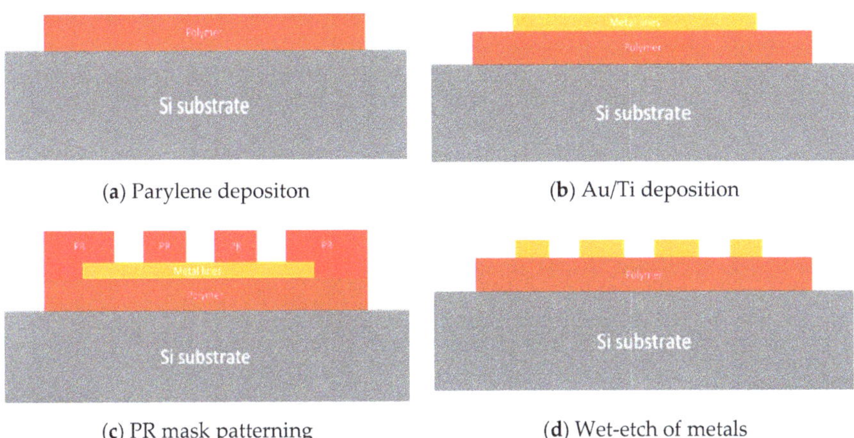

(**a**) Parylene depositon

(**b**) Au/Ti deposition

(**c**) PR mask patterning

(**d**) Wet-etch of metals

Figure 2. Test sample fabrication process.

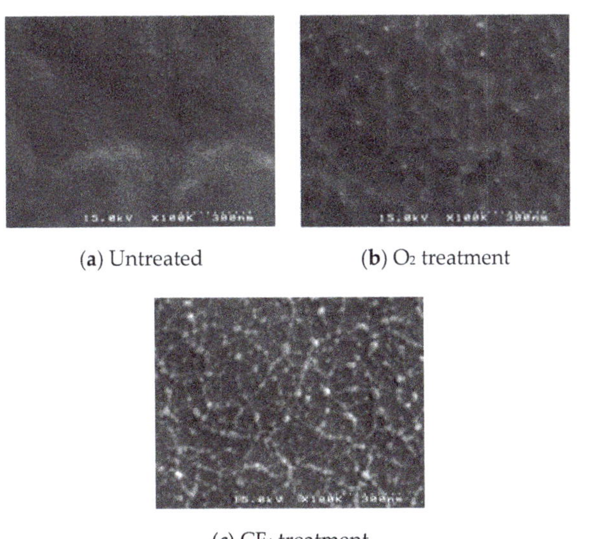

(**a**) Untreated

(**b**) O_2 treatment

(**c**) CF_4 treatment

Figure 3. Parylene surface modification.

Table 1. Photoresist etch mask process conditions.

Step	Conditions
PR coating	10 s @1500 rpm
Soft bake	60 s @ 95 °C
Exposure	10 s @10 mW
Develop	10 s @developer
Post exposure bake	30 s @110 °C
Thickness	1.2 µm

Resultant test samples on 4-inch silicon wafer is shown in Figure 4. All of metal lines were successfully implemented on the parylene substrate without any delamination during etching process. Note that the metal lines were fully delaminated from the parylene substrate during metal wet-etch step.

Figure 4. Test samples fabricated on 4-inch silicon substrate.

3. Scotch Tape Test for Metal Adhesion to Parylene Substrate

The prepared test samples underwent the scotch tape test to evaluate adhesion strength of the metal electrodes to the parylene substrate. The machine used for the scotch tape test was Shimadzu EZ-S machine (Shimadzu, Kyoto, Japan) dedicated for tensile testing. Figure 5 shows a photo of scotch tape attached on the test sample and schematics of the scotch tape test, respectively. The scotch tape is 3 M transparent tape 550. It has thickness of 50 µm and 12 mm width, and it provides adhesion to steel of 1.8 N/cm (or 0.18 N/mm).

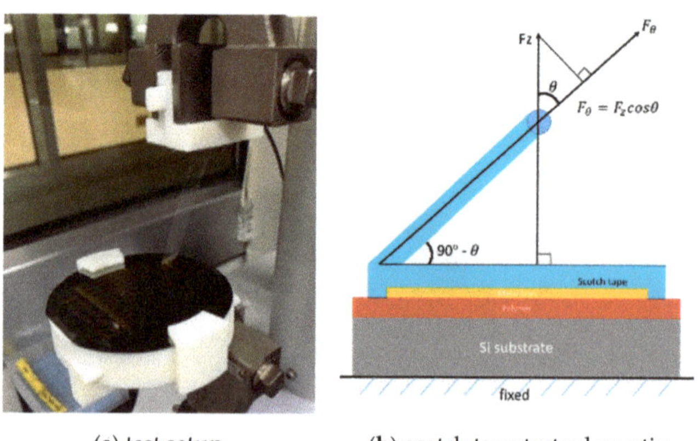

(**a**) test setup   (**b**) scotch tape test schematic

Figure 5. Scotch tape test for metal adhesion to parylene substrates.

Referring to Figure 5b, the scotch tape test was carried out in the following way; the machine applies stroke (unit: mm/min) into one end of the scotch tape, and then it measures force F_z (unit: N). As force of interest is F_θ, relationship between F_z and F_θ can be calculated as Equation (1).

$$F_\theta = F_z \cos \theta. \tag{1}$$

During the scotch tape test, the angle $(90° - \theta)$ was changed from 60° to 90°; thus, $F_\theta = 0.5\,F_z$ at 45°, and $F_\theta = F_z$ at 90°. For simplicity, we used the measured F_z from the scotch tape test to extract the adhesion strength.

Metal adhesion to parylene substrate was then measured with the scotch tape test. Scotch tape was attached to the parylene surface, slightly away from the left-end metal line to the right-end of the metal. After that, two different strokes (10 mm/min and 1 mm/min) were applied to the scotch tape, and corresponding force was measured as shown in Figure 6. All metal lines were debonded from the parylene substrate for all the cases. It was found that the sample with CF_4 treatment needs more force than that without CF_4 treatment, which means a parylene surface with CF_4 treatment sticks better to scotch tape. This is a proof that parylene surface energy can be increased with only CF_4 treatment without O_2.

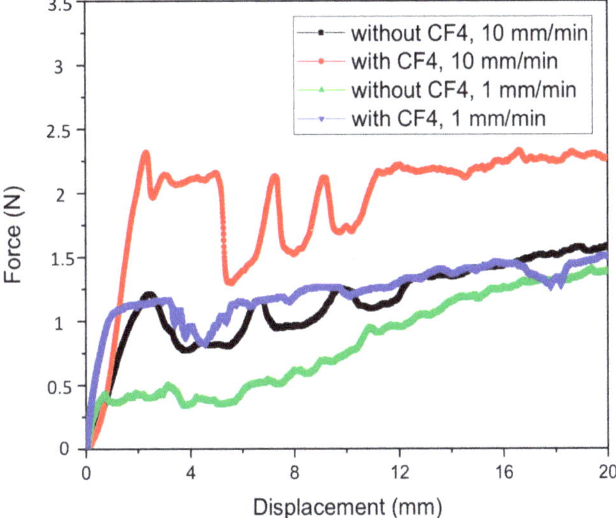

Figure 6. Force-displacement curves from scotch tape test with metal lines on parylene substrate.

The first peak in each measured force was due to initiation of metal debonding, which makes abrupt drop of force. Minimal adhesion force is found when a stroke of 1 mm/min was applied to the parylene test sample without CF_4 treatment (green line). It can be said that the metal adhesion had lower than the adhesion value estimated from the first peak. The adhesion can be calculated as follows; (0.5 N/12 mm) × (1.2 mm/12 mm) = 4.2 N/m. As all metal lines were debonded, the metal adhesion should have had lower than 4.2 J/m^2. Thus, a lower stoke of 0.1 mm/min was applied to find metal adhesion to the parylene substrate. In this case, the scotch tape was attached only to the narrow metal lines of CF_4 treated parylene and then force was recoded while stroke of 0.1 mm/min is applied. The measured force-displacement curve was compared with the previous results of 1 mm/min and 10 mm/min, as shown in Figure 7. As remarked in Figure 7, the scotch tape was debonded up to 0.9 N without metal line delamination, and one metal line started to debond from 0.91 N. Therefore, metal adhesion could be extracted from this peak force; (0.91 N/12 mm) × (0.2 mm/12 mm) = (76 N/m) ×

$(0.017) = 1.29$ N/m $= 1.29$ J/m^2. Note that inset shows transferred metal lines on the scotch tape. Scotch tape strokes of 1 mm/min and 10 mm/min introduced large force fluctuation, which would result from relatively large applied force compared with interfacial energy.

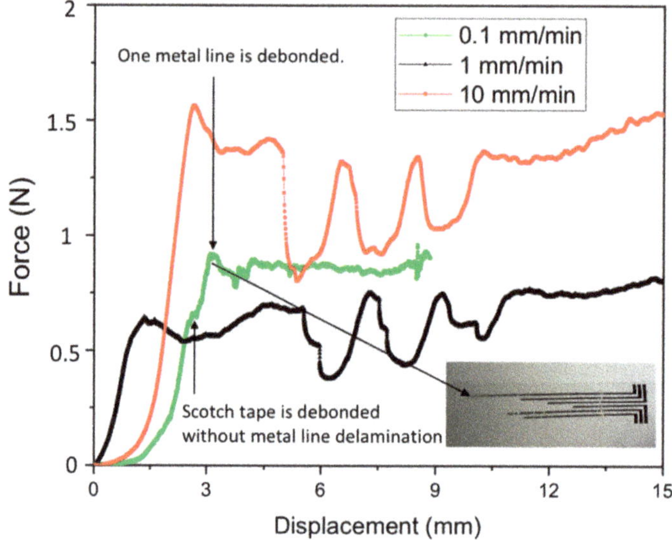

Figure 7. Force-displacement curves from scotch tape test with metal lines on parylene substrate with CF$_4$ treatment.

Table 2 summarizes adhesion force between parylene and metal layer of previous reports compared with this work. The majority of the studies on metal-parylene adhesion force deal with parylene layer deposited on metal film with inteface treatment. Thus, they can provide from tens N/m to hundreds N/m of adhesion force. A thin metal layer, like Al thin film, on PET(Polyethylene terephthalate) layer has 46.7 N/m of adhesion. The thin gold metal line of this work had 1.29 N/m of adhesion force, which may result from the relatively narrow metal line of 200 μm.

Table 2. Comparison of other research results with scotch tape test.

Test Method	Test Sample Structure	Interface Treatment	Environment	Remarks
		Adhesion Force		
Peel 90° [18]	Parylene (10 μm)/sputtered Ti (up to 100 nm)/Au (300 nm)/glass substrate	Gorham process before parylene deposition to make adhesion promoting layer	48 h in PBS solution	metal width = 10 mm
		35.5 mN/mm		
Peel 180° [19]	Al thermal evaporated metal (60 nm)/PET (12 μm)/Al substrate (1 mm)	-	RT	metal width = 15 mm
		0.7 N/15 mm		
Tensile [15]	Au (200 nm)/Cr (13 nm)/Parylene (10 μm)/glass substrate	O$_2$ RIE etching	RT	metal width = 15 mm
		2.13 ± 0.12 MPa		
Peel test [22]	Parylene (9–20 μm)/Ti substrate	Plasma Polymerized Ethane (PPE)	RT	metal width = NA
		0.34 N/mm		
Peel test [This work]	Au(300 nm)/Ti(30 nm)/ Parylene(5 μm)/Si substrate	CF$_4$ RIE	RT	metal width = 200 μm
		1.29 N/m		

4. FEM Modeling and Simulation

FEM modeling and simulation is very useful to understand stress effect and corresponding deformation of MEMS package, debonding characteristics of a transfer packaging, and mechanical behaviors related with delamination [23–27]. Especially, debonding of a substrate and film delamination can be studied by adopting a CZM (Cohesive Zone model) to represent the interface of interest [28–30].

For FEM modeling, material properties of each element are important to get good simulation results. Required material properties in this modeling are Young's moduli and poisson ratios of scotch tape and parylene and strain energy release rate of interface between the scotch tape and parylene. Young's modulus of the scotch tape is extracted from tensile test result, as shown in Figure 8. The Young's modulus of the scotch tape is 6.9 MPa in the elastic region, and the maximum applied force in the elastic region is 7.6 N when the applied strain is 2.2% (2.2 mm elongation as test scotch tape length is 100 mm). From the tensile test result, scotch tape in the metal adhesion test would be in the elastic region as the applied force is less than 2 N in all the cases. Poisson ratio of the scotch tape is assumed to be 0.4 as other polymer materials. Young's modulus and poisson ratio of parylene are 2.67 GPa, as extracted in previous work, and 0.4, respectively [23]. Table 3 summarizes material properties for the FEM model. Note that interface material properties were defined with critical strain energy release rate. The value for critical strain energy release rate was the measured adhesion force as explained earlier.

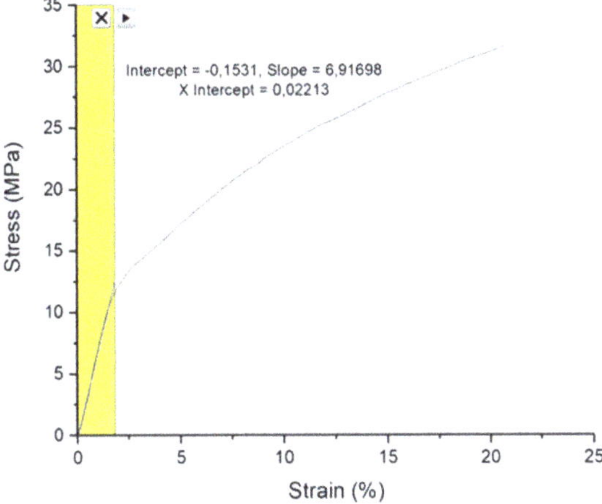

Figure 8. Tensile test result of scotch tape.

Table 3. Material properties of materials and interface.

Material Properties	Scotch Tape	Parylene
Young's modulus (Pa)	6.9×10^6	2.7×10^9
Poisson ratio	0.4	0.4
-		Interface
Critical mode I energy release rate (J/m^2)		1.3
Critical mode II energy release rate (J/m^2)		1.3
Critical mode III energy release rate (J/m^2)		1.3

Given with material properties, modeling and simulation of the scotch tape test was performed in two steps: 1) crack propagation behavior of the interface between the scotch tape and parylene substrate 2) debonding of the scotch tape from parylene substrate based on CZM.

A 2D FEM model for crack propagation was built, as shown in Figure 9a. The length of this model was 1000 µm, and thickness was 50 µm for scotch tape and 5 µm for parylene polymer. The following boundary conditions were applied: bottom line is fixed and displacement load is applied to left-top end, having 50 µm width. Note that 2D model behavior was defined as plane strain. Total deformation of the model when displacement of 100 µm in y direction was applied to the scotch tape is presented in Figure 9b. As is in the scotch tape test, delamination of the interface between scotch tape and parylene occurred, and crack propagated in x-direction.

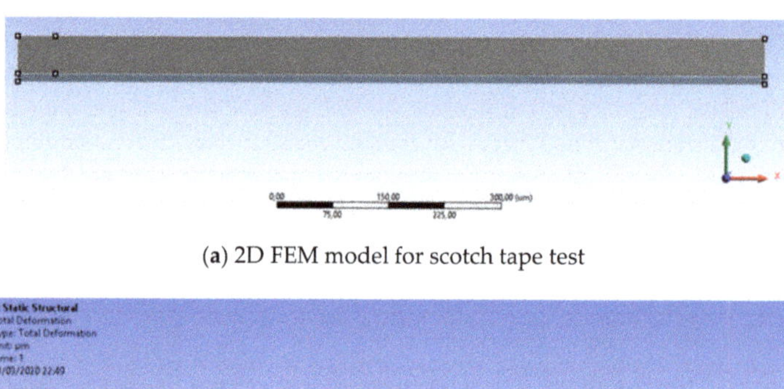

(**a**) 2D FEM model for scotch tape test

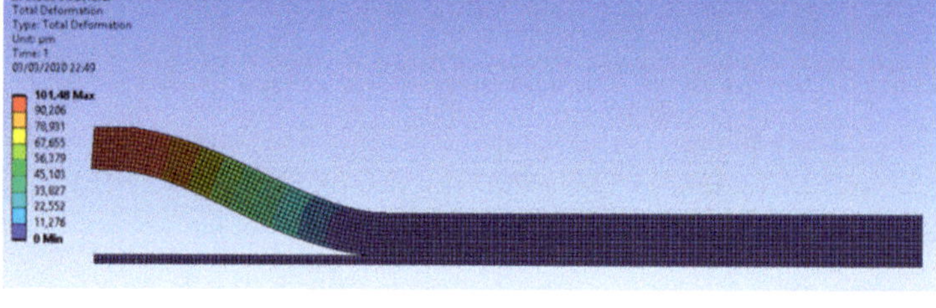

(**b**) Total deformation

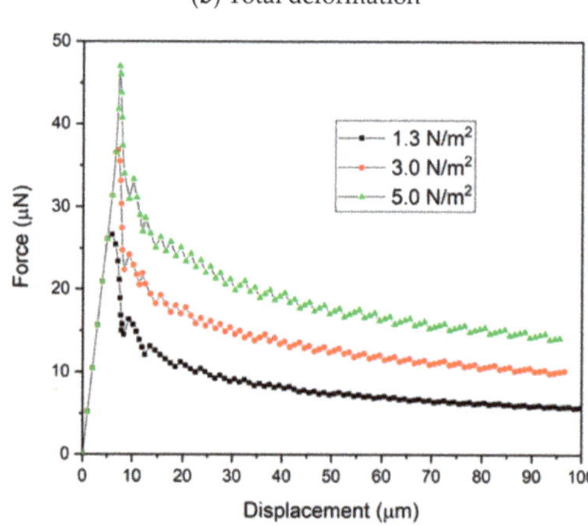

(**c**) Force-Displacement curves as function of interface energy

Figure 9. Crack propagation model and simulation results.

Force-displacement was investigated as function of interface adhesion energy, as shown in Figure 9c. The required force for crack initiation was increased as interface adhesion energy increased, as expected. The force magnitude smaller than the measurement would have been due to thickness effect in 2D simulation. An important parameter in this graph is minimal displacements for crack initiation: 5 μm, 6.7 μm, and 7.6 μm for 1.3 N/m^2, 3.0 N/m^2, and 5.0 N/m^2, respectively. These parameters are included in the following 3D interface delamination as a part of CZM parameters. From the simulation results, strain energy release rate (SERR) for mode 1 referring to VCCT (Virtual Crack Closure Technique) (G1), SERR for mode 2 from VCCT (G2), and SERR for mode 3 from VCCT (G3) were found 0.5 J/m^2, 0.75 J/m^2 and 0 J/m^2, respectively. The total amount of VCCTs corresponded to the interface energy of 1.3 N/m^2. Principal modes of the fracture of the delamination was from mode 1 and mode 2.

Figure 10a shows 3D model for the CZM interface delamination. As indicated in Figure 10b, there were two different regions in this model: pre-cracked (interface I) and CZM-modeled (interface II). CZM is a useful way to simulate interface delamination, which is frequently used for thin film delamination and transfer packaging technique [24,25]. The interface II, which is of interest for the adhesion, is modeled with CZM (Cohesive Zone Model) parameters, as explained in Figure 10c. As the bilinear CZM model needs at least two parameters, maximum normal traction and normal displacement at debonding was defined, as presented in Table 4. The minimal gaps for the fracture initiation found from the previous crack propagation simulation were included for 3D CZM simulation to estimate applied force to initiate the interface crack.

Table 4. Cohesive Zone model (CZM) parameters.

Parameter Name	Value
Maximum normal traction	0.5 MPa
Normal displacement jump at completion of debonding	5 μm
Maximum tangential traction	0.5 MPa
Tangential displacement jump at completion of debonding	5 μm

A displacement load was applied to one-end of the scotch tape, and then the force-displacement was extracted from the simulation. Referring to bilinear CZM model, critical strain energy release rate was calculated 1.25 J/m^2. Initial width of the 3D model was 200 μm as was the fabricated metal electrode width. As in the 2D case, displacement load was applied to left-tip end. Extracted force-displacement curve at the loading place is presented in Figure 11. Minimal force for debonding of scotch tape was estimated 1.2 N/m, while measured one was 1.29 N/m. Adhesion force of the simulation had a good agreement with the measurement. Width of metal electrode could have been increased to get larger interface adhesion, as shown in Figure 11 When wider metal electrode is used to achieve larger metal-parylene adhesion, metal line impedance for neural signal acquisition should be taken into account.

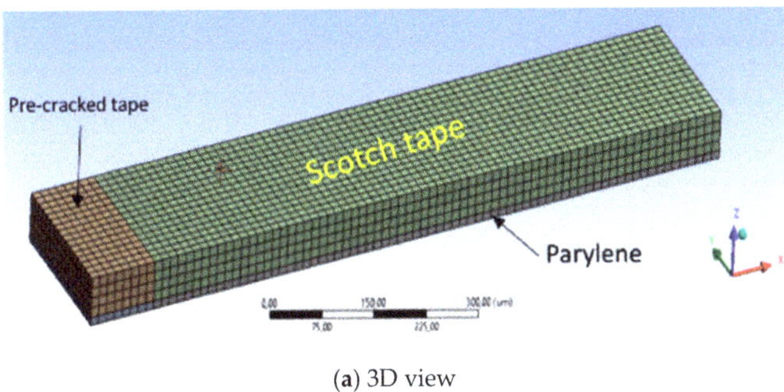

(a) 3D view

Figure 10. *Cont.*

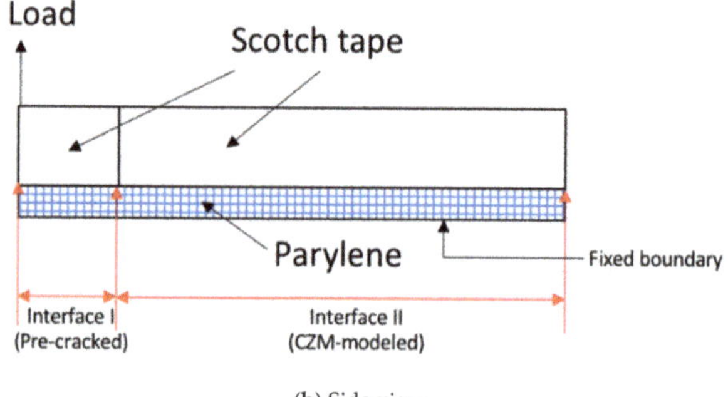

(**b**) Side view

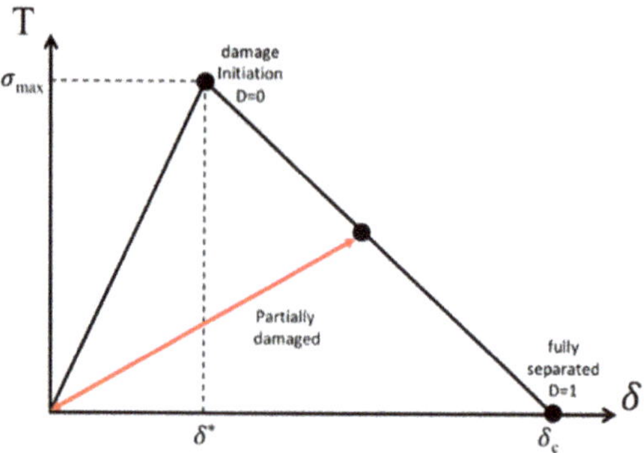

Tmax = maximum normal cohesive traction (σmax)

δ^* = normal displacement jump at maximum normal cohesive traction

δc = normal displacement jump at the completion of debonding

(**c**) Bilinear CZM model parameters

Figure 10. Three-dimensional Finite Element Method (FEM) model for Cohesive Zone model (CZM) interface delamination.

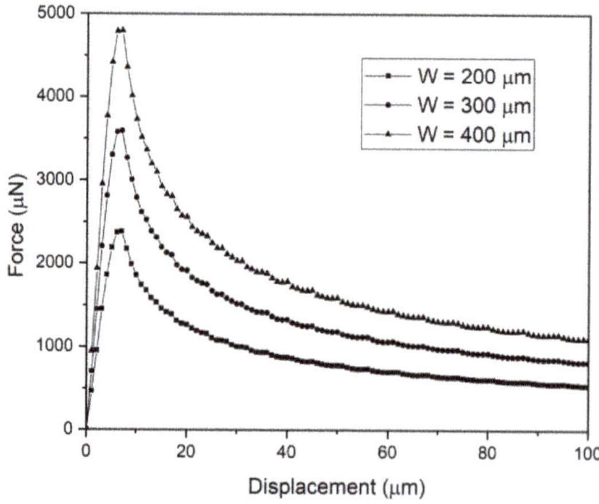

Figure 11. Force-displacement curves as function of metal width.

5. Conclusions

Thin film flexible polymeric devices, such as a parylene-metal-parylene system, are being used for measuring nerve impulse from the central or peripheral nerve systems. Such thin-film polymeric devices provide advantages of flexibility and biocompatibility, but they are prone to delamination and carry concerns about their mechanical robustness. Therefore, metal adhesion strength to polymer substrate is important. The adhesion of metal electrodes to parylene substrate was measured by the scotch tape test. Thin and long metal electrodes was patterned on a parylene substrate in which the surface was modified by CF_4 plasma etch before the metal deposition through e-beam evaporation. Metal adhesion strength was estimated by measuring force-displacement curve of the scotch tape test. The estimated metal adhesion was 1.3 J/m^2. Experiment result was verified through FEM modeling of the scotch tape test. The proposed modeling method provided adhesion force having good agreement with experimental result. Although a thin-film parylene-based device can provide excellent short-term reliability, there exists one significant drawback of poor adhesion to metallic layer. The failure of the metal electrode on the parylene substrate is accelerated in the wet environment of a human body and under mechanical forces originating from body movement. Therefore, mechanical integrity in conditions of a human body implant or movement will be performed to assess long-term reliability of the parylene-metal devices, along with the biocompatibility of the parylene-base neural probe.

Author Contributions: Conceptualization, S.S. and J.K.; methodology, H.P.; software, H.P.; validation, S.S., H.P. and J.K.; formal analysis, S.S.; investigation, H.P.; resources, H.P.; data curation, H.P.; writing—original draft preparation, S.S.; writing—review and editing, S.S.; visualization, S.S.; supervision, J.K.; project administration, J.K.; funding acquisition, J.K. All authors have read and agreed to the published version of the manuscript.

Funding: This research was supported by the convergence technology development program for bionic arm through the National Research Foundation of Korea (NRF) funded by the Ministry of Science & ICT (No. 2017M3C1B2085292). This work was also supported by the Korea Institute of Science and Technology institutional program (2E30090 and 2V08590).

Conflicts of Interest: The authors declare no conflict of interest.

References

1. Anderson, D.J.; Najafi, K.; Tanghe, S.J.; Evans, D.A.; Levy, K.L.; Hetke, J.F.; Xue, X.; Zappia, J.J.; Wise, K.D. Batch-fabricated thin-film electrodes for stimulation of the central auditory system. *IEEE Trans. Biomed. Eng.* **1989**, *36*, 693–704. [CrossRef] [PubMed]

2. Campbell, P.K.; Normann, R.A.; Horch, K.W.; Stensaas, S.S. A chronic intracortical electrode array: Preliminary results. *J. Biomed. Mater. Res. Appl. Biomat.* **1989**, *23*, 245–259.

3. Eichenbaum, H.; Kuperstein, M. Extracellular neural recording with multichannel microelectrodes. *J. Electrophysiol. Tech.* **1986**, *13*, 189–209.

4. Najafi, K.; Ji, J.; Wise, K.D. Scaling limitation on silicon multichannel recoding probe. *IEEE Trans. Biomed. Eng.* **1990**, *37*, 1–11. [CrossRef]

5. Campbell, P.K.; Jones, K.E.; Huber, R.J.; Horch, K.W.; Normann, R.A. A silicon-based, three-dimensional neural interface: Manufacturing processes for an intracortical electrode array. *IEEE Trans. Biomed. Eng.* **1991**, *38*, 758–768. [CrossRef]

6. Takeuchi, S.; Suzuki, T.; Mabuchi, K.; Fujita, H. 3D flexible multichannel neural probe array. *J. Micromechanics Microengineering* **2004**, *14*, 104–107. [CrossRef]

7. Kim, O.; Choi, W.; Jung, W.; Jung, S.; Park, H.; Park, J.W.; Kim, J. Novel Neural Interface Electrode Array for the Peripheral Nerve. In Proceedings of the International Conference on Rehabilitation Robotics (ICORR), London, UK, 17–20 July 2017.

8. Castagnola, V.; Descamps, E.; Lecestre, A.; Dahan, L.; Remaud, J.; Nowak, L.G.; Bergaud, C. Parylene-based flexible neural probes with PEDOT coated surface for brain stimulation and recording. *Biosens. Bioelectron.* **2015**, *67*, 450–457. [CrossRef]

9. Cheung, K.C.; Renaud, P.; Tanila, H.; Djupsund, K. Flexible Polyimide Microelectrode Array for in Vivo Recordings and Current Source Density Analysis. *Biosens. Bioelectron.* **2007**, *22*, 1783–1790. [CrossRef]

10. Stieglitz, T. Development of a micromachined epiretinal vision prosthesis. *J. Neural Eng.* **2009**, *6*, 065005. [CrossRef]

11. Ordonez, J.; Schuettler, M.; Boehler, C.; Boretius, T.; Stieglitz, T. Thin films and microelectrode arrays for neuroprosthetics. *MRS Bull.* **2012**, *37*, 590–598. [CrossRef]

12. Chen, Z.; Ryzhik, L.; Palanker, D. Current Distribution on Capacitive Electrode-Electrolyte Interfaces. *Phys. Rev. Appl.* **2020**, *13*. [CrossRef]

13. Onzález, C.; Rodríguez, M. A flexible perforated microelectrode array probe for action potential recording in nerve and muscle tissues. *J. Neurosci. Methods* **1997**, *72*, 189–195. [CrossRef]

14. Lee, C.D.; Meng, E. Mechanical properties of thin-film Parylene- metal-Parylen devices. *Front. Mech. Eng.* **2015**, *1*, 10. [CrossRef]

15. Xie, Y.; Pei, W.; Guo, D.; Zhang, L.; Zhang, H.; Guo, X.; Xing, X.; Yang, X.; Wang, F.; Gui, Q.; et al. Improving adhesion strength between layers of an implantable parylene-C electrode. *Sens. Actuators A Phys.* **2017**, *260*, 117–123. [CrossRef]

16. Oliva, N.; Mueller, M.; Stieglitz, T.; Navarro, X.; del Valle, J. On the use of Parylene C polymer as substrate for periphral nerve electrode. *Sci. Rep.* **2018**, *8*, 1–12.

17. Chang, J.H.; Lu, B.; Tai, Y.C. Adhesion-enhancing surface treatments for parylene deposition. In Proceedings of the International Solid-State Sensors, Actuators and Microsystems Conference (Transducers 2011), Beijing, China, 5–9 June 2011.

18. Jesdinszki, M.; Struller, C.; Rodler, N.; Blondin, D.; Cassio, V.; Kucukpinar, E.; Langowski, H.-C. Evaluation of Adhesion Strength Between Thin Aluminum Layer and Poly(ethylene terephthalate) Substrate by Peel Tests—A Practical Approach for the Packaging Industry. *J. Adhes. Sci. Technol.* **2012**, *26*, 2357–2380. [CrossRef]

19. Tsai, L.-C.; Rezaee, M.; Haider, M.I.; Yazdi, A.; Salowitz, N.P. QUANTITATIVE MEASUREMENT OF THIN FILM ADHESION FORCE. In Proceedings of the ASME 2019 Conference on Smart Materials, Adaptive Structures and Intelligent Systems SMASIS 2019, Louisville, KY, USA, 9–11 September 2019.

20. Min, K.; Rammohan, A.R.; Lee, H.S.; Shin, J.; Lee, S.H.; Goyal, S.; Park, H.; Mauro, J.C.; Stewart, R.; Botu, V.; et al. Computational approaches for investigating interfacial adhesion phenomena of polyimide on silica glass. *Sci. Rep.* **2017**, *7*, 10475. [CrossRef]

21. ASTM-D-3359. *Standard Test Methods for Measuring Adhesion by Tape Test*; ASTM International: West Conshohocken, PA, USA, 2010.

22. Yamagishi, F.G. Investigations of plasma-polymerized films as primers for Parylene-C coatings on neural prosthesis materials. *Thin Solid Film.* **1991**, *202*, 39–50. [CrossRef]

23. Seok, S.; Park, H.D.; Choi, W.; Kim, O.; Kim, J. Analysis of Thin Film Parylene-Metal-Parylene Device Based on Mechanical Tensile Strength Measurement. In Proceedings of the DTIP 2019, Berlin, Germany, 8–12 September 2019.

24. Lishchynska, M.; O'Mahony, C.; Slattery, O.; Wittler, O.; Walter, H. Evaluation of Packaging Effect on MEMS Performance: Simulation and Experimental Study. *IEEE Trans. Adv. Packag.* **2007**, *30*, 629–635. [CrossRef]

25. Seok, S.; Rolland, N.; Rolland, P.-A. A Theoretical and Experimental Study of BCB Thin-Film Cap Zero-Level Package Based on FEM Simulations. *J. Micromechanics Microengineering* **2010**. [CrossRef]

26. Seok, S. A Theoretical Study on Post-it-like Debonding Process for BCB Cap Transfer Packaging Based on FEM Simulation. *IEEE Trans. Compon. Packag. Manuf. Technol. CPMT* **2015**, *5*, 1417–1422. [CrossRef]

27. Seok, S. Fabrication and Modeling of Nitride Thin Film Encapsulation Based on Anti-Adhesion-Assisted Transfer Technique and Nitride/BCB Bilayer Wrinkling. *IEEE Trans. Compon. Packag. Manuf. Technol. CPMT* **2016**, *6*, 1301–1307. [CrossRef]

28. Yan, Y.; Shang, F. Cohesive zone modeling of interfacial delamination in PZT thin films. *Int. J. Solids Struct.* **2009**, *46*, 2739–2749. [CrossRef]

29. Samimi, M.; van Dommelen, J.A.W.; Geers, M.G.D. A three-dimensional self-adaptive cohesive zone model for interfacial delamination. *Comput. Methods Appl. Mech. Eng.* **2011**, *200*, 3540–3553. [CrossRef]

30. Moslemi, M.; Khoshravan, M. Cohesive Zone Parameters Selection for Mode-I Prediction of Interfacial Delamination. *J. Mech. Eng.* **2015**, *61*, 507–516. [CrossRef]

 *micromachines*

Article

On the Feasibility of Fan-Out Wafer-Level Packaging of Capacitive Micromachined Ultrasound Transducers (CMUT) by Using Inkjet-Printed Redistribution Layers

Ali Roshanghias [1,*], **Marc Dreissigacker** [2], **Christina Scherf** [3,4], **Christian Bretthauer** [5], **Lukas Rauter** [1], **Johanna Zikulnig** [1], **Tanja Braun** [2,6], **Karl-F. Becker** [2,6], **Sven Rzepka** [3,4] and **Martin Schneider-Ramelow** [2,6]

[1] Silicon Austria Labs GmbH, Europastrasse 12, 9524 Villach, Austria; lukas.rauter@silicon-austria.com (L.R.); johanna.zikulnig@silicon-austria.com (J.Z.)
[2] Microperipheric Center, Technical University Berlin, 13355 Berlin, Germany; marc.b.dreissigacker@tu-berlin.de (M.D.); Tanja.Braun@izm.fraunhofer.de (T.B.); Karl-Friedrich.Becker@izm.fraunhofer.de (K.-F.B.); martin.schneider-ramelow@izm.fraunhofer.de (M.S.-R.)
[3] Materials and Reliability of Microsystems, Chemnitz University of Technology, 09111 Chemnitz, Germany; christina.scherf@enas.fraunhofer.de (C.S.); Sven.Rzepka@enas.fraunhofer.de (S.R.)
[4] Micro Material Center, Fraunhofer Institute for Electronic Nano Systems, 09126 Chemnitz, Germany
[5] Infineon Technologies AG, 85579 Neubiberg, Germany; christian.bretthauer@infineon.com
[6] Fraunhofer-Institut für Zuverlässigkeit und Mikrointegration (IZM), 13355 Berlin, Germany
* Correspondence: ali.roshanghias@silicon-austria.com

Received: 4 May 2020; Accepted: 30 May 2020; Published: 31 May 2020

Abstract: Fan-out wafer-level packaging (FOWLP) is an interesting platform for Microelectromechanical systems (MEMS) sensor packaging. Employing FOWLP for MEMS sensor packaging has some unique challenges, while some originate merely from the fabrication of redistribution layers (RDL). For instance, it is crucial to protect the delicate structures and fragile membranes during RDL formation. Thus, additive manufacturing (AM) for RDL formation seems to be an auspicious approach, as those challenges are conquered by principle. In this study, by exploiting the benefits of AM, RDLs for fan-out packaging of capacitive micromachined ultrasound transducers (CMUT) were realized via drop-on-demand inkjet printing technology. The long-term reliability of the printed tracks was assessed via temperature cycling tests. The effects of multilayering and implementation of an insulating ramp on the reliability of the conductive tracks were identified. Packaging-induced stresses on CMUT dies were further investigated via laser-Doppler vibrometry (LDV) measurements and the corresponding resonance frequency shift. Conclusively, the bottlenecks of the inkjet-printed RDLs for FOWLP were discussed in detail.

Keywords: microelectromechanical systems (MEMS) packaging; inkjet printing; redistribution layers; capacitive micromachined ultrasound transducers (CMUT); fan-out wafer-level packaging (FOWLP)

---

1. Introduction

Fan-out wafer-level packaging (FOWLP) has spurred increasing interest due to significant cost advantages over competitive technologies, increased interconnect density, as well as enhanced electrical and thermal package performance. With its roots in integrated circuit (IC) manufacturing technology, FOWLP has also recently gained a lot of attention for microelectromechanical systems (MEMS) and sensors packaging. Some examples for FOWLP of sensors including MEMS-based acceleration and pressure sensors, capacitive micromachined ultrasonic transducers (CMUTs), gas sensors and biomedical sensors were recently realized and reported [1–7].

Employing FOWLP for MEMS sensor packaging creates some unique challenges [2]. For instance, the thin and sensitive parts of MEMS components are incompatible with FOWLP processes such as laminating, molding, back-grinding and dicing. This leads to deterioration of component performance, i.e., a resonance frequency shift of a MEMS microphone, or membrane rupture, which are issues with mold encapsulation per se [7].

Moreover, some issues are emerging from the typical fabrication processes of redistribution layers (RDL) for fan-out packages, a combination of sputtering, photolithographic processes, etching, as well as electroplating. RDLs are typically metal interconnects used to provide power supply and route signals within the package and towards its periphery [8]. To ensure proper functionality of MEMS, it is essential to temporarily protect delicate sensing areas during RDL processing and ensure that temporary protection is entirely removed afterward. This complex procedure for the protection of thin film is also known as keep-out-zones (KOZ) processing on RDLs [2].

In this study, an alternative FOWLP concept by implementing additively manufactured RDLs for CMUT array packaging was proposed. The printed RDLs served as an interconnect between capacitive microphones and speakers, operating in the ultrasonic domain, with corresponding application-specific integrated circuits (ASICs), which allow features such as touchless activation or control using gestures [9,10]. As schematically illustrated in Figure 1, metallic and dielectric structures can be deposited selectively and in a controlled volume via a drop-on-demand printing technology (e.g., inkjet, aerosol, electrostatic, electrohydrodynamic, etc.). The concept of inkjet-printed RDLs for FOWLP was introduced in our previous work [11]. Inkjet-printed circuitry was also evaluated for the fabrication of low-cost silicon [12] and organic interposers [13], which showed great potential for rapid prototyping and signal probing. Aerosol printed RDLs for 3D smart devices were also recently reported by Serpelloni et al. [14] as well as screen printed RDLs by Chia-Yen et al. [15].

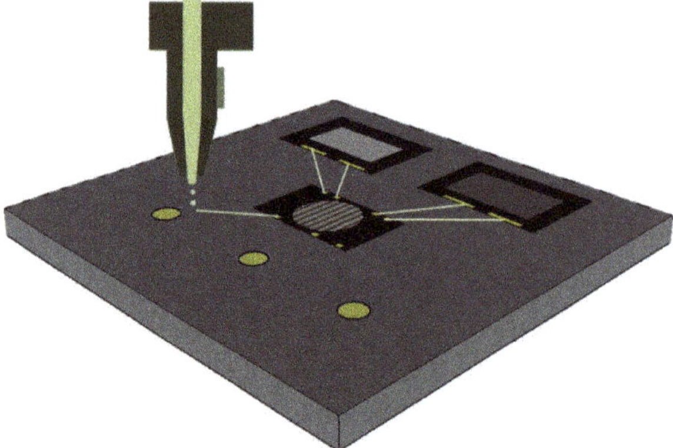

Figure 1. Schematic illustration of the fabrication process of redistribution layers in FOWLP by using drop-on-demand inkjet-printing.

Inkjet printing technology avoids long lithography procedures including global resist coating and sputtering, thus lower thermo-mechanical stresses are expected to be applied to the sensitive MEMS structures which will be assessed here. Additionally, long-term reliability of these printed RDLs for FOWLP will also be investigated and discussed.

2. Materials and Methods

In this study, an advanced R&D inkjet printer (PiXDRO LP50, Meyer Burger Technology AG, Gwatt, Switzerland) equipped with an industrial inkjet print-head (Spectra SE-128, Fujifilm Dimatix

Inc., Santa Clara, CA, USA) was used. A commercial nanoparticle silver (Ag)-ink (Sicrys 115-TM 119, PV Nano Cell Ltd., HaZafon, Israel) with 50 wt.% metal loading and an average particle size of 120 nm (d90) was deposited at the operational jetting voltage of 120 V, printing frequency of 1000 Hz and at carefully adjusted jetting pulse duration profile. Consequently, droplets with an average volume of 20 picoliters and a velocity of 2 m/s were generated. The printing of Ag was performed at room temperature, while the substrate was heated up to 50 °C to facilitate the evaporation of solvents. After printing, the Ag lines were thermally sintered at 150 °C for 30 min. The electrical resistivity of the tracks after thermal sintering at 150 °C was 5.61 E-8 Ωm corresponding to ~33% of the conductivity of bulk Ag. For multilayering, several UV-curable dielectric inks were examined [16,17].

The morphology of the printed RDLs was characterized using scanning electron microscopy (SEM, Helios, Thermo Fisher Scientific Inc., Waltham, MA, USA) and a mechanical stylus profilometer (Dektak XT-A, Brucker, Elk Grove Village, IL, USA). In addition, the surface roughness of the substrates was measured using a white light interferometer (MSA 500, Polytec GmbH, Waldbronn, Germany) and calculated as defined in ISO 25,178 [18].

Capacitive acoustic sensor chips (Infineon Technologies AG, Neubiberg, Germany) with a size of 1.6 mm × 1.6 mm were utilized in this study [19,20]. The chip possesses a circular polysilicon membrane with a diameter of 0.9 mm and gold pads with a diameter of 0.1 mm. Silicon dummy chips with a size of 5 mm × 5 mm were also employed for reliability analysis. Chip placement was done using a Datacon 2200evo, while encapsulation by means of compression molding was done using a TOWA Y-120 to form an 8″ wafer. The chips were assembled facedown onto a temporary carrier with laminated thermal release tape. Subsequently, compression molding, post-curing and release from the temporary carrier were performed, resulting in mold-embedded components into an 8″ mold-wafer ready for inkjet printing. The liquid epoxy molding compound (EMC) had a filler load of 89 wt.% with a top filler cut of 75 µm, resulting in an overall coefficient of thermal expansion (CTE) of 7 ppm/K ($T < T_g$).

The reliability performance of the printed tracks was evaluated by resistance measurements (Süss Microtec PM5, SUSS MicroTec, Garching, Germany). A temperature cycling test (−40 to 125 °C, 130 min per cycle; CTS CSR 60/600-5) was conducted for reliability analysis. Measurements were done at defined intervals during accelerated aging. Here two sets of experiments were planned to examine the difference between the single-layer and double-layer Ag printed lines as well as to explore the influence of the insulating layer beneath the Ag printed lines.

To investigate the introduced stresses due to the packaging process, laser-Doppler vibrometry (LDV) measurements were performed with both bare-die and mold-embedded CMUTs [21]. For the LDV measurements, a micromotion analyzer (MSA 400, MSA Safety, Cranberry Township, PA, USA) was used to conclude the first resonance frequency. The system provides a real-time velocity and displacement signal of vibration frequencies up to 1.5 MHz at maximal velocity amplitudes of 10 m/s. The excitation of the CMUT resonators was triggered by electrostatic forces. An electrostatic probe was connected to a high voltage excitation signal of up to 400 Volt and placed nearby the CMUT resonator at a distance of approximately 100 µm. The fringing field electrodes provided a strong electrostatic force; therefore, it was not necessary to electrically connect the samples to any potential.

3. Results and Discussion

3.1. Microfabrication

In Figures 2 and 3, two realizations of inkjet-printed RDLs for CMUT FOWLP are depicted. As shown in Figure 2, a system-in-package (SiP) layout for FOWLP of CMUT consisting of two sensors and one ASIC was realized by inkjet printing. Here two Ag layers were ink-jetted over the metallic pads of components and EMC in a drop-on-demand manner. The dielectric layer was inkjet-printed between the two Ag tracks, which enabled multilayering. Depending on the number of input/output (I/Os), pad-diameter, and -pitch of the components, various line-widths for the Ag tracks (between

50 µm to 500 µm) were tailored. In fact, inkjet printing enabled us to customize RDL geometries and realize complex circuitries within a short time.

Since inkjet printing of particle-loaded inks is limited to inks with low viscosity (5–20 mPa·s [22]) and the major fraction of the ink is evaporated during curing, single layer printing usually yields a thin layer. It is noteworthy to state that the metal loading of the ink in this study was 50 wt.%. The average height of a single-layer Ag track was <1 µm. As a comparison, double-layer Ag lines were also prepared. Concerning the sequential processing of double-layer samples, the first layer was only dried before applying the second layer without an intermediate sintering step [23]. Here the average thickness reached ~2 µm, however, the line-width was also increased by up to ~20%. As an example, in Figure 4 the surface morphologies of the single-layer and double-layer Ag lines are compared. Here, Ag tracks with the line-width of 275 µm were aimed. As implied by this figure, the line-width of the double-layer extended to ~325 µm. The calculated area (under the surface profile) of the single-layer and double-layer Ag tracks is compared in Figure 4c. The results for 3 samples per kind are plotted here and averaged. It can be inferred that double-layer Ag tracks have on average double the area of the Ag materials when compared to single-layer tracks.

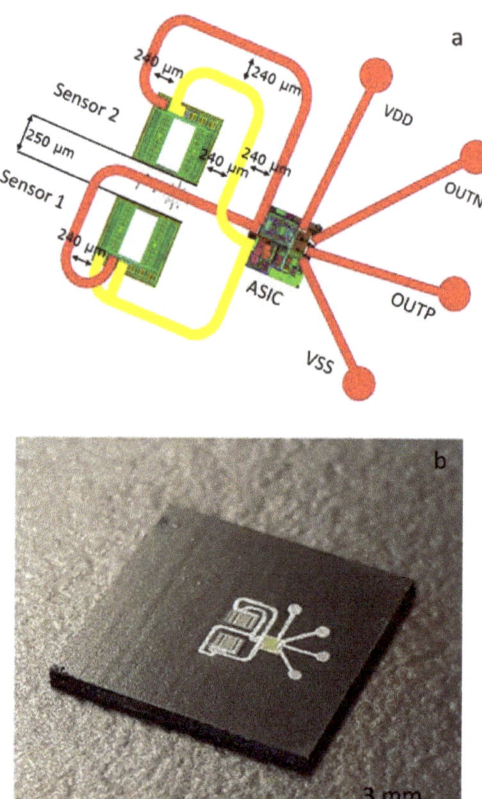

Figure 2. An example for the layout (**a**) and the final demonstrated FOWLP of two capacitive micromachined ultrasonic transducers (CMUTs) and an application-specific integrated circuit (ASIC) chip with inkjet-printed redistribution layers (RDLs) (Comprised of 2 layers of Ag printed lines (red and yellow) and an intermediate insulating printed layer (not shown)) (**b**).

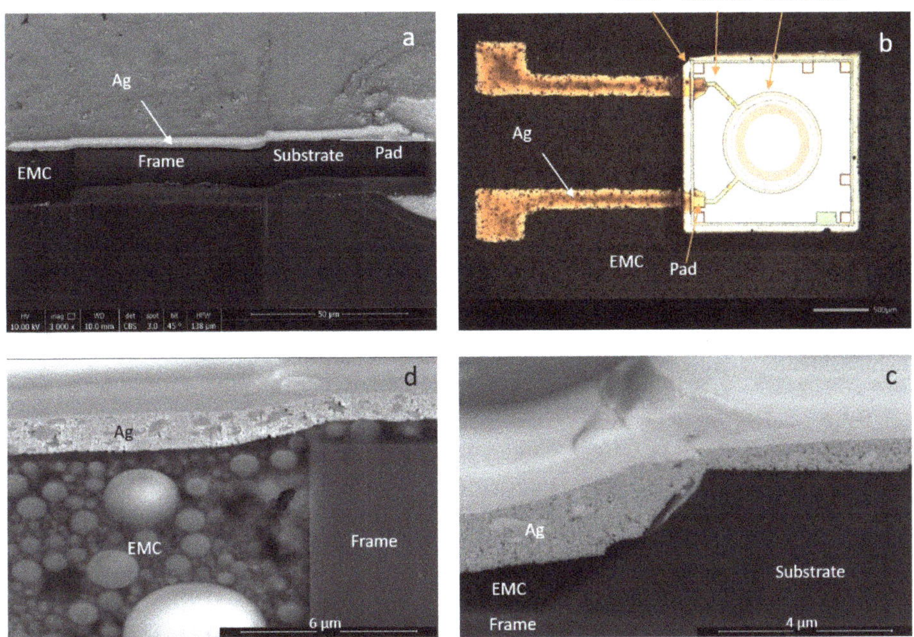

Figure 3. The cross-sectional (**a**) and the planar (**b**) view of the test vehicle comprised of a CMUT microphone with a sensitive membrane and two interconnecting Ag tracks. The critical interfaces between the CMUT substrate-frame (**c**) and CMUT frame-epoxy molding compound (EMC) (**d**) are shown in higher magnifications.

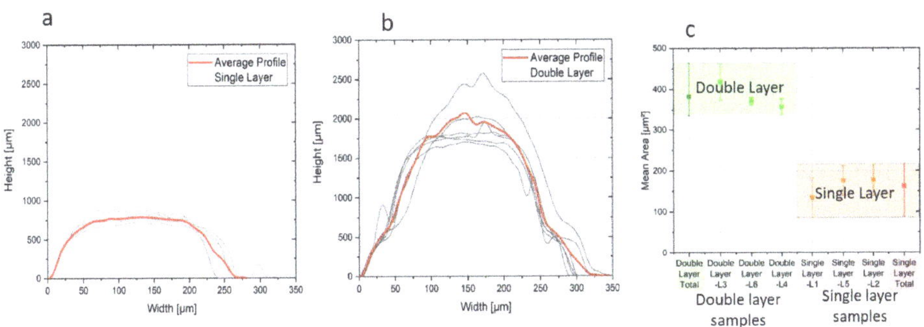

Figure 4. Surface profiles of the single-layer (**a**) and double-layer Ag tracks (**b**). The calculated areas under the profiles are compared in (**c**).

Figure 3 shows a rather simplified example of two inkjet-printed Ag tracks, which was selected as the test-vehicle for the performance analysis and fast signal probing. Here, the CMUT microphone possesses a sensitive membrane and the packaging-induced stress could be extracted by the resonance frequency shift. By observing cross-sectional images of the samples, it was observed that the chip surface is ~3–6 µm higher than the surface of the EMC. This sharp step was predicted to be the weak-spot for the long-term reliability of the RDLs. Figure 3c,d revealed two critical steps at the interfaces between CMUT frame/ CMUT substrate and EMC/ CMUT frame, respectively. In an attempt to smoothen the steps, dielectric inks were selectively inkjet printed on the edges of the components, forming a ramp. In Figure 5, one example of the printed dielectric ramp is depicted. By employing this approach, the Z-offset was smoothened and the sharp step between the chip and mold surface was diminished.

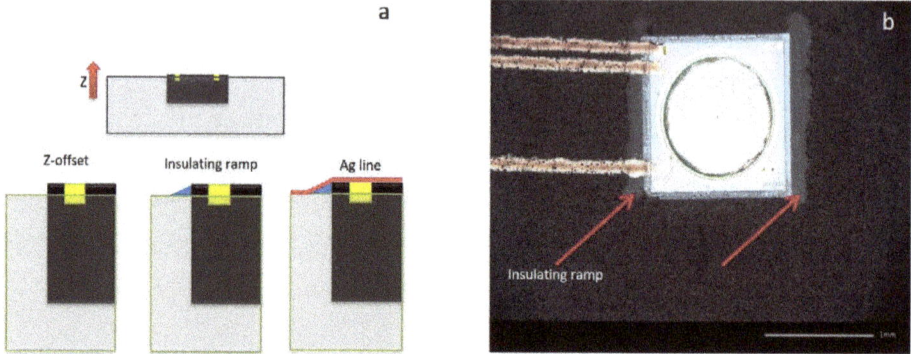

Figure 5. The schematic demonstration (**a**) of the inkjet-printed insulating ramp approach to smoothen the Z-offset and the corresponding optical image (**b**).

3.2. LDV Analysis

As aforementioned, the mechanical stresses upon FOWLP of CMUTs can lead to either physical damage to the sensitive membrane or shifts of the acoustical properties of the CMUT due to the change of membrane stiffness. Accordingly, LDV measurements were performed on both bare-die and packaged die to quantify the shifts of the first resonance frequency of the CMUT.

Figure 6 shows the result of the average LDV-measurement for the bare dies. The response curves of the membranes were monitored and averaged over several measuring points, which were positioned as an overlaid grid. The speed frequency response with its real (blue) and imaginary (red) part is shown in the graphs. When the imaginary part had an extremum and the real component had a turning point, a characteristic mode was formed. The averaged LDV measurement result of the packaged dies is also shown and compared with the bare-dies in Figure 7. The first resonance frequency of the membrane was measured after packaging with a value of 78 kHz. As inferred from Figures 6 and 7, there was a shift in the first resonance frequency caused by the packaging of ca. 16 kHz. This shift was attributed to the compressive stresses due to the chemical and thermal shrinkage (CTE mismatch) of the encapsulation material after cooling down from molding temperature to room temperature. It was postulated that the inkjet printing of RDLs could not have a remarkable influence on the induced stress, since the curing temperature of the inks is identical to the post-mold curing temperature of 150 °C. Besides, neither pressure, aggressive chemical treatment nor a physical contact to the CMUT membranes was imposed during inkjet printing. It is noteworthy to mention that in the course of FOWLP, after compression molding, two temperature-assisted processes were exerted, i.e., post-mold curing at 150 °C and removal of the release tape at 180 °C.

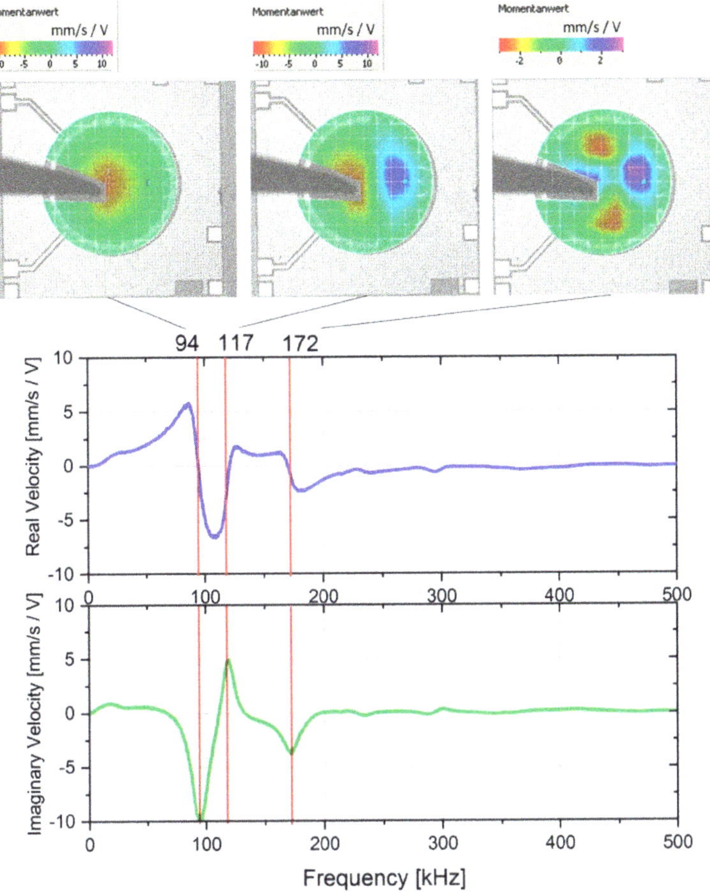

Figure 6. Averaged laser-Doppler vibrometry (LDV) measurement results of the bare CMUT dies with the images of the modes shown above at the characteristic frequency.

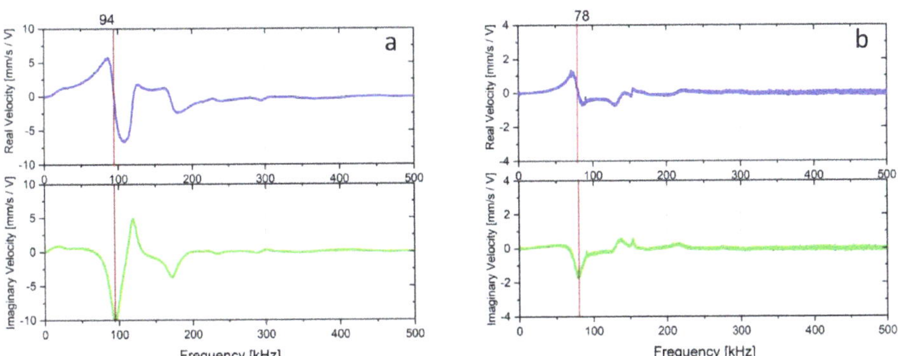

Figure 7. A comparison between the averaged LDV-Measurements of CMUT die before (**a**) and after (**b**) packaging indicating a shift in the first resonance frequency of ca. 16 kHz.

3.3. Reliability Analysis

Accelerated tests are often used to get a deeper understanding of the reliability of components and the collaboration of those within a system. Since the proposed technology (developed within the Silense project [8]) was aimed for mobile and automotive applications, temperature cycling according to the automotive electronics council (AEC-Q100) Grade 1 standard was selected as the main verification methodology. Consequently, the temperature profile was selected according to the standard (−40 to 125 °C) with a time course of a cycle being 130 min with a 30 min holding phase at each peak temperature [24]. Here, a new set of samples was designed and fabricated. The simplified design and final configuration of the test samples for reliability analysis are shown in Figure 8. As seen, the test samples comprised embedded Si chip arrays with 6 printed Ag tracks per chip. Electrical characterization via a two-wire-method was conducted beforehand and at defined intervals during testing at room temperature.

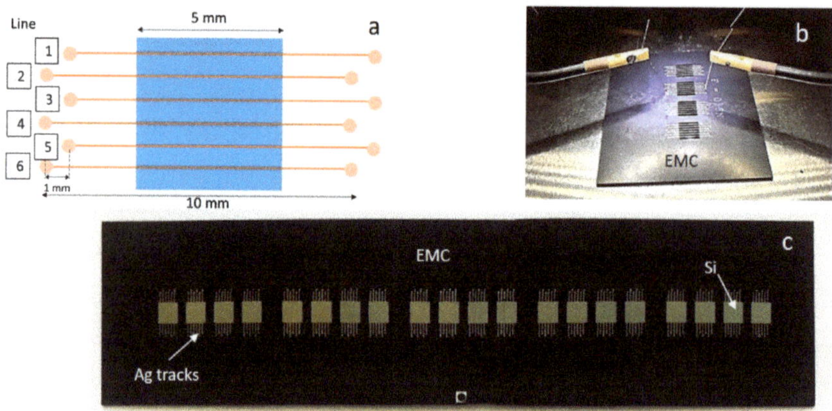

Figure 8. Schematic (**a**) and experimental setup (**b**) for temperature cycling test. An example of the test sample arrays for temperature cycling test is shown in (**c**).

In Figure 9, the measured resistances of the single and double layer printed conductors are compared. It can be seen that the average electrical resistances of the double layer ones were lower than that of the single layer. Moreover, in the course of thermal cycling, the double-layered lines exhibited more consistency compared to the single-layered lines. As highlighted in the graphs, several open circuits emerged during thermal cycles of single layers.

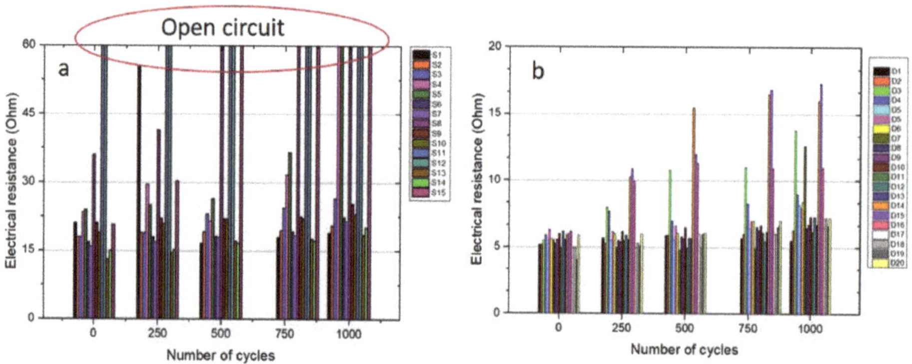

Figure 9. Temperature cycling test results of single-layered (**a**) and double-layered (**b**) Ag tracks.

It was postulated that the thin layers are prone to higher reliability issues since the thin single layer cannot accommodate the thermomechanical stresses due to the CTE mismatch between EMC (~7 ppm/K) and sintered ink (~19 ppm/K) during the thermal cycle tests. There was also a risk that a single-layer Ag could not fully cover the 3–6 μm step-height between the chip and the EMC. The surface roughness (arithmetic mean of the measured absolute height, Sa) of the EMC was measured to be in the range of 450–603 nm, given that a double-layered Ag layer with a thickness of 2 μm seemed to be a more reliable approach to provide a homogeneous layer all over the surface compared to a single layer with a thickness of less than 1 μm. Conclusively, the double-layer printing of Ag lines was proposed to be the best compromise between the reliability, process speed and the final line-width of the RDLs.

In the second set of experiments, the effect of the dielectric ramp on the reliability of the printed interface between die and EMC was investigated. The results of the thermal cycle tests of the samples with and without the insulating ramp are compared in Figure 10. In contrast to our expectation, the insulating ramp did not improve the performance of the printed lines, but rather increased the total electrical resistance of the tracks. There were also more cases of open circuits found. A possible explanation for this observation was the higher CTE of the dielectric polymer inks, which induced additional thermomechanical stresses during thermal cycling. In fact, as to keep the viscosity of the ink low, the inkjettable dielectric materials usually do not contain any fillers and thus possess high CTEs [16,17,25]. In another study, SU8 dielectric ink was inkjet-printed as an insulating ramp to generate 3D interconnects for a millimeter-wave system-on-package [26]. The used SU8 had a CTE of about 52 ppm/K [27]. The inks employed in the current study were also supposed to have similar CTEs, although these values were not provided by the material suppliers.

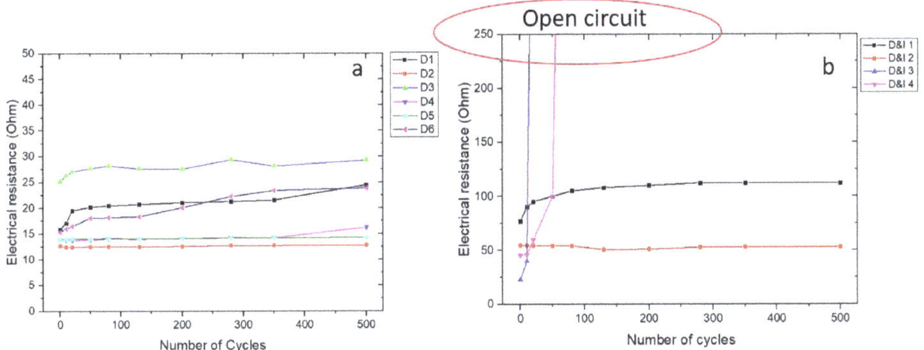

Figure 10. Temperature cycling test results of double-layer Ag lines without (**a**) and with (**b**) insulating ramp.

The cross-sectional images of two failed samples with and without insulating ramp after 500 cycles are presented in Figure 11. It is evident that the fracture took place at the interface between the die and EMC. This interface was subjected to the fusion of thermomechanical stresses due to CTE mismatches between different materials, i.e., Ag/EMC/Si or Ag/EMC/dielectric ramp/Si. It can be seen that the gap was broader in the case of the insulating ramp approach, which is consistent with the measured electrical resistance results.

Conclusively, it can be deduced that the double-layer Ag layer yielded more reliable interconnects compared to single-layered interconnects or interconnects with the insulating ramp. It was also found that the step-height between the chip and EMC led to reliability issues and should be minimized. The magnitude of this height difference is assumed to be dependent on the CTE mismatch between the EMC and the dies, thus also on the thermal budget throughout the manufacturing process. Chemical shrinkage of the EMC impacts this further, as well as the choice of temporary adhesive (thermal release tape) and possibly the placement force. The resulting steps, even though only a few μm high, was identified as a potential bottleneck for the printed lines, especially considering reliability performance. Additionally, the observed delamination in Figure 11a implied a degraded adhesion between the Ag and EMC upon temperature cycling. By comparing Figure 11a to Figure 3d, one can deduce that the delamination emerged during the thermal cycling test due to the CTE mismatch.

3.4. Barriers to Overcome

This study sheds light on challenges and opportunities in FOWLP of CMUT arrays by using inkjet-printed RDLs. It can be inferred that inkjet printing is a cost-effective, powerful and rapid way to form RDLs, especially in comparison to conventional lithography- and electrochemical-based formation of the RDLs. It is well-suited for MEMS packaging, as the additive manufacturing of the RDL eliminates all sorts of challenges with delicate sensing surfaces (i.e., fragile membranes). MEMS also have typically few I/Os; thus, a low-density FOWLP with larger line-width within the resolution of inkjet printing (30–100 μm) can be feasible.

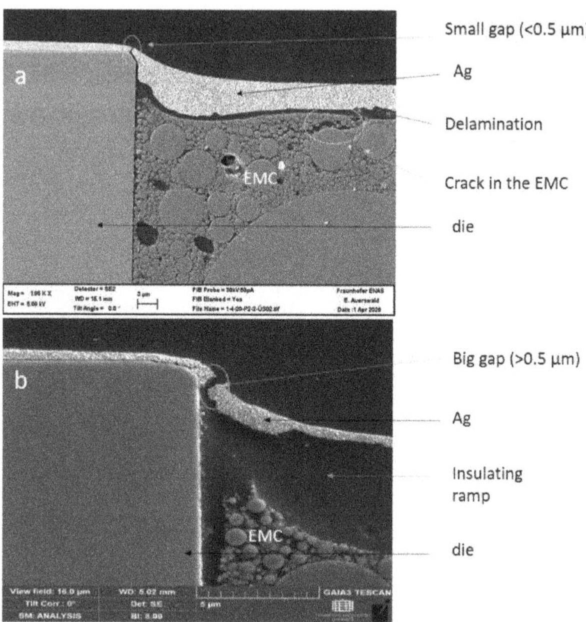

Figure 11. Cross-sectional scanning electron microscopy (SEM) images of the broken Ag tracks at the interface of EMC/die after thermal cycling manifesting the hotspots of the interconnects; (**a**) the sample with double-layered Ag line (**b**) the sample with double-layered Ag line and insulating ramp.

The challenges can be divided into inkjet-related and FOWLP-stemmed issues. There are still some crucial issues that can hinder the full implementation of inkjet printing for FOWLP packaging, such as the poor conductivity of the metallic inks, limitations in resolution for high-density FOWLP, and signal integrity for high frequencies as well as the reliability issues. One of the major limitations of inkjet-printing lies in the strict rheological requirements of the inks, i.e., small range of viscosity and surface tension. Additionally, the coffee-ring effects due to the uneven drying of the inks are still an issue with inkjet-printed structures [28]. Despite the current limitations, with the advancement of additive-manufacturing processes and improved materials, digital printing can certainly overcome the existing challenges. For instance, the current line-width limitations and rheological requirements can be overcome using other printing approaches such as electrohydrodynamic inkjet printing [29,30].

Concerning FOWLP, the encapsulation of delicate CMUT components for FOWLP was found to be challenging, since stress management and protective concepts for sensing areas were determining overall system performance. In addition, the Z-offset issue, which was found to be the reason for the reduced long-term stability, should be further investigated and mitigated upon the consequent process and material optimizations. Future work can be also devoted to improving the adhesion of the Ag ink to EMC by employing different surface pretreatments.

4. Conclusions

In this study, an innovative approach for FOWLP of CMUT sensors by using inkjet-printed redistribution layers was pursued. Two realizations of the proposed FOWLP were shown and the performance of the packaged sensors with sensitive membranes was compared to bare dies performance. The reliability of the printed RDLs was assessed by using a thermal cycling test. The effects of multilayering and incorporating an insulating ramp between the die and mold were investigated. The cross-sectional analysis of the failed samples manifested the bottlenecks of the printed lines. Consequently, the challenges and opportunities of printed RDLs were addressed. The proposed approach for FOWLP of MEMS by using inkjet printing could eventually lead to a new platform for cost-effective heterogeneous integration.

Author Contributions: Investigation: A.R., M.D., C.S., L.R. and J.Z.; Methodology: A.R., M.D., C.B.; Writing—original draft: A.R., M.D., C.S.; Writing—review & editing: A.R., M.D., C.S., J.Z. and K.-F.B.; Supervision: M.S.-R., S.R., T.B., and K.-F.B. All authors have read and agreed to the published version of the manuscript

Funding: This project has received funding from the Electronic Component Systems for European Leadership Joint Undertaking under grant agreement No 737487. This Joint Undertaking receives support from the European Union's Horizon 2020 research and innovation program and Belgium, Netherlands, France, Germany, Austria, Spain, Czech Republic, Norway.

Acknowledgments: Credit for the LDV measurements goes to Steffen Kurth.

Conflicts of Interest: The authors declare no conflict of interest. The funders had no role in the design of the study; in the collection, analyses, or interpretation of data; in the writing of the manuscript, or in the decision to publish the results.

References

1. Braun, T.; Becker, K.F.; Hoelck, O.; Voges, S.; Kahle, R.; Graap, P.; Wöhrmann, M.; Aschenbrenner, R.; Dreissigacker, M.; Schneider-Ramelow, M.; et al. Fan-Out Wafer Level Packaging-A Platform for Advanced Sensor Packaging. In Proceedings of the 2019 IEEE 69th Electronic Components and Technology Conference (ECTC), Las Vegas, NV, USA, 28–31 May 2019; pp. 861–867.
2. Cardoso, A.; Kroehnert, S.; Pinto, R.; Fernandes, E.; Barros, I. Integration of MEMS/Sensors in Fan-Out wafer-level packaging technology based system-in-package (WLSiP). In Proceedings of the 2016 IEEE 18th Electronics Packaging Technology Conference (EPTC), Singapore, 30 November–3 December 2016; pp. 801–807.
3. Lau, J.H. *Fan-Out Wafer-Level Packaging*; Springer: Singapore, 2018.
4. Kuisma, H.; Cardoso, A.; Braun, T. Fan-out wafer-level packaging as packaging technology for MEMS. In *Handbook of Silicon Based MEMS Materials and Technologies*; Elsevier: Amsterdam, The Netherlands, 2020; pp. 707–720.
5. Kuisma, H.; Cardoso, A.; Mäntyoja, N.; Rosenkrantz, R.; Nurmi, S.; Gall, M. FO-WLP multi-DOF inertial sensor for automotive applications. In Proceedings of the 7th Electronic System-Integration Technology Conference (ESTC), Dresden, Germany, 18–21 September 2018; pp. 1–7.

6. Martins, A.; Pinheiro, M.; Ferreira, A.F.; Almeida, R.; Matos, F.; Oliveira, J.; Silva, R.P.; Santos, H.; Monteiro, M.; Gamboa, H. Heterogeneous integration challenges within wafer level fan-out SiP for wearables and IoT. In Proceedings of the 2018 IEEE 68th Electronic Components and Technology Conference (ECTC), San Diego, CA, USA, 29 May–1 June 2018; pp. 1485–1492.

7. Theuss, H.; Geissler, C.; Muehlbauer, F.X.; von Waechter, C.; Kilger, T.; Wagner, J.; Fischer, T.; Bartl, U.; Helbig, S.; Sigl, A.; et al. A MEMS Microphone in a FOWLP. In Proceedings of the 2019 IEEE 69th Electronic Components and Technology Conference (ECTC), Las Vegas, NV, USA, 28–31 May 2019; pp. 855–860.

8. Flack, W.W.; Hsieh, R.; Nguyen, H.A.; Slabbekoorn, J.; Lorant, C.; Miller, A. One micron redistribution for fan-out wafer level packaging. In Proceedings of the 2017 IEEE 19th Electronics Packaging Technology Conference (EPTC), Singapore, 6–9 December 2017; pp. 1–7.

9. Anzinger, S.; Lickert, F.; Fusco, A.; Bosetti, G.; Tumpold, D.; Bretthauer, C.; Dehé, A. Low Power Capacitive Ultrasonic Transceiver Array for Airborne Object Detection. In Proceedings of the 2020 IEEE 33rd International Conference on Micro Electro Mechanical Systems (MEMS), Vancouver, BC, Canada, 18–22 January 2020; pp. 853–856.

10. Available online: https://silense.eu/ (accessed on 4 May 2020).

11. Roshanghias, A.; Ma, Y.; Dreissigacker, M.; Braun, T.; Bretthauer, C.; Becker, K.F.; Schneider-Ramelow, M. The Realization of Redistribution Layers for FOWLP by Inkjet Printing. *Proceedings* **2018**, *2*, 703. [CrossRef]

12. Laurila, M.M.; Khorramdel, B.; Mäntysalo, M. Combination of E-jet and inkjet printing for additive fabrication of multilayer high-density RDL of silicon interposer. *IEEE Trans. Electron Devices* **2017**, *64*, 1217–1224. [CrossRef]

13. Roshanghias, A.; Krivec, M.; Bardong, J.; Binder, A. Additive-manufactured organic interposers. *J. Electron. Packag.* **2020**, *142*, 014501. [CrossRef]

14. Serpelloni, M.; Cantù, E.; Borghetti, M.; Sardini, E. Printed Smart Devices on Cellulose-Based Materials by means of Aerosol-Jet Printing and Photonic Curing. *Sensors* **2020**, *20*, 841. [CrossRef] [PubMed]

15. Chia-Yen, L.; Tsai, H. Printing Method for Redistribution Layer and Filling of through Silicon Vias Using Sintering Silver Paste. In Proceedings of the 2014 9th International Microsystems, Packaging, Assembly and Circuits Technology Conference (IMPACT), Taipei, Taiwan, 22–24 October 2014. [CrossRef]

16. Roshanghias, A.; Krivec, M.; Binder, A. Digital micro-dispension of non-conductive adhesives (NCA) by inkjet printer. In Proceedings of the 2017 IEEE 19th Electronics Packaging Technology Conference (EPTC), Singapore, 6–9 December 2017; pp. 1–3.

17. Roshanghias, A.; Ma, Y.; Gaumont, E.; Neumaier, L. Inkjet printed adhesives for advanced M(O)EMS packaging. *J. Mater. Sci. Mater. Electron.* **2019**, *30*, 20285–20291. [CrossRef]

18. ISO—International Organization of Standardization. *ISO 25178 Geometrical Product Specifications (GPS)—Surface Texture: Areal*; International Organization of Standardization: Geneva, Switzerland, 2016.

19. Dehé, A.; Wurzer, M.; Füldner, M.; Krumbein, U. A4. 3-The infineon silicon MEMS microphone. *Proc. Sens.* **2013**, *2013*, 95–99.

20. Dehe, A.; Froemel, A.; Infineon Technologies AG. MEMS Microphone with Low Pressure Region between Diaphragm and Counter Electrode. U.S. Patent 9,181,080, 10 November 2015.

21. Rembe, C.; Siegmund, G.; Steger, H.; Wörtge, M. Measuring MEMS in motion by laser-Doppler vibrometry. In *Optical Inspection of Microsystems*; CRC Press: Boca Raton, FL, USA, 2006; pp. 245–292.

22. Bhushan, B.; Luo, D.; Schricker, S.R.; Sigmund, W.; Zauscher, S. (Eds.) *Handbook of Nanomaterials Properties*; Springer Science & Business Media: Berlin/Heidelberg, Germany, 2014; p. 202.

23. Nilsson, H.E.; Unander, T.; Siden, J.; Andersson, H.; Manuilskiy, A.; Hummelgard, M.; Gulliksson, M. System integration of electronic functions in smart packaging applications. *IEEE Trans. Compon. Packag. Manuf. Technol.* **2012**, *2*, 1723–1734. [CrossRef]

24. AEC-Q100-REV-H. *Failure Mechanism Based Stress Test Qualification for Integrated Circuits*; Automotive Electronics Council: Denver, CO, USA, 2014.

25. De Gans, B.J.; Duineveld, P.C.; Schubert, U.S. Inkjet printing of polymers: State of the art and future developments. *Adv. Mater.* **2004**, *16*, 203–213. [CrossRef]

26. Tehrani, B.K.; Cook, B.S.; Tentzeris, M.M. Inkjet-printed 3D interconnects for millimeter-wave system-on-package solutions. In Proceedings of the 2016 IEEE MTT-S International Microwave Symposium (IMS), San Francisco, CA, USA, 22–27 May 2016; pp. 1–4.

27. SU8 Datasheet, Microchem. Available online: http://web.mit.edu/3.042/team1_08f/documents/SU8-2050.pdf (accessed on 4 May 2020).
28. Sundriyal, P.; Bhattacharya, S. Inkjet-printed sensors on flexible substrates. In *Environmental, Chemical and Medical Sensors*; Springer: Singapore, 2018; pp. 89–113.
29. Han, Y.; Dong, J. Electrohydrodynamic printing for advanced micro/nanomanufacturing: Current progresses, opportunities, and challenges. *J. Micro Nano Manuf.* **2018**, *6*, 040802. [CrossRef]
30. Zhou, P.; Yu, H.; Zou, W.; Wang, Z.; Liu, L. High-Resolution and Controllable Nanodeposition Pattern of Ag Nanoparticles by Electrohydrodynamic Jet Printing Combined with Coffee Ring Effect. *Adv. Mater. Interfaces* **2019**, *6*, 1900912. [CrossRef]

Article

Effect of Au Film Thickness and Surface Roughness on Room-Temperature Wafer Bonding and Wafer-Scale Vacuum Sealing by Au-Au Surface Activated Bonding

Michitaka Yamamoto [1,2], **Takashi Matsumae** [2], **Yuichi Kurashima** [2], **Hideki Takagi** [2], **Tadatomo Suga** [3], **Seiichi Takamatsu** [1], **Toshihiro Itoh** [1] **and Eiji Higurashi** [2,*]

1   The University of Tokyo, 5-1-5 Kashiwanoha, Kashiwa-shi, Chiba 277-8563, Japan; myamamoto@s.h.k.u-tokyo.ac.jp (M.Y.); seiichi-takamatsu@edu.k.u-tokyo.ac.jp (S.T.); toshihiro-itoh@edu.k.u-tokyo.ac.jp (T.I.)
2   National Institute of Advanced Industrial Science and Technology (AIST), 1-2-1 Namiki, Tsukuba, Ibaraki 305-8564, Japan; t.matsumae@aist.go.jp (T.M.); y-kurashima@aist.go.jp (Y.K.); takagi.hideki@aist.go.jp (H.T.)
3   Meisei University, 2-1-1 Hodokubo, Hino, Tokyo 191-8506, Japan; suga@gakushikai.jp
*   Correspondence: eiji.higurashi@aist.go.jp

Received: 20 March 2020; Accepted: 24 April 2020; Published: 27 April 2020

Abstract: Au-Au surface activated bonding (SAB) using ultrathin Au films is effective for room-temperature pressureless wafer bonding. This paper reports the effect of the film thickness (15–500 nm) and surface roughness (0.3–1.6 nm) on room-temperature pressureless wafer bonding and sealing. The root-mean-square surface roughness and grain size of sputtered Au thin films on Si and glass wafers increased with the film thickness. The bonded area was more than 85% of the total wafer area when the film thickness was 100 nm or less and decreased as the thickness increased. Room-temperature wafer-scale vacuum sealing was achieved when Au thin films with a thickness of 50 nm or less were used. These results suggest that Au-Au SAB using ultrathin Au films is useful in achieving room-temperature wafer-level hermetic and vacuum packaging of microelectromechanical systems and optoelectronic devices.

Keywords: heterogeneous integration; wafer bonding; wafer sealing; room-temperature bonding; Au-Au bonding; surface activated bonding; Au film thickness; surface roughness

---

1. Introduction

Sealing techniques are essential to protect the sensitive elements of microelectromechanical systems (MEMS) and optoelectronic devices from the environment [1–3]. An effective way to achieve sealing is to bond cap wafers to device wafers. Many types of bonding techniques such as anodic bonding [4], thermocompression bonding [5–7], solder bonding [8,9], and eutectic bonding [10] have been used as sealing techniques. However, these techniques require high bonding temperature, which causes problems such as thermally induced mechanical stress due to thermal expansion mismatch. Therefore, low-temperature bonding using metal intermediate layers is becoming increasingly attractive because of the high bonding strength and good reliability that can now be achieved. Research on bonding using Au intermediate layers has been increasing [11–35], because Au has several highly desirable properties such as high resistance to oxidation and corrosion.

Au-Au surface activated bonding (SAB) [17–35] is a promising technique for low-temperature bonding. In Au-Au SAB, the Au surfaces are activated by plasma treatment and then brought into contact at low temperature (<150 °C). Au-Au SAB has also been applied to hermetic sealing, as well as the integration of different materials [17,24]. An advantage of Au-Au SAB is that the Au films can be patterned using photolithography before bonding, enabling high transparency to be achieved by using glass wafers [35]. For example, chip-scale hermetic sealing in air has been achieved at low temperature (150 °C). However, high bonding pressure (300 MPa) was necessary because thick Au films (thickness: 300–500 nm) with rough surfaces (root mean square (RMS) surface roughness: 4.0 nm) were used as sealing rings [24]. Various methods such as thermal-imprint [25], lift-off [26], and direct-transfer [31] have been investigated to reduce the bonding pressure required for sealing by using Au-Au SAB. However, high bonding pressure (>100 MPa) is still required to achieve sealing [28].

Room-temperature pressureless wafer bonding was recently achieved with Au-Au SAB using ultrathin Au films (thickness <50 nm) with small grains, and thus, smooth surfaces (RMS surface roughness: <0.5 nm) [30,34]. Furthermore, room-temperature pressureless wafer-scale hermetic sealing in both air and vacuum was achieved using Au-Au SAB with ultrathin Au films (thickness: 15 nm) [35]. However, the effect of the film thickness on Au-Au bondability and sealing quality has not been investigated quantitatively.

This paper reports on the use of Au thin films with different film thicknesses in room-temperature pressureless wafer bonding and vacuum sealing processes. We also investigate the effect of film thickness and surface roughness on wafer bonding and vacuum sealing quality.

2. Experimental Methods

2.1. Room-Temperature Pressureless Wafer Bonding in Ambient Air

In the first experiment, Au thin films with different thicknesses (15, 50, 100, 300, 500 nm) and Ti thin films with a thickness of 5 nm as adhesion layers were deposited on 4-inch Si wafers by DC sputtering (JSP-8000, ULVAC, Inc., Chigasaki, Japan). The sputtering was performed at a chamber pressure of 0.15 Pa and a sputtering power of 200 W for the Ti films and 100 W for the Au films. The surface roughness of the deposited films was measured with an atomic force microscope (AFM; L-trace, Hitachi High-Tech Science Corporation, Tokyo, Japan), with a scanning area of 500 nm × 500 nm. The average grain size of the Au films was calculated from the observed AFM data using the watershed algorithm [36]. To investigate the stress in the deposited films, we measured the curvature radius of the wafers before and after film deposition using a thin-film stress measurement system (FLX-2320-S, TOHO Inc., Nagoya, Japan). The film stress σ_f was calculated using the Stoney equation [37]:

$$\sigma_f = \frac{E_s t_s^2}{6(1 - v_s)t_f} \cdot \left(\frac{1}{R_1} - \frac{1}{R_0}\right) \tag{1}$$

where E_s, v_s, and t_s, are Young's modulus, Poisson's ratio, and substrate thickness, respectively, R_0 and R_1 are the curvature radii of the wafer before and after film deposition, and t_f is the deposited film thickness. In this work, Si was assumed to be isotropic, and Young's modulus, Poisson's ratio, and substrate thickness were set to 169 GPa, 0.06, and 525 μm, respectively [38,39]. The t_f was calculated as the total thickness of the Au and Ti thin films.

The bonding was performed by placing two wafers with the Au sides facing each other in ambient air and squeezing their centers together with tweezers once with an estimated applied force of <10 N. Before bonding, Ar plasma treatment (RF power: 200 W, operating pressure: 60 Pa, treatment time: 60 s) was performed for surface activation using the plasma equipment installed in the bonding system (WAP-1000, Bondtech Co., Ltd., Kyoto, Japan). The treatment time (60 s) was short enough not to affect the surface roughness of the Au surfaces [34]. The bonded area was observed with a surface acoustic microscope (SAM; SAM 300, PVA TePla Analytical Systems, Westhausen, Germany), and the percentage of the bonded area was calculated using ImageJ software [40]. The bonding strength was evaluated using the razor blade test, which is also known as the crack opening method [41]. The crack length caused by inserting a blade was observed with the SAM.

2.2. Room-Temperature Wafer Sealing in Vacuum

In the second experiment, Au thin films with different film thicknesses (15, 50, 100, 300 nm) and Ti thin films with a thickness of 5 nm as adhesion layers were deposited on Si wafers (4-inch diameter) with cavities and on alkali-free ultrathin glass wafers (80 mm square and 50 μm thick, G-Leaf, Nippon Electric Glass Co., Ltd., Otsu, Japan) by DC sputtering. More than 100 cavities with lateral dimensions of 2 mm × 2 mm, a depth of 100 μm, and a pitch of 3 mm were fabricated in the middle of the wafers by wet chemical etching. A schematic of a bonded wafer pair is shown in Figure 1a, and a cross-sectional schematic of a vacuum-sealed sample is shown in Figure 1b. The surface roughness of each wafer was measured with the AFM, and the average grain size was calculated using the watershed algorithm.

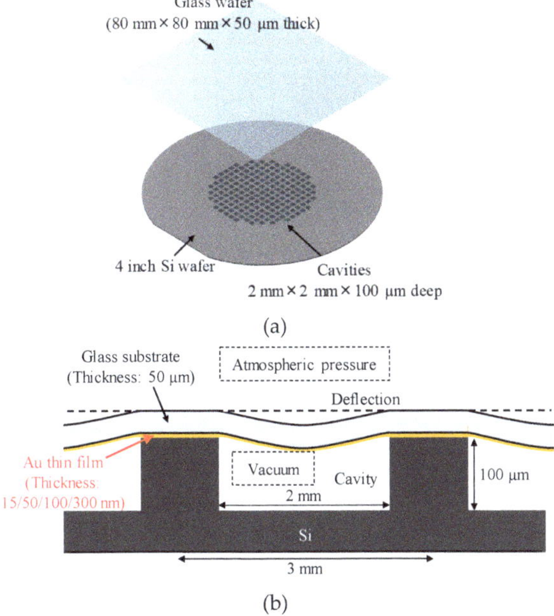

Figure 1. Schematics of bonded wafer pair and bonded structure: (**a**) Wafer pair (thin glass wafer and Si wafer with cavities) deposited with Au thin films (**b**) Cross-sectional schematic of vacuum-sealed sample. Glass substrate exhibited deflection due to pressure difference between sealed vacuum cavity and ambient atmosphere.

Room-temperature vacuum sealing was performed using the bonding system (WAP-1000, Bondtech Co., Ltd.). Two wafers were bonded in a vacuum chamber (~10^{-2} Pa) at room temperature and a contact load of 2000 N. Before bonding, the Au surfaces were activated by Ar plasma (RF power: 200 W, operating pressure: 60 Pa, treatment time: 60 s). The applied contact load (2000 N) corresponded to less than 1.6 MPa for the bonded samples.

The sealing quality of the vacuum-sealed samples was evaluated by visually checking the number of cavities with cap deflection. Since the glass wafers were thin (thickness: 50 μm), the glass caps on the vacuum-sealed cavities exhibited deflection after bonding due to the pressure difference between the sealed vacuum cavities and the ambient atmosphere, as shown in Figure 1b. Furthermore, microstructure observation of the bonded interface was performed with a transmission electron microscope (TEM; H-9500, Hitachi High-Tech Science Co., Tokyo, Japan).

3. Results

3.1. Room-Temperature Pressureless Wafer Bonding in Ambient Air

A measured AFM image of a Si wafer before Au thin film deposition is shown in Figure 2a, and the images of Au thin films with thicknesses of 15, 50, 100, 300, and 500 nm deposited on Si wafers are shown in Figure 2b–f, respectively. Before film deposition, the RMS surface roughness was 0.3 nm. The grain geometry of each Au thin film (Figure 3) was determined using the watershed algorithm. The effect of the film thickness on the surface roughness and average grain size deposited on the Si wafers is illustrated in Figure 4. Both increased exponentially with the thickness, which is consistent with the results of previous studies [42].

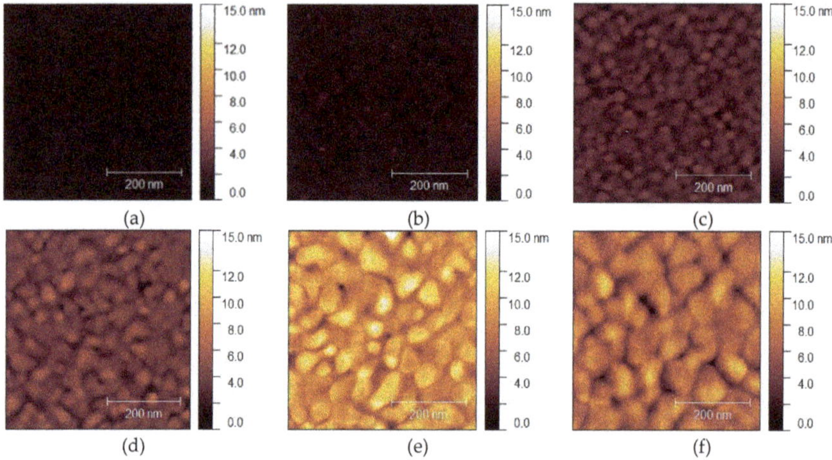

Figure 2. Typical atomic force microscope (AFM) images of Au thin films deposited on Si wafers: (**a**) before deposition; (**b–f**) deposited films with thicknesses of 15, 50, 100, 300, and 500 nm, respectively.

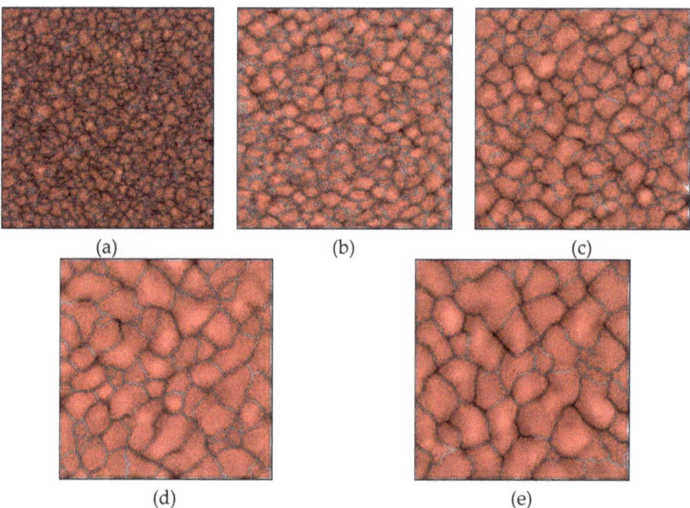

Figure 3. Typical AFM images of Au thin films deposited on Si wafers with grains segmented using the watershed algorithm: (**a–e**) films with thicknesses of 15, 50, 100, 300, and 500 nm, respectively.

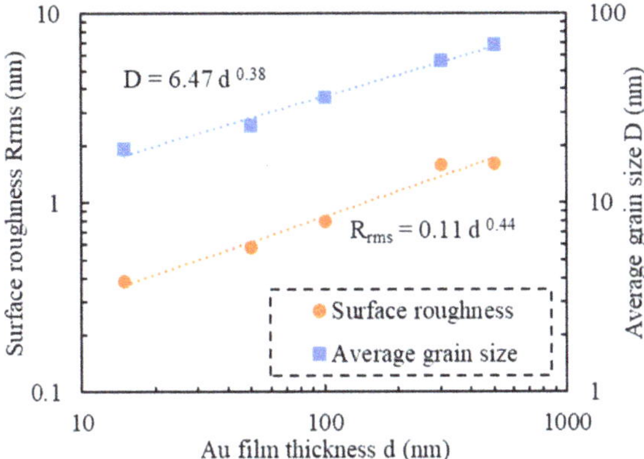

Figure 4. Effect of thickness of Au films deposited on Si wafers on surface roughness and average grain size.

In a previous study [43], the effect of surface roughness on spontaneous bonding was discussed in terms of elastic deformation and energy gain due to bond formation. If bonding is to be achieved, the elastic energy must be smaller than the work of adhesion, W_A, i.e., the energy gain due to bond formation at the interface. If the surface profile is assumed to be a sinusoidal curve, the surface is assumed to be elastic, and wavelength λ is assumed to correspond to the average grain size, the necessary surface roughness for pressureless bonding can be estimated using

$$\frac{R_{rms}^2}{\lambda} < \frac{2\left(1 - v^2\right)}{\pi E} \cdot W_A \tag{2}$$

where R_{rms} and λ are the RMS surface roughness and wavelength of the bonding surface. E and v are Young's modulus and Poisson's ratio. The relationship between average grain size and surface roughness calculated with this formula is shown in Figure 5. The measured average grain size and surface roughness of Au thin films with different thicknesses are also plotted. A thickness of 100 nm or less satisfied the above assumptions, and pressureless bonding should thus be achieved.

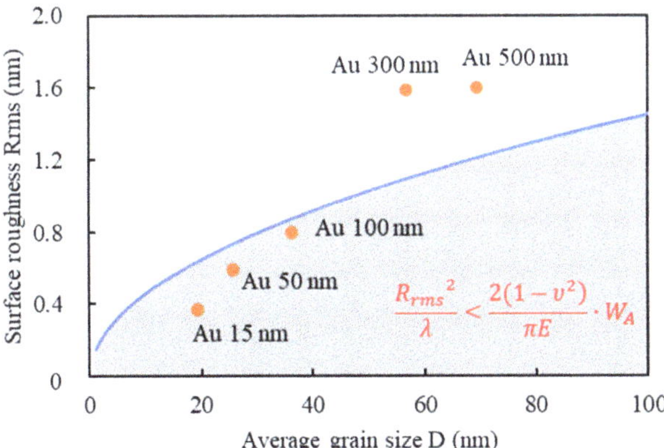

Figure 5. Relationship between surface roughness calculated using Equation (2) and average grain size. Measured average grain size and surface roughness of Au thin films with different thicknesses are also plotted.

The measured film stress is plotted in Figure 6. Previous studies reported that the residual stress strongly depended on the sputtering parameters, especially the chamber pressure [44–46]. In this experiment, the film stress was compressive for all film thicknesses, and the compressive residual stress decreased to −20 MPa when the film thickness was increased to 500 nm. Moreover, the change in the wafer bow after film deposition was less than 1 μm. This means that film stress and wafer bow due to residual stress in Au thin films should not affect bonding.

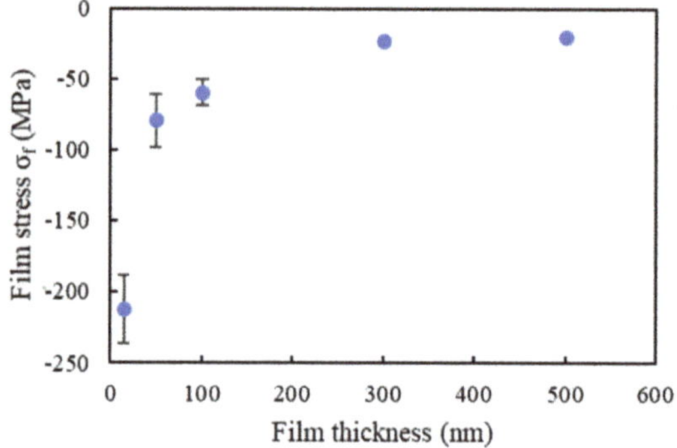

Figure 6. Residual stress in Au thin films as a function of film thickness.

The bonded area of the room-temperature pressureless bonded wafers was observed with the SAM. Typical SAM measurement results are shown in Figure 7. Most of the wafer, except for the particles, was bonded successfully when Au thin films with a thickness of 100 nm or less were used. As shown in Figure 8, the bonded area was inversely proportional to the Au film thickness. When the film thickness was 15, 50, 100, or 300 nm, there was a sufficient bonding area for a razor blade test. Sufficient bonding strength over the surface energy of bulk Si (2.5 J/m^2) [47] was obtained using Au thin films with a thickness of less than or equal to 300 nm although the entire wafer was not bonded when the thickness was 300 nm.

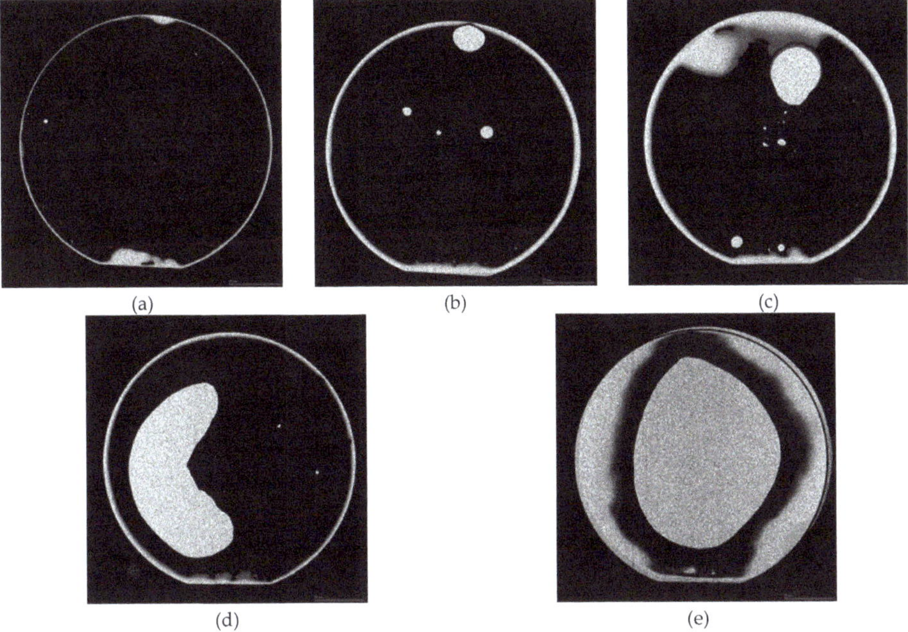

Figure 7. Typical surface acoustic microscope (SAM) images of room-temperature pressureless wafer-scale bonding with Au film thicknesses of (**a–e**) 15, 50, 100, 300, and 500 nm, respectively.

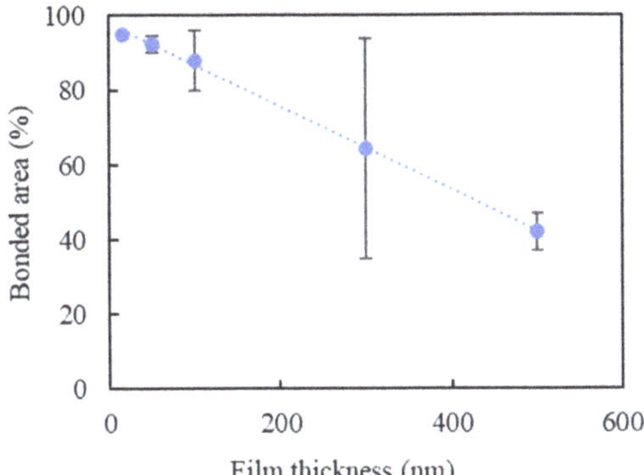

Figure 8. Relationship between bonded area and Au film thickness.

3.2. Room-Temperature Wafer Sealing in Vacuum

Measured AFM images of Au thin films with thicknesses of 15, 50, 100, 300 nm deposited on Si wafers with cavities and glass wafers are shown in Figures 9 and 10. The RMS surface roughness of the Si and glass wafers before film deposition was 0.2 nm. The AFM measurement results showed that the change in surface roughness due to wet chemical etching was small, and thus, had little effect on bonding. The grain geometry of each Au thin film was determined using the watershed algorithm. AFM images of Au thin films deposited on Si wafers with cavities with the grains segmented are shown in Figures 11 and 12. The relationships between Au film thickness, surface roughness, and average grain size are plotted in Figure 13. The surface roughness and grain size increased exponentially as the film thickness was increased, which was consistent with the results when Au thin films were deposited on Si wafers (Figure 4).

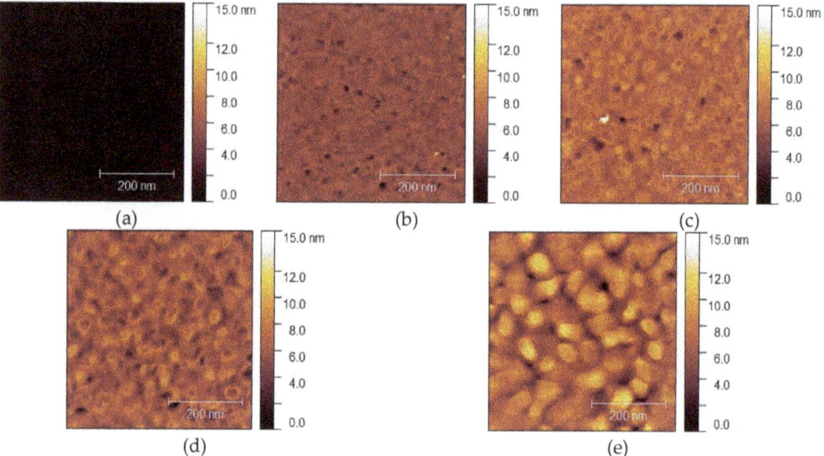

Figure 9. Typical AFM images of Au thin films deposited on Si wafers with cavities: (**a**) before deposition; (**b–e**) after deposition of films with thicknesses of 15, 50, 100, and 300 nm, respectively.

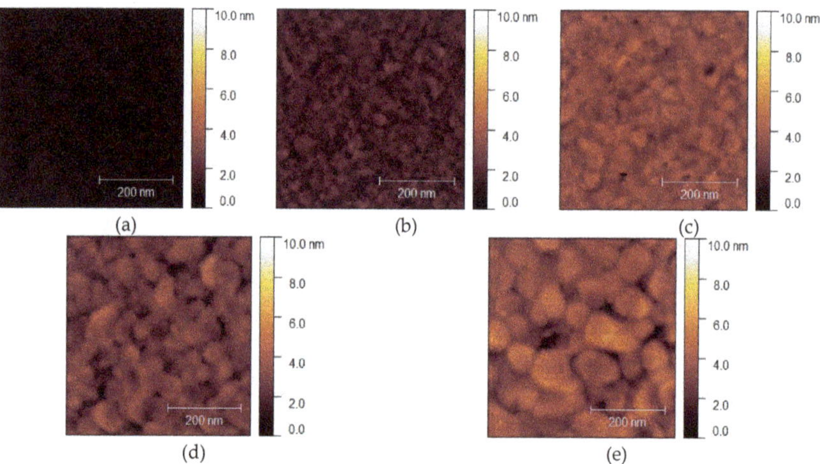

Figure 10. Typical AFM images of Au thin films deposited on glass wafers: (**a**) before deposition; (**b–e**) after deposition of films with thicknesses of 15, 50, 100, 300 nm, respectively.

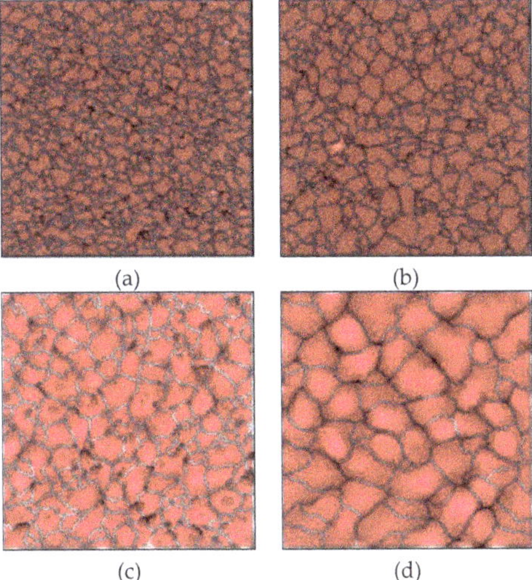

Figure 11. AFM images of Au thin films deposited on Si wafers with cavities with grains segmented using the watershed algorithm: (**a–d**) films with thicknesses of 15, 50, 100, 300 nm, respectively.

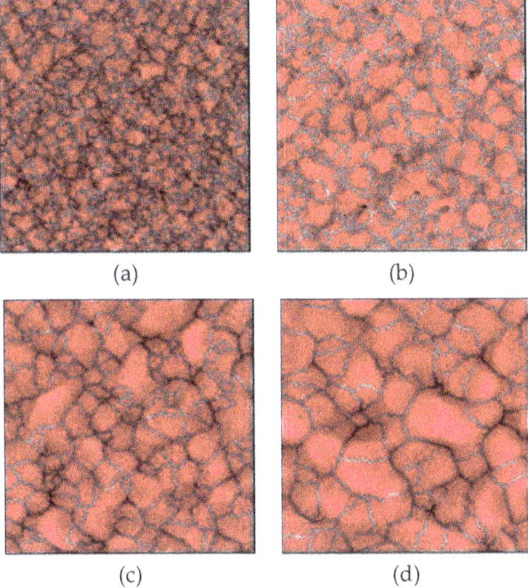

Figure 12. AFM images of Au thin films deposited on glass wafers with grains segmented using the watershed algorithm: (**a–d**) films with thicknesses of 15, 50, 100, 300 nm, respectively.

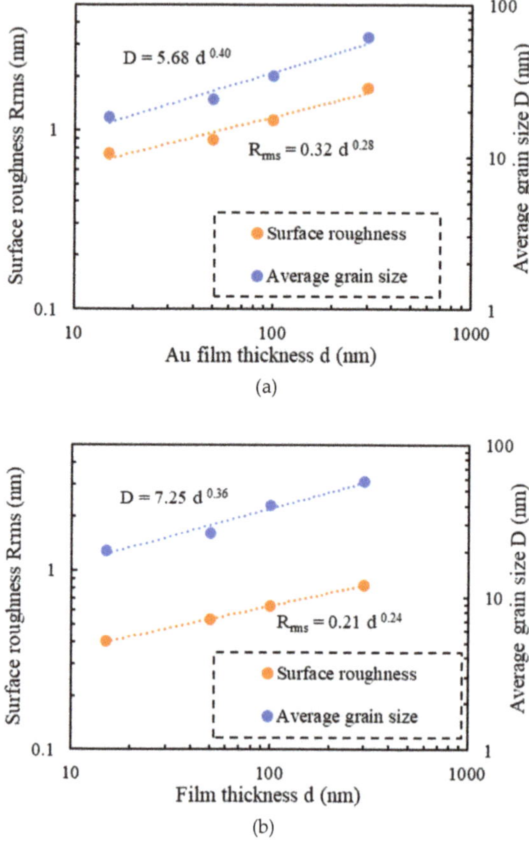

Figure 13. Effect of Au film thickness on surface roughness and average grain size: (**a**) films deposited on Si wafers with cavities; (**b**) films deposited on glass wafers.

The success or failure of the vacuum sealing was determined by observing the deflection of the glass caps caused by the differential pressure between the vacuum-sealed cavities and the ambient atmosphere. The measurement results are plotted in Figure 14. When the Au thin films were 100 nm thick or more, the deflection was observed only in the center region of the wafer immediately after bonding, and the deflection disappeared as the bonded wafer pairs were exposed to air. When the films were 50 nm thick or less, the deflection did not change even after 150 days of air exposure. These results indicated that Au thin films with a thickness of 50 nm or less could be used effectively for wafer bonding, especially for vacuum sealing.

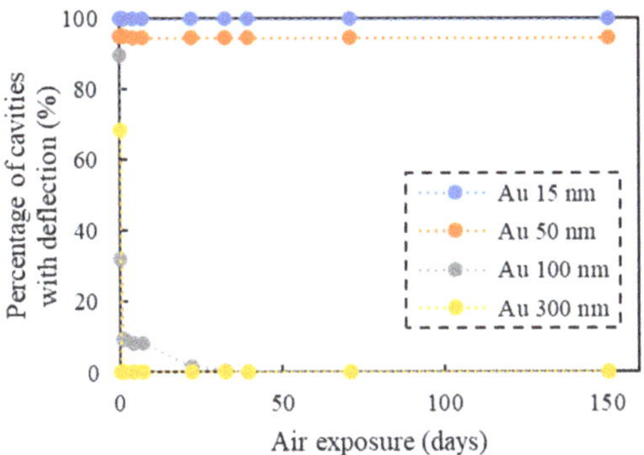

Figure 14. Vacuum sealing results: percentage of cavities with a deflection after air exposure.

Moreover, the air leakage of samples vacuum sealed using Au thin films with a thickness of 15 nm measured by the time dependence of the deflection of the thin glass caps was less than 1.3×10^{-14} Pa m^3/s [35]. This satisfied the reject limit defined by MIL-STD-883 K, method 1014 $(5.0 \times 10^{-9}$ Pa m^3/s).

Cross-sectional TEM observation of the Au-Au bonded interface with 15-nm-thick Au films was performed to investigate its microstructure. The example TEM image in Figure 15 shows that bonding was achieved at the atomic level and that Au atoms diffused around the grain boundaries. This indicated that good sealing could be obtained using 15-nm-thick Au films.

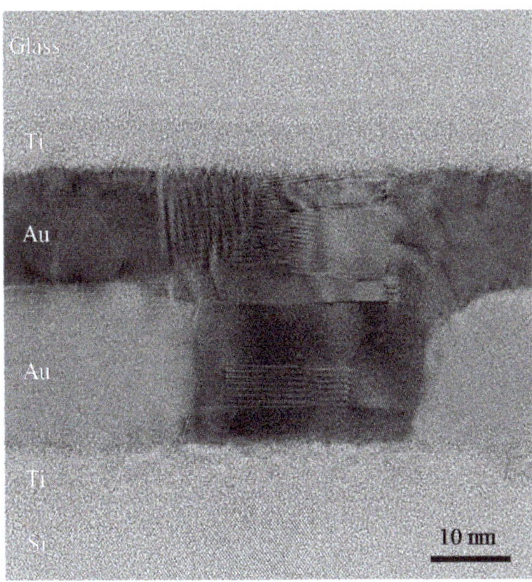

Figure 15. Transmission electron microscope (TEM) image of the bonded sample with a 15-nm-thick Au film.

4. Conclusions

We investigated the effect of Au film thickness (15–500 nm) and surface roughness on room-temperature pressureless wafer bonding and sealing by Au-Au surface activated bonding. The RMS surface roughness and grain size of Au thin films sputtered on Si wafers, Si wafers with cavities, and glass wafers increased with the film thickness. When the film thickness was 100 nm or less, most of the wafer was bonded; the bonded area decreased as the Au film thickness was increased.

Room-temperature wafer-scale vacuum sealing was achieved using Au thin films with a thickness of 50 nm or less. These results suggest that Au-Au surface activated bonding using ultrathin Au films is useful in achieving room-temperature wafer-level hermetic and vacuum packaging of MEMS and optoelectronic devices.

Author Contributions: Investigation: M.Y., T.M., Y.K., H.T., T.S., S.T., T.I., and E.H.; Methodology: M.Y. and E.H.; Validation: M.Y.; Data curation: M.Y.; Writing—original draft: M.Y.: Writing—review & editing: T.M., Y.K., H.T., T.S., S.T., T.I., and E.H.; Funding acquisition: T.H. and E.H.; Resources: H.T., T.S., S.T., T.I., and E.H.; Supervision: T.S., T.I., and E.H.; Project administration: E.H. All authors have read and agreed to the published version of the manuscript.

Funding: This research received no external funding.

Acknowledgments: The authors would like to thank Nippon Electric Glass Co., Ltd. for providing the glass wafers used in the wafer-sealing experiment.

Conflicts of Interest: The authors declare no conflicts of interest.

References

1. Higurashi, E. Heterogeneous integration based on low-temperature bonding for advanced optoelectronic devices. *Jpn. J. Appl. Phys.* **2018**, *57*, 04FA02. [CrossRef]
2. Higurashi, E.; Suga, T. Review of Low-Temperature Bonding Technologies and Their Application in Optoelectronic Devices. *Electron. Commun. Jpn.* **2016**, *99*, 63–71. [CrossRef]
3. Higurashi, E.; Sawada, R.; Suga, T. Optical Microsensors Integration Technologies for Biomedical Applications. *IEICE Trans. Electron.* **2009**, 231–238. [CrossRef]
4. Tanaka, S. Wafer-level hermetic MEMS packaging by anodic bonding and its reliability issues. *Microelectron. Reliab.* **2014**, *54*, 875–881. [CrossRef]
5. Tan, C.S.; Fan, J.; Lim, D.F.; Chong, G.Y.; Li, H. Low temperature wafer-level bonding for hermetic packaging of 3D microsystems. *J. Micromech. Microeng.* **2011**, *21*, 75006. [CrossRef]
6. Henmi, H.; Shoji, S.; Shoji, Y.; Yoshimi, K.; Esashi, M. Vacuum packaging for microsensors by glass-silicon anodic bonding. *Sens. Actuators A Phys.* **1994**, *43*, 243–248. [CrossRef]
7. Tanaka, K.; Hirano, H.; Kumano, M.; Froemel, J.; Tanaka, S. Bonding-Based Wafer-Level Vacuum Packaging Using Atomic Hydrogen Pre-Treated Cu Bonding Frames. *Micromachines* **2018**, *9*, 181. [CrossRef] [PubMed]
8. Higurashi, E.; Kawai, H.; Suga, T.; Okada, S.; Hagihara, T. Low-Temperature Solid-State Bonding Using Hydrogen Radical Treated Solder for Optoelectronic and MEMS Packaging. *ECS Trans.* **2014**, *64*, 267–274. [CrossRef]
9. Goyal, A.; Cheong, J.; Tadigadapa, S. Tin-based solder bonding for MEMS fabrication and packaging applications. *J. Micromech. Microeng.* **2004**, *14*, 819–825. [CrossRef]
10. Wang, Q.; Choa, S.-H.; Kim, W.; Hwang, J.; Ham, S.; Moon, C. Application of Au-Sn eutectic bonding in hermetic radio-frequency microelectromechanical system wafer level packaging. *J. Electron. Mater.* **2006**, *35*, 425–432. [CrossRef]
11. Tsau, C.; Spearing, S.; Schmidt, M.A. Characterization of Wafer-Level Thermocompression Bonds. *J. Microelectromech. Syst.* **2004**, *13*, 963–971. [CrossRef]
12. Taklo, M.M.V.; Storås, P.; Schjølberg-Henriksen, K.; Hasting, H.K.; Jakobsen, H. Strong, high-yield and low-temperature thermocompression silicon wafer-level bonding with gold. *J. Micromech. Microeng.* **2004**, *14*, 884–890. [CrossRef]
13. Park, G.-S.; Kim, Y.-K.; Paek, K.-K.; Kim, J.-S.; Lee, J.-H.; Ju, B.-K. Low-Temperature Silicon Wafer-Scale Thermocompression Bonding Using Electroplated Gold Layers in Hermetic Packaging. *Electrochem. Solid State Lett.* **2005**, *8*, G330. [CrossRef]
14. Shimatsu, T.; Uomoto, M.; Oba, K.; Furukata, Y. Atomic diffusion bonding of wafers in air with thin Au films and its application to optical devices fabrication. In Proceedings of the 2012 3rd IEEE International Workshop on Low Temperature Bonding for 3D Integration, Tokyo, Japan, 22–23 May 2012; p. 103. [CrossRef]
15. Goorsky, M.; Schjølberg-Henriksen, K.; Beekley, B.; Bai, T.; Mani, K.; Ambhore, P.; Bajwa, A.; Malik, N.; Iyer, S.S. Characterization of interfacial morphology of low temperature, low pressure Au–Au thermocompression bonding. *Jpn. J. Appl. Phys.* **2017**, *57*, 02BC03. [CrossRef]

16. Matsumae, T.; Kurashima, Y.; Umezawa, H.; Mokuno, Y.; Takagi, H. Room-temperature bonding of single-crystal diamond and Si using Au/Au atomic diffusion bonding in atmospheric air. *Microelectron. Eng.* **2018**, *195*, 68–73. [CrossRef]

17. Okada, H.; Itoh, T.; Frömel, J.; Gessner, T.; Suga, T. Room temperature vacuum sealing using surfaced activated bonding with Au thin films. In Proceedings of the 13th International Conference on Solid-State Sensors, Actuators and Microsystems, 2005. Digest of Technical Papers. TRANSDUCERS '05, Seoul, Korea, 5–9 June 2005; Volume 1, pp. 932–935. [CrossRef]

18. Higurashi, E.; Imamura, T.; Suga, T.; Sawada, R. Low-Temperature Bonding of Laser Diode Chips on Silicon Substrates Using Plasma Activation of Au Films. *IEEE Photonics Technol. Lett.* **2007**, *19*, 1994–1996. [CrossRef]

19. Takigawa, R.; Higurashi, E.; Suga, T.; Shinada, S.; Kawanishi, T. Low-temperature Au-to-Au Bonding for LiNbO$_3$/Si Structure Achieved in Ambient Air. *IEICE Trans. Electron.* **2007**, *E90-C*, 145–146. [CrossRef]

20. Takigawa, R.; Higurashi, E.; Suga, T.; Sawada, R. Room-Temperature Bonding of Vertical-Cavity Surface-Emitting Laser Chips on Si Substrates Using Au Microbumps in Ambient Air. *Appl. Phys. Express* **2008**, *1*, 112201. [CrossRef]

21. Higurashi, E.; Chino, D.; Suga, T.; Sawada, R. Au–Au Surface-Activated Bonding and Its Application to Optical Microsensors With 3-D Structure. *IEEE J. Sel. Top. Quantum Electron.* **2009**, *15*, 1500–1505. [CrossRef]

22. Takigawa, R.; Higurashi, E.; Suga, T.; Kawanishi, T. Passive Alignment and Mounting of LiNbO$_3$ Waveguide Chips on Si Substrates by Low-Temperature Solid-State Bonding of Au. *IEEE J. Sel. Top. Quantum Electron.* **2011**, *17*, 652–658. [CrossRef]

23. Higurashi, E.; Fukunaga, T.; Suga, T. Low-Temperature Bonding of GaN on Si Using a Nonalloyed Metal Ohmic Contact Layer for GaN-Based Heterogeneous Devices. *IEEE J. Quantum Electron.* **2011**, *48*, 182–186. [CrossRef]

24. Yamamoto, S.-I.; Higurashi, E.; Suga, T.; Sawada, R. Low-temperature hermetic packaging for microsystems using Au–Au surface-activated bonding at atmospheric pressure. *J. Micromech. Microeng.* **2012**, *22*, 55026. [CrossRef]

25. Kurashima, Y.; Maeda, A.; Takigawa, R.; Takagi, H. Room temperature wafer bonding of metal films using flattening by thermal imprint process. *Microelectron. Eng.* **2013**, *112*, 52–56. [CrossRef]

26. Kurashima, Y.; Maeda, A.; Takagi, H. Replication of atomically smooth surface shape onto electroplated Au patterns by lift-off process and room-temperature Au–Au bonding in atmospheric air. *Microelectron. Eng.* **2014**, *129*, 1–4. [CrossRef]

27. Goto, M.; Hagiwara, K.; Iguchi, Y.; Ohtake, H.; Saraya, T.; Higurashi, E.; Toshiyoshi, H.; Hiramoto, T. 3-D Silicon-on-Insulator Integrated Circuits With NFET and PFET on Separate Layers Using Au/SiO$_2$ Hybrid Bonding. *IEEE Trans. Electron Devices* **2014**, *61*, 2886–2892. [CrossRef]

28. Kurashima, Y.; Maeda, A.; Takagi, H. Room-temperature wafer scale bonding using smoothed Au seal ring surfaces for hermetic sealing. *Jpn. J. Appl. Phys.* **2015**, *55*, 16701. [CrossRef]

29. Higurashi, E.; Yamamoto, M.; Sato, T.; Suga, T.; Sawada, R. Room-Temperature Gold-Gold Bonding Method Based on Argon and Hydrogen Gas Mixture Atmospheric-Pressure Plasma Treatment for Optoelectronic Device Integration. *IEICE Trans. Electron.* **2016**, 339–345. [CrossRef]

30. Higurashi, E.; Okumura, K.; Kunimune, Y.; Suga, T.; Hagiwara, K. Room-Temperature Bonding of Wafers with Smooth Au Thin Films in Ambient Air Using a Surface-Activated Bonding Method. *IEICE Trans. Electron.* **2017**, 156–160. [CrossRef]

31. Kurashima, Y.; Matsumae, T.; Takagi, H. Room-temperature Au–Au bonding in atmospheric air using direct transferred atomically smooth Au film on electroplated patterns. *Microelectron. Eng.* **2018**, *189*, 1–5. [CrossRef]

32. Shiratori, Y.; Hoshi, T.; Ida, M.; Higurashi, E.; Matsuzaki, H. High-Speed InP/InGaAsSb DHBT on High-Thermal-Conductivity SiC Substrate. *IEEE Electron Device Lett.* **2018**, *39*, 807–810. [CrossRef]

33. Yamamoto, M.; Higurashi, E.; Suga, T.; Sawada, R.; Itoh, T. Properties of various plasma surface treatments for low-temperature Au–Au bonding. *Jpn. J. Appl. Phys.* **2018**, *57*, 04FC12. [CrossRef]

34. Yamamoto, M.; Matsumae, T.; Kurashima, Y.; Takagi, H.; Suga, T.; Itoh, T.; Higurashi, E. Comparison of Argon and Oxygen Plasma Treatments for Ambient Room-Temperature Wafer-Scale Au–Au Bonding Using Ultrathin Au Films. *Micromachines* **2019**, *10*, 119. [CrossRef] [PubMed]

35. Yamamoto, M.; Kunimune, Y.; Matsumae, T.; Kurashima, Y.; Takagi, H.; Iguchi, Y.; Honda, Y.; Suga, T.; Itoh, T.; Higurashi, E. Room-temperature pressureless wafer-scale hermetic sealing in air and vacuum using surface activated bonding with ultrathin Au films. *Jpn. J. Appl. Phys.* **2019**, *59*, SBBB01. [CrossRef]
36. Klapetek, P.; Ohlídal, I.; Montaigne-Ramil, A.; Bonanni, A.; Stifter, D.; Sitter, H. Atomic Force Microscopy Characterization of ZnTe Epitaxial Thin Films. *Jpn. J. Appl. Phys.* **2003**, *42*, 4706–4709. [CrossRef]
37. Stoney, G.G. The tension of metallic films deposited by electrolysis. *Proc. R. Soc. London. Ser. A Math. Phys. Sci.* **1909**, *82*, 172–175. [CrossRef]
38. Fournel, F.; Continni, L.; Morales, C.; Da Fonseca, J.; Moriceau, H.; Rieutord, F.; Barthélémy, A.; Radu, I. Measurement of bonding energy in an anhydrous nitrogen atmosphere and its application to silicon direct bonding technology. *J. Appl. Phys.* **2012**, *111*, 104907. [CrossRef]
39. Brantley, W.A. Calculated elastic constants for stress problems associated with semiconductor devices. *J. Appl. Phys.* **1973**, *44*, 534. [CrossRef]
40. Available online: http://imagej.nih.gov/ij/ (accessed on 25 April 2020).
41. Maszara, W.P.; Goetz, G.; Caviglia, A.; McKitterick, J.B. Bonding of Silicon Wafers for Silicon-on-Insulator. *J. Appl. Phys.* **1988**, *64*, 4943–4950. [CrossRef]
42. Zhang, X.; Song, X.H.; Zhang, D.L. Thickness Dependence of Grain Size and Surface Roughness for DC Magnetron Sputtered Au Films. *Chin. Phys. B* **2010**, *19*, 086802. [CrossRef]
43. Takagi, H.; Maeda, R.; Chung, T.R.; Hosoda, N.; Suga, T. Effect of Surface Roughness on Room-Temperature Wafer Bonding by Ar Beam Surface Activation. *Jpn. J. Appl. Phys.* **1998**, *37*, 4197–4203. [CrossRef]
44. Tan, C.W.; Miao, J. Optimization of Sputtered Cr/Au Thin Film for Diaphragm-Based MEMS Applications. *Thin Solid Films* **2009**, *517*, 4921–4925. [CrossRef]
45. Abadias, G.; Chason, E.; Keckes, J.; Sebastiani, M.; Thompson, G.B.; Barthel, E.; Doll, G.L.; Murray, C.E.; Stoessel, C.H.; Martinu, L. Review Article: Stress in Thin Films and Coatings: Current Status, Challenges, and Prospects. *J. Vac. Sci. Technol. A Vac. Surf. Film* **2018**, *36*, 020801. [CrossRef]
46. Hoffman, D.W.; Thornton, J.A. Internal Stresses in Cr, Mo, Ta, and Pt Films Deposited by Sputtering from a Planar Magnetron Source. *J. Vac. Sci. Technol.* **1982**, *20*, 355–358. [CrossRef]
47. Tong, Q.-Y.; Kim, W.J.; Lee, T.; Gösele, U. Low Vacuum Wafer Bonding. *Electrochem. Solid-State Lett.* **1999**, *1*, 52–53. [CrossRef]

MDPI

St. Alban-Anlage 66

4052 Basel

Switzerland

Tel. +41 61 683 77 34

Fax +41 61 302 89 18

www.mdpi.com

Micromachines Editorial Office

E-mail: micromachines@mdpi.com

www.mdpi.com/journal/micromachines